Benchmark Papers
in Animal Behavior

Series Editor: Martin W. Schein
West Virginia University

PUBLISHED VOLUMES

HORMONES AND SEXUAL BEHAVIOR / *Carol Sue Carter*
TERRITORY / *Allen W. Stokes*
SOCIAL HIERARCHY AND DOMINANCE / *Martin W. Schein*
EXTERNAL CONSTRUCTION BY ANIMALS / *Nicholas E. Collias and
    Elsie C. Collias*
PSYCHOPHYSIOLOGY / *Stephen W. Porges and M. G. H. Coles*
SOUND RECEPTION IN FISHES / *William N. Tavolga*
SOUND PRODUCTION IN FISHES / *William N. Tavolga*
IMPRINTING / *E. H. Hess and Slobodan B. Petrovich*
VERTEBRATE SOCIAL ORGANIZATION / *Edwin M. Banks*
EVOLUTION OF PLAY BEHAVIOR / *D. Müller-Schwarze*
PARENTAL BEHAVIOR IN BIRDS / *Rae Silver*

RELATED TITLES IN OTHER BENCHMARK SERIES

BEHAVIOR AS AN ECOLOGICAL FACTOR / *D. E. Davis*
CYCLES OF ESSENTIAL ELEMENTS / *L. R. Pomeroy*
NICHE: Theory and Application / *R. H. Whittaker and S. A. Levin*
ECOLOGICAL ENERGETICS / *R. G. Wiegert*
ECOLOGICAL SUCCESSION / *F. B. Golley*
CONCEPTS OF SPECIES / *C. N. Slobodchikoff*

Benchmark Papers
in Animal Behavior / 11

A BENCHMARK® Books Series

PARENTAL BEHAVIOR
IN BIRDS

Edited by

RAE SILVER

**Barnard College
of Columbia University
and
The American Museum
of Natural History**

Dowden, Hutchinson & Ross, Inc.

STROUDSBURG, PENNSYLVANIA

LIBRARY OF CONGRESS CATALOGING IN PUBLICATION DATA
Main entry under title:
Parental behavior in birds.
    (Benchmark papers in animal behavior ; 11)
    Includes indexes.
    1. Birds—Behavior—Addresses, essays, lectures. 2.   Parental behavior
in animals—Addresses, essays, lectures. I.   Silver, Rae.
    QL698.3.P37   598.2′5′6   77-8495
    ISBN 0-87933-289-1

Exclusive Distributor: Halsted Press
A Division of John Wiley & Sons, Inc.
ISBN: 0-470-99237-9

SERIES EDITOR'S FOREWORD

It was not too many years ago that virtually all research publications dealing with animal behavior could be housed within the covers of a very few hard-bound volumes which were easily accessible to the few workers in the field. Times have changed! The present day students of animal behavior have all they can do to keep abreast of developments within their own area of special interest, let alone in the field as a whole; and of course we have long since given up attempts to maintain more than a superficial awareness of what is happening "in biology," "in psychology," "in sociology," or in any of the broad fields touching upon or encompassing the behavioral sciences.

It was even fewer years ago that those who taught animal behavior courses could easily choose a suitable textbook from among the very few that were available; all "covered" the field, according to the bias of the author. Students working on a special project used *the* text and *the* journal as reference sources, and for the most part successfully covered their assigned topics. Times have changed! The present day teacher of animal behavior is confronted with a bewildering array of books to choose among, some purported to be all-encompassing, others confessing to strictly delimited coverage, and still others being simply collections of recent and profound writings.

In response to the problem of the steadily increasing and overwhelming volume of information in the area, the Benchmark Papers in Animal Behavior was launched as a series of single-topic volumes designed to be some things to some people. Each volume contains a collection of what an expert considers to be *the* significant research papers in a given topic area. Each volume, then, serves several purposes. To teachers, a Benchmark volume serves as a supplement to other written materials assigned to students; it permits in-depth consideration of a particular topic while at the same time confronting students (often for the first time) with original research papers of outstanding quality. To researchers, a Benchmark volume serves to save countless hours digging through the various journals to find *the* basic articles in their area of interest; often the journals are not easily available. To students, a Benchmark volume provides a readily accessible set of original papers on the topic, a set that forms the core of the more extensive bibliography that they are likely to compile; it also permits them to see at first hand what an "expert" thinks is important in the area, and to react accordingly. Finally, to librarians, a Benchmark volume represents a collection of important papers from many diverse sources, thus making readily available materials that might otherwise not be economically possible to obtain or physically possible to keep in stock.

The choice of topics to be covered in this series is no small matter. Each of us could come up with a long list of possible topics and then search for potential volume editors. Alternatively, we could draw up long lists of recognized and prominent scholars and try to persuade them to do a volume on a topic of their choice. For the most part, I have followed a mix of both approaches: match a distinguished researcher with a desired topic, and the results should be outstanding. And so it is with the present volume.

Dr. Silver has been studying avian reproduction and parental behavior virtually all of her professional life. With a background that includes work in Daniel Lehrman's laboratory at Rutgers University, the American Museum of Natural History in New York, and subsequent research at Barnard College of Columbia University, she is appropriately knowledgeable to organize the present volume. I am grateful that she has agreed to take on the job; all of us are the beneficiaries.

MARTIN W. SCHEIN

PREFACE

The papers in this volume were chosen to highlight advances in the understanding of the parental behavior of birds. Some papers are primarily descriptive, and show how each phase of the reproductive cycle is important in the production of offspring. Others focus on questions about parental behavior such as underlying causal mechanisms, development, adaptive significance, and evolution.

A decision had to be made on whether to organize the papers around theoretical problems in animal behavior, or around the stages of the reproductive cycle. To focus attention on the animals themselves, the latter route was chosen. The papers are arranged in the following categories; the initiation of breeding, pair formation and nest building, egg laying, post-laying behavior, and post-hatching behavior. Other topics, such as mating and sexual behavior, territoriality, "imprinting" of young on their parents, are not considered. The former topic is more closely linked to sexual than to parental aspects of reproduction. The latter two topics are treated in other books in this series (Stokes, 1974; Hess and Petrovich, 1977). The theoretical significance of the papers emerge in both the introductory editorial comments and in the discussion by the original author(s).

The decision on which aspects of reproduction to include in the category of parental care was difficult to make. The special responses that are directly parental in that they function to further the development of young are not always clearly differentiated from behaviors that serve other functions. Many behaviors which are a part of parental behavior are also a part of nonparental behavior. For example, when the fourteen- to fifteen-day-old young Ring Dove (Streptopelia risoria) begs for food, the response of the adult is often to peck and eat grain. Parental eating of grain in turn stimulates pecking of grain by the squab. Parental pecking in response to squab begging provides nutrition for the adult and at the same time stimulates independent feeding by the young.

Another point to consider is that responses to the young are not always clearly separable within the behavioral repertoire of the animal from responses to the mate, nest, or self. For example, the motor movements of "courtship feeding" when the male Ring Dove feeds the female are indistinguishable from the movements involved in parental feeding of young. The mechanisms regulating feeding by the male may be identical in both cases, though the function differs in the two instances. Although functional categories are arbitrary, they are convenient, and the papers have been organized around functional aspects of the reproductive cycle.

In preparing this volume, problems of categorization arose. Some papers were selected because they represent significant theoretical contributions to the study of animal behavior in general but do not fit any single functional category of parental behavior. Others deal with more than one functional aspect of parental behavior. Finally, many important papers have already been printed in other volumes in this series. The chart on pages, xii and xiii provides a cross-referencing of information. By examining the chart one can find that a particular paper deals with evolution of nest building, or physiological mechanisms underlying incubation, and also emphasizes another aspect of parental care. In addition, relevant papers in the series Benchmark Papers in Animal Behavior and Benchmark Papers in Ecology are indicated. In this way landmark papers that set the direction of future studies are available to the individual interested in the parental behavior in birds and to the reader interested in comparative aspects of parental care.

ACKNOWLEDGMENTS

I wish to express thanks to Colin Beer, Barry Komisaruk, and Jay Rosenblatt for suggestions regarding the papers included in this volume and Mike Leon, Martin W. Schein, and H. Phillip Zeigler for comments on the editor's introductory material. In addition, I am indebted to Martin W. Schein for editorial assistance and to Geoffrey and Leonard Silver for support during the preparation of this volume.

CONTENTS

PART I: THE INITIATION OF BREEDING

PART II: PAIR FORMATION AND NEST BUILDING

Contents

PART V: POST-HATCHING BEHAVIOR

	Initiation of Breeding	Nesting
Stimulus and Environmental Factors	Rowan, p. 28[1] Bissonnette, p. 30 Lack, p. 31 Marshall & Disney, p. 37 Lehrman, BPAB, v. 4[2]	Coulson, p. 104 Lehrman et al., p. 129 Frith, p. 201 Smith, p. 218 Collias & Collias, BPAB, v. 4 Frith, BPAB, v. 4 White et al., BPAB, v. 4 White & Kinney, BPAB, v. 4
Hormonal Determinants	Lack, p. 31	Lehrman, BPAB, v. 4
Neural Determinants		
Learning, Experience, and Development	Coulson, p. 104	Dilger, p. 54 Collias & Collias, BPAB, v. 4
Evolution, Adaptation, and Natural History	Lack, p. 31 Lack, BPAB, v. 2 Verner, BPAB, v. 2 Wolf & Stiles, BPAB, v. 2 Miller, BPEco, v. 2[3]	von Haartman, p. 45 Dilger, p. 54 Cullen, p. 76 Smith, p. 218 Thomson, p. 393 Crook, BPAB, v. 2 Lack, BPAB, v. 2 Collias, BPAB, v. 4 Davis, BPEco, v. 2

[1]Author's name followed by page number indicates a paper in this volume.

[2]BPAB followed by volume number indicates a paper included in a volume in the Benchmark Papers in Animal Behavior series.

[3]BPEco followed by volume number indicates a paper included in a volume in the Benchmark Papers in Ecology series.

Egg Laying	Incubation	Brooding and Post Hatching Behavior
Craig, p. 118 Matthews, p. 125 Phillips, p. 139	Lehrman et al., p. 129 Frith, p. 201 Hinde & Steel, p. 240 Lorenz & Tinbergen, p. 247	Smith, p. 218 Schleidt et al., p. 290 Tinbergen et al., p. 309
	Lienhart, p. 230 Komisaruk, p. 234 Höhn, p. 245 Lehrman, BPAB, v. 4	Goodale, p. 303 Lehrman, BPAB, v. 4
	Komisaruk, p. 234	
Cullen, p. 76 Coulson, p. 104 Lack, p. 140 Friedmann, p. 212	New, p. 223 Lorenz & Tinbergen, p. 247 Tinbergen et al., p. 309	Schleidt et al., p. 290 Tinbergen et al., p. 309 Smith, p. 355 Norton-Griffiths, p. 359 Ramsay, BPAB, v. 5 Hess, BPAB, v. 5 Hinde, BPAB, v. 5
Craig & Toth, BPAB, v. 3	Swanberg, p. 160 Skutch, p. 176 Lorenz & Tinbergen, p. 247 Lack, BPAB, v. 2 Davis, BPEco, v. 2	Cullen, p. 76 Lack, p. 140 Tinbergen et al., p. 309 Norton-Griffiths, p. 359 Cade & MacLean, p. 372 Thomson, p. 393 Skutch, p. 399 Crook, BPAB, v. 2 Lack, BPAB, v. 2 Davis, BPEco, v. 2 Hess, BPAB, v. 5

CONTENTS BY AUTHOR

PARENTAL BEHAVIOR
IN BIRDS

INTRODUCTION

Why do people study the parental behavior of birds? The description and analysis of the parental behavior of birds has been of interest to amateurs and scientists of many different disciplines including ornithology, ethology, and psychology. The earliest studies were primarily descriptive and focused on the collection of eggs and nests and not on the behavior of the animals (see Paper 12). A series of twenty-one books by A. C. Bent on the life history of North American Birds, published between 1919 and 1961 by the U.S. National Museum Bulletins documents the contribution of these naturalists and is part of the heritage of ornithology.

The lack of attention to the behavior of birds at the nest was not due to lack of interest, but rather to the absence of suitable techniques and equipment for following, identifying, and observing birds at a distance. For example, John J. Audubon (1780–1851), a devoted student of birds, in his sketches and paintings often posed his subjects at the nest with eggs or young. But he had to kill his subjects or find dead specimens for use as models. With the advent of binoculars, photographic equipment, and hides for observation, more detailed analysis of bird behavior became feasible. Interest in birds by scientists and amateurs has not waned since Audubon's time. The number of books published about birds is enormous. In 1964 there were approximately 215 ornithological journals and about 245 additional publications containing substantial amounts of information on avian biology (Baldwin and Oehlerts, 1964).

At the turn of the century, field naturalists and nest hunters were viewed as mere amateurs, while scientific ornithology (that branch of zoology that deals with birds) focused on problems of taxonomy

1

(classification of organisms) and on morphology (form and structure) rather than behavior. By 1900 undescribed avian species were rare. As a result birds are better known taxonomically than any other class of animals. Also, more is known about the vital statistics of avian populations living in their natural environments than about any other animal populations. Taken together, these factors had important consequences for subsequent research on the behavior of birds.

The interest in avian taxonomy at the turn of the century proved to be an important factor in the development of a new scientific discipline, ethology. Ethology is the scientific study of the behavior of organisms, especially in their natural environment. The disciplines of ornithology and taxonomy came together in the work of several students of bird behavior, forming the background of ethology. Oskar Heinroth, of the Berlin Zoological Gardens, and Charles Otis Whitman and his student Wallace Craig, of the University of Chicago, were ornithologists with a background in zoology and an interest in taxonomy and the evolution of animals. They enjoyed watching animals in nature.

In the course of rearing and studying many species of birds, both Heinroth and Whitman realized that behavior could be used as a taxonomic indicator and could provide clues to the classification of organisms in much the same way that organs and structural characters were used. The notion that behavior could be used as a taxonomic tool was given much impetus by Konrad Lorenz, a student of Heinroth. According to Lorenz (1950, 1974), it was the discovery of inherited units of behavior, or fixed action patterns characteristic of each species, that started the field of ethology. Fixed action patterns are distinguished from other behaviors in several ways: (1) they are stereotyped, (2) they are shown by all members of a species, and (3) they are elicited by simple but specific external stimuli. As an outgrowth of ethological writings, many comparative studies were undertaken by ornithologists to find clues to the phylogeny (or evolutionary history) of behavior.

As indicated above, ornithologists at the turn of the century considered field naturalists to be unscientific and mere amateurs. A number of forces brought a change in this attitude. First, with the writings of the European ethologists behavior became a respectable subject for analysis by scientists interested in the classification of organisms. Second, field observations yielded information on the adaptive significance (function and survival value) of behavior, which was useful to taxonomists. Third, the publication of Nice's (1937, 1943) monographs on the Song Sparrow (*Melospiza melodia*) and Hann's (1937) study of the Ovenbird (*Seiurus aurocapillus*) stimulated the interests of both scientists and bird watchers around the world. As Skutch (1976) points out in his survey of the parental behavior of birds, most of the detailed, accurate knowledge of avian breeding has been gathered in the present century by an increasing number of dedicated amateurs and professionals who spent long, patient hours in observation. By this century, the previously distinct methods of taxonomy and field studies were beginning to merge.

Ethologists became interested in behavior as a taxonomic tool, in its adaptive and evolutionary aspects. Another group of scientists, the psychologists, were interested in universal underlying causal mechanisms and in the development of behavior. Early psychologists studied only humans. They were interested primarily in human behavior and by studying animals they hoped to find laws which applied to all species. Comparative psychology became a respectable discipline when Darwin pointed to a continuity between the behavior of humans and animals in his *Origin of Species* (1859) and again in *Expressions of Emotions in Man and Animals* (1872). Comparative psychologists often viewed animals as conveniently simplified versions of man, thereby suited to experimental analysis. Increasingly, biological concepts were used to explain both animal and human behavior. In this context psychologists viewed "instincts" as agents which moved organisms to purposive goals. Many psychologists made use of the "instincts" notion in discussing similarities and differences between animal and human behavior. The complex and highly stereotyped parental behavior of birds was among the behaviors believed to be instinctive.

The concept of instinct as an explanation for behavior was widely acceptable to the founders of psychology. William James, author of the first textbook of psychology (*Principles of Psychology*, 1890), wrote that man has more instincts than any other animal. James Watson, who proclaimed in 1913 that "psychology as the behaviorist views it is a purely objective experimental branch of natural science," relied on the concepts of instinct and learning to account for much of behavior. In his view, instincts developed during the evolution of a species, and were "a combination of congenital responses unfolding serially under appropriate stimulation." From the beginning of his career, Watson was enthusiastic about working with animals. Karl Lashley (1915), one of his most brilliant and famous students, followed up on some of Watson's observations on the choice of nest sites by Sooty Terns (*Sterna fuscata*) and Noddy Terns (*Anoüs Stolidus*). This work formed some of the data base for his later important theoretical writings on the experimental analysis of instincts (1938).

By the turn of the century the concept of instinct had become quite popular in psychology. However, in 1919 Dunlap published a general attack challenging the usefulness of the instinct notion. Watson (1919) altered his previous position and argued that the "instincts" were a substitute for true explanation of observed activity. Kuo (1924) and Lashley (1938) made important arguments on the need for further experimental analysis of instincts and for studies on the development of behavior. Experimental psychologists working primarily with rats had long been arguing these issues (Beach, 1950) when Lorenz and Tinbergen reintroduced the concept of instinct to explain species typical behavior. Although Lorenz had been publishing since 1931, the early writings were in German, and were not read by psychologists in America. Also, a great part of the material on which Lorenz based his hypotheses came from studies of bird behavior which were known to ornithologists but not to psychologists, who worked primarily with mammals. When the ideas of ethologists and psychologists met, there was often heated debate.

Daniel S. Lehrman, an avid bird watcher and student of avian behavior, launched his scientific career with a paper entitled, "A Critique of Konrad Lorenz's Theory of Instinctive Behavior" (1953). Others, including Donald Hebb (1953), Frank Beach (1955) and Theodore Schnierla (1956), challenged the value of distinguishing innate and instinctive classes of behavior from learned and acquired behaviors. One outcome of this controversy was that investigators began to analyze instinctive activities experimentally.

In the last thirty years there has been an increase in communication between different disciplines. Those whose primary interest is in understanding the behavior of animals themselves, and those who study the behavior of animals to understand man, the organization of the central nervous system, or evolution and ecological pressures, have come to see that each of these problems is interrelated. Today it is generally accepted that an understanding of behavior requires multifaceted analysis by workers in different areas. The problems of interest to taxonomists, field naturalists, ornithologists, ethologists, psychologists, physiologists, and endocrinologists increasingly overlap. It has become evident that the description and analysis of naturally occurring behavior both in the field and laboratory is a first step towards identifying significant problems for further study. There is increased integration and overlap of previously distinct areas of research in behavior.

Birds have been important in the scientific understanding of behavior. There are several reasons for the key role played by ornithological research. Birds have a number of highly specialized motor and sensory capacities and display a wide variety of adaptations to a great range of environments. They possess a great variety of highly stereotyped response which vary dramatically from species to species, thereby often providing clear-cut and striking examples of whole classes of phenomena. They are well known taxonomically. Much is known about naturally occurring avian populations. They are relatively easy to observe in the field.

There are several basic questions which can be asked about behavior: What are the underlying mechanisms? How does behavior develop within the lifespan of the individual? What function does behavior serve? How did a given behavior evolve in the phylogenetic history of the species? These questions have been fruitfully asked about avian parental behavior.

MECHANISMS UNDERLYING PARENTAL CARE

Many early experimenters directed attention to the interaction of stimulus, endocrine, and response factors in the parental behavior of birds (Maier and Schnierla, 1935; Armstrong, 1942, 1964). A number of features make avian reproductive behavior particularly suited to this type of analysis. There are striking changes in the state of both primary and secondary sex structures associated with an array of stimulus factors such as seasonal changes in light, temperature, rainfall, and availability of nest, food, and mate. In the absence of more sophisticated chemical techniques, these target organ changes provide good indirect measures of changes in hormonal state. The parental behaviors of birds are interesting to observe and are sufficiently elaborate and repetitive to yield to simple description and categorization. Often the participation and cooperation of both male and female partners over an entire reproductive cycle and even over a lifetime is involved. The intricate coordination between behavior (of male, female, and young), endocrine state, and stimulus conditions during reproduction presents a complex model well suited to the study of mechanisms underlying behavior.

One might conceive of parental behavior as involving only nest building, incubation of eggs, and care of young. Conceptually and functionally, these phases are distinct from each other and from the preceding phases of reproduction including migration, courtship, and mating. However, these apparently different behaviors may be mediated by related underlying causal mechanisms.

Rowan (Paper 1) was the first to experimentally show a link between increasing day length, migration and breeding, and seasonal gonadal growth. His finding opened the way for several new lines of research. Is day length but one of many environmental cues which influence the onset of breeding seasons? How does information about day length (or other cues) reach the nervous system? Which neural pathways and which photoreceptors are involved? Is there a fluctuation in responsiveness to environmental cues? Or, is there evidence of an endogenous "biological clock" with a circadian or a circannual periodicity? Finally, is there a causal relationship among hormone secretion, migration, and breeding?

At some level, many of these problems were known well before

Rowan's work. Men have realized since time immemorial that reproduction is regulated in various ways by seasonal changes. It was also known that once breeding condition was achieved, the timing of subsequent stages was influenced by external factors. For example, Francis Herrick (1901, 1911) found that nest building in birds varied according to the weather, the nature of the nest site, and the condition of the bird. Craig (1913) found that ovulation in pigeons (*Columba livia*) was delayed if the birds were disturbed. On the other hand, a mirror in the cage, or the experimenter stroking the bird's head induced ovulation. Taibell (1928) successfully induced incubation in a male domestic turkey (*Meleagris gallopavo*) by fastening it down to a clutch of eggs. Clearly, stimuli in the environment influence the display of parental behavior.

That internal factors have an important influence on behavior was also well established. Berthold (1849) had shown that testicular secretions could support male sexual behavior in chickens (*Gallus gallus*). Goodale (1918) found that a capon (castrated male) would brood young almost like a hen. These experiments clearly implicated hormones in sexual and parental behavior. Thus, the phenomena were known, but the physiological mechanisms whereby stimulation from a nest or young, or from environmental factors such as day length or availability of food and nesting material could act to affect hormone secretion and behavior were not known.

In 1955 Harris published his classic monograph demonstrating conclusively that the link between the hypothalamus and the pituitary was of neurovasculatory (involving neural and vascular systems) nature. Here was the missing link between the environment, the endocrine system, the nervous system, and behavior. External stimuli affect receptors and reach appropriate hypothalamic brain areas to modify the production of neurohormones (hormones synthesized by nerves) by the brain. Neurohormones reach the pituitary by a local blood supply and influence the production of pituitary trophic hormones which act on the gonads. Gonadal hormones in turn influence behavior in ways which had long been established. Harris' work made understandable the ways in which external stimuli and stimulation which results from a behavioral response can in turn modify hormone secretion and further change behavior.

The effectiveness of social and sexual stimuli in integrating the behavior of the reproductive cycle, which was pointed out by Craig (Paper 9), Matthews (Paper 10), and others, can be more broadly interpreted in view of what is now known about the neuroendocrine system. There is a complex feedback system where the environment modulates endocrine secretion, and the endocrine system modulates responsiveness to the environment. In some ways this conception formed the starting point for the important research programs on avian parental behavior by Lehrman and Hinde. Lehrman, working with Ring Doves, and Hinde, working with canaries (*Fringilla serinus*), have shown that stimuli from the male and the nest induce hormone changes which influence sexual responsiveness, nest building, and incubation behavior. In doves stimuli from the eggs induce prolactin secretion, which in turn influences parental care of eggs and of young, which in turn increase prolactin secretion. Fragmentary evidence on hormonal-behavioral integration of the breeding cycle is available on many species, but the understanding of the parental behavior of doves and canaries is most complete. (See book by Collias and Collias, 1976, in this series for review article by Lehrman.)

Before modern neuroendocrinology developed, how were the findings of Berthold, Craig, Herrick, and others interpreted, and what implications for behavioral organization were seen? Social and sexual stimulation were seen as contributing to generalized emotion or excitement. Selous (1909–1910) concluded that greyhens (*Tetrao tetrix*) come to the lek (display grounds where birds that court in assemblies gather) for the purpose of being aroused sexually, and in the absence of sufficient stimulation, they leave without mating. He further argued that the origin of nest building lies in the movements of the birds in sexual frenzy during copulation (Selous, 1901–1902, 1095, 1916). Selous was not alone in postulating the existence of excitement or energy-cumulating mechanisms. McDougall (1923), a founder of psychology, believed that external stimuli liberate energy on the afferent side of the nervous system, and "sluice gates" hold back the energy. Only when appropriate stimuli are received does the appropriate gate open, and all available energy flows to this response. Over the years, the terminology has changed, but the same problems attract experimenter's attention. Cognate terms, such as emotion, are avoided and "motivation" or simply description of the behavior is cautiously substituted. Selous' theory on the origin of nest-building movements is seen as simplified; however his main point, that ritualization or change in function of a response pattern occurs in evolution, is an accepted notion.

The problem attacked by Selous, on the nature and patterning of complex sequences of responses, continues as a central theme in the study of behavior. One advance has been in the study of neuroendocrine integration. A breakthrough came with the work of Konrad Lorenz and Niko Tinbergen in the mid-1930s. Lorenz and Tinbergen developed energy models of behavior very similar to that of McDougall. Lorenz postulated the existence of "reaction specific energy," analogous to gas being pumped into a container (1937) or liquid in a reservoir (1950). Only under appropriate conditions could this energy be released, and the behavior expressed. Tinbergen (1951) pictured a hierarchy of nervous centers each of which has properties similar to the reservoir in Lorenz's model. The implied locus of Lorenz's (1935) energy storing mechanism, and the locus at which hormones act to induce stereotyped behavior was in the central nervous system. As techniques became available for direct studies of the brain, the energy models of behavior were tested.

Neural Basis of Parental Behavior

Relatively little work has been done on the role of the brain in the regulation of parental behavior. The commonly used technique of destroying selected portions of the brain by lesioning has not been applied in avian research on parental behavior. Another technique for studying neural tissue involves electrical stimulation of the brain (ESB). Von Holst and St. Paul (1963) pioneered in the use of ESB to study social behavior of the chicken. They reported that with electrical stimulation of the brain ". . . nearly all known movements can be elicited, serving orientation and the needs of the body directed towards enemies, rivals, the sex partner, and the young, with all the associated calls." Phillips and Youngren (1971) sampled 2,600 neural loci in chickens and ducks and found that ESB usually elicited only fragments of more complex behavior. They suggest that complex sequences of behavior depend on the presence of appropriate environmental stimuli along with ESB.

The above studies emphasize the neural control of motor responses, or the efferent component. Another aspect of neural control, and one which has received a good deal of attention, is the afferent component whereby environmental stimuli reach the nervous system to influence the display of species typical behavior. This kind of question, on how stimuli in the natural environment are encoded and how they elicit species typical behavior, received a major new emphasis with developments in ethology. Although ethologists were concerned with a naturalistic approach to the study of behavior, a large part of classical ethology was concerned with the physiological mechanisms of behavior. Lorenz pointed out that many animals recognize biologically significant objects in their environment, such as mate, young, prey, predator, by using a small set of sensory cues. Under suitable conditions, these stimuli release the behavior appropriate for each situation. The basic ideas of ethological theory are demonstrated by the classic study of egg-retrieving behavior in the Greylag Goose (*Anser anser*) described by Lorenz and Tinbergen (Paper 24). The goose lays her eggs in a shallow nest on the ground. If one egg rolls out of the nest, the goose responds in a characteristic way. First she looks at the egg and then looks away repeatedly. Then she stretches her neck as far as possible towards the egg. If she cannot reach the egg, she eventually gets out of the nest, neck still outstretched, and walks toward the egg. She touches the egg with lower beak and rolls the egg back to the nest in a stereotyped way. The neck muscles quiver and the neck contracts and at the same time the beak moves from side to side and the goose walks slowly backwards. In this way the egg is rolled in a fairly straight line, back to the nest. If the egg slips away en route to the nest, the goose continues the back and forth movements of the neck all the way back to the nest, as though the egg were still there. In the case of the Greylag Goose, egg-rolling behavior is released by the sight of the egg outside of the nest. Lorenz (1935) postulated that a releasing stimulus (egg out of the nest) triggers an "innate releasing mechanism" (IRM) which releases a fixed action pattern (egg rolling). A greylag goose recognizes an egg outside the nest because the stimulus it presents matches the internal release image of the egg. The site of the IRM mechanism was thought to be in the central nervous system (Schleidt, 1962; Konishi, 1971). In 1935, when Lorenz put forth his theory, there were no supporting neurophysiological data, although generally neurophysiologists believed that the brain analyzed sensory information which passed the sense organs unprocessed. Critics of Lorenz argued that there was no need to invoke a special central mechanism if the behavior could be explained by the specific response of the sense organ (Schnierla, 1952; Lehrman, 1953).

There is a wealth of supporting data, largely on insects, indicating that peripheral sense organ sensitivity can account for behavioral responses to specific stimuli (Marler, 1961). In the analysis of the parental behavior of birds, behavioral and physiological data have not yet been examined in conjunction to see whether there is a match between peripheral sensitivity and behavioral response. Some advances have been made. For example, Baerends (1957) analyzed the characteristics of the stimulus which is effective in eliciting egg rolling. He used wooden dummy eggs with different egg characteristics to see which were important in releasing egg-retrieving in the herring gull (*Larus argentatus*). He found that large eggs were more effective than small eggs, especially if the size of the model was larger than the normal egg. Speckling was an important feature, and eggs bearing small darks spots were most effective. The color and shape of the egg were less important than speckling and size. He suggests that the concept "releasing mechanism" encompasses a complex of factors localized in diverse parts of the nervous system and not in one locus of the brain. Thus far, the characteristics of the effective stimuli from the egg have been identified, but the sensitivity of the visual system to these stimulus characteristics is not known. Even if a match was found between the peripheral sensitivity of the eye and the animal's response, this would not account fully for the behavior. Beer (1962) showed that over the course of a breeding cycle there are changes in the readiness to roll an egg back into the nest. Changes in responsiveness to constant stimuli can not be accounted for by peripheral processing mechanisms alone. Processing of sensory input probably occurs at many stages. How sensory input leads eventually to a motor response is a continuing empirical problem for analysis. The naturalistic observations of the ethologists, and physiological mechanisms they inferred, have generated widespread interest and enthusiasm and have proven to be useful models, facilitating the understanding of neurobehavioral mechanisms in development (Konishi, 1971).

To conclude, the neural basis of avian parental behavior is largely unexplored. In recent years, stereotaxic atlases of the brain of the pigeon (*Columba Livia*; Karten and Hodos, 1967), the canary (*Serinus canaria*; Stokes et al., 1974) and the Ring Dove (*Streptopelia risoria*; Vowles et al., 1975) have been published. Given the fact that avian parental behavior depends primarily on visual, auditory, and hormonal cues which are readily accessible to the experimenter, this is likely to be an area of increased research activity in the future.

Hormonal Basis of Parental Behavior

Another way to study the role of the brain in regulating behavior is to analyze the neural effect of hormones. What are the sites of action in the brain of hormones which initiate and maintain parental behavior? Komisaruk (Paper 21) showed that hypothalamic progesterone implants initiate incubation behavior in (gonadally intact) male and female doves. Stern and Lehrman (1969), and Stern (1974) showed that castrated males had to be treated with either estrogen or testosterone along with progesterone in order to induce incubation behavior. By removing known endogenous sources of progesterone (Silver and Buntin, 1973; Silver and Feder, 1973; Cheng and Silver, 1975), and by measuring blood levels of progesterone throughout a breeding cycle (Silver et al., 1974), it has been shown that progesterone normally plays a role in the onset of incubation in the female but not in the male dove. While it is of some interest that a given neural manipulation, such as implantation of progesterone, can influence the display of a particular behavior, this fact in isolation is not proof of the neural basis of the *naturally* occurring behavior. To clarify the mechanisms underlying behavior, many different types of experiments must be done, as described below.

The earlier literature on the role of hormones in mediating parental behavior of birds has been extensively and thoroughly reviewed by Lehrman (1961) and Eisner (1960). A more recent review by Lofts and Murton (1973) on reproduction in birds emphasizes gonadal physiology and endocrinology, and briefly covers parental behavior.

10

These articles are widely available and no purpose would be served in re-reviewing the topic here. However, some of the conclusions reached in the older literature may not be valid. First, many of the experiments using exogenous gonadal hormones to study parental behaviors were performed on intact rather than on gonadectomized animals. Second, much of the evidence is based on correlational studies which relate gross seasonal behavioral and endocrine changes. Finally, three lines of evidence are required to ascertain that a particular hormonal change normally plays a role in the display of a particular behavior. When the known source of the hormone is removed, the behavior should not be displayed (or be displayed at a lower frequency). When the hormone is replaced in a gonadectomized or hypophysectomized animal, the behavior should reappear. During a reproductive cycle, a change in hormone levels should occur, as determined by assay, at about the time a change in the behavior occurs. Even when all three lines of evidence implicate hormones in producing behavioral changes, it must not be assumed that the only site of action of hormones is in the central nervous system.

The earlier literature implied that apparently complex behaviors of animals were the outcome of many simple stimulus-response connections. Hormones could act by changing the stimuli coming from the body surface thereby altering responses. For example, Maier and Schnierla (1935), commenting on Whitman's description of mating in the pigeon, write the following: ". . . the tail raising, so important for the actual mating act, is specific to this period and may depend upon a local irritation of cloacal surfaces for which the condition of the female's genital tissues is responsible." In other words, the peripheral irritating effect of hormones on the cloaca may stimulate mating. An argument of this type on the relative importance of central and peripheral factors is central to the debate between Lehrman and Riddle on the role of prolactin as the hormone of parental behavior. Riddle noted that the pituitary principle that induced lactation in mammals was the same as that which caused growth of the pigeon's crop gland. This discovery provided a useful bioassay (measure of hormone level by examination of a biological reaction it produces) for prolactin and hastened its chemical isolation. In fact, prolactin was the first pituitary hormone to be isolated as a pure protein.

After the pioneering work of Riddle and his colleagues (Riddle and Bates, 1939; Riddle and Lahr, 1944) it was widely believed that prolactin was crucial in the control of avian and mammalian parental behavior (Riddle, 1935). Riddle et al. (1935) showed that prolactin is effective in inducing incubation in the hen while progesterone is not. Riddle and Lahr (1944) also found that Ring Doves bearing subcutaneous progesterone implants would incubate. At autopsy after fourteen days of incubation the doves had enlarged crops indicating prolactin secretion. This led Riddle and Lahr to argue that progesterone had stimulated prolactin secretion which in turn induced incubation. However, Lehrman (1961), Lehrman and Brody (1961), Lehrman et al. (Paper 11) reported that exogenously administered progesterone is more effective than prolactin in inducing incubation; and furthermore, there is no evidence that prolactin is secreted in doves at the onset of incubation (as indicated by crop growth). Crop weight, a sensitive index of prolactin secretion, increases from about 0.5 grams to about 3.0 grams from day 7 of incubation to the day of hatching on day 15 of incubation; a further growth to about 4.5 grams occurs by day 5 of brooding (Hansen, 1966; Silver and Barbiere, 1977). Lehrman and Brody (1964) suggested that prolactin may play a role in the maintenance but not in the initiation of incubation. However, birds treated with pharmacological agents which block crop growth by preventing prolactin secretion continue to incubate (Silver and Buntin, 1973), negating a necessary role for prolactin in the maintenance of incubation. Certainly prolactin is secreted by doves and pigeons and by many other animals at the same time that parental behavior is displayed. How prolactin influences parental behavior is not yet clear.

The hormonal basis of incubation behavior in birds is a problem of continuing interest for several reasons. First, when hens become "incubatory" or "broody" they cease to lay eggs. It would save the poultry industry a great deal of money to eliminate this parental response in egg-producing hens. From a scientific perspective, the types of parental behavior shown by the orders and families of birds of the world are extremely varied. Are the underlying endocrine mechanisms and relevant hormones equally varied, or do a few hormones produce this diversity of response? In the regulation of mating behavior, it seems that testosterone in the male and estrogen or estrogen and progesterone in the female produce the particular species typical behavior of the avian and mammalian species which have been studied. As further research is done, we should have greater insight into the evolution of hormone-behavior relationships in the regulation of parental behavior.

DEVELOPMENT OF PARENTAL BEHAVIOR

Why do weaver birds (Ploceidae) build elaborate communal nests? Why do pigeons and doves (Columbidae) regurgitate food to their young? Observations of parent birds with their young often give the viewer the strong impression that the behavior is performed correctly and in the appropriate context the first time it occurs. Probably because of this powerful initial impression, remarkably few experiments and remarkably many hypotheses have been presented on the development of avian parental behavior.

At one time, it was believed that all complex behavior was composed of arrays of simple reflexes. A reflex is a stereotyped and fixed response to a stimulus and involves a simple neural mechanism. In this view, nest building, care of eggs and young, and even learning and thinking were considered to be nothing but sequences of simple reflexes. Each component of the behavior had an invariable relationship between stimulus and response. For example, Bruckner (1933) studied the reaction of the domestic fowl to the distress call of the chick. If a chick was placed under a glass bowl so that it could be seen but not heard, it was ignored by the mother. If, however, it was placed behind a fence so that it could be heard but not seen, the hen quickly responded. Although the hen seems to be protecting the young and seems to be aware of danger to her brood, the defense response is not actually elicited by the young in danger, but is based on a simple stimulus (distress call)-response relationship.

One might think that any behavior could be analyzed into its component reflexes, but this approach has not proven successful. Responses are variable even in a fixed stimulus situation and the influence of more complex neural processes than reflexes is evident. For example, Ring Dove parents which have gone through a breeding cycle including nest building and incubation will feed their own young, or foster young of the same approximate age. However, if a male or female at some other stage of the cycle is presented with squab begging for food, it will ignore or even kill them. Parental feeding can not be explained in simple stimulus (young begging)-response (parent feeding) terms.

Another attempt to explain complex and apparently purposive behavior involved the establishment of a category called "instinctive behavior" defined by certain criteria. Instincts were (1) unlearned, (2) characteristic of the species, and (3) adaptive. According to Lorenz (1950) species-typical behaviors were instinctive (according to the above criteria) and were distinct from all other behavior. Tinbergen referred to these stereotyped behaviors as *fixed action patterns*. In the early 1950s the development of species-typical adaptive behavior became the focal point of debate between ethologists and comparative psychologists (Lehrman, 1953). The debate was directed to the question "Is the behavior innate or is it learned?" Ethologists argued that species-typical fixed action patterns were innate, occurred when a particular stimulus released an innate releasing mechanism in the brain, and that fixed action patterns developed centrally without any control by feedback from the environment to the senses. Comparative psychologists argued that sensory input from the environment and learning are essential for the development of species-typical behavior. Both groups looked to the same types of experiments to prove their point, and often drew different conclusions from the same results. These studies included deprivation experiments, behavioral genetics, and experiments testing central versus peripheral control of behavior.

It was assumed that if an organism had no previous experience in performing a given behavior, and that if that behavior was performed correctly when the opportunity first arose, then the behavior must be innate or unlearned. Hinde (1958) found that a chaffinch (*Fringilla coelebs*) kept in a cage without nesting material plucked its own feathers to use as nesting material. However, in the following year when it was kept in an aviary with plenty of nesting material, it still plucked its own feathers. Previous experience without nesting material had modified later nest-building behavior. In the same paper Hinde reported that naive canaries that had never had an opportunity to manipulate nesting material would, when given grass for the first time, pick it up and carry it to the nest pan within a minute or so. In this case, nest building was not altered by previous deprivation of nest material.

A slightly different approach to the question of how behavior develops involves studying experienced and inexperienced animals. Lehrman (1955) reported that the latency of doves to feed their young after hatching was longer in the first breeding than in the second. Craig (1918) reported that a female Ring Dove which had no previous experience with a nest, nesting material, or a mate will lay an egg on the floor, but an experienced dove will withhold the egg until a nest is available.

These experiments on the effect of restrictive environment and on the role of previous breeding experience provide merely a first step forward in the analysis of which factors do and which factors do not influence the development of parental behavior. They do not address the problem of whether the behavior is innate or learned. Other approaches yield different kinds of clues and force a reevaluation of the innate-learned dichotomy.

Studies indicating that genetic factors influence behavior have been taken as proof for the inheritance of behavior. It has long been known that strains of domestic hens differ with respect to broodiness. Goodale (Paper 26) noted that only 2 to 3 percent of White Leghorn hens ever became broody compared to 93 percent of Rhode Island Red hens. Goodale et al. (1920) showed that a nonbroody strain could be developed from a broody one by selective breeding. What is (are) the mechanism(s) underlying strain differences? Riddle et al. (1935) found that prolactin injected into laying hens induced incubation behavior more readily in a broody than in a nonbroody race. Byerly and Burrows (1936) found that the pituitary glands of genetically broody hens contained more prolactin than did those of nonbroody birds. Apparently broody and nonbroody races differ in responsiveness to prolactin, and also in prolactin production.

One of the most vivid examples of a genetic contribution to a parental response comes from the work of Dilger (1962) with lovebirds (Agapornis). Some lovebirds transport nest material back to the nest by tucking several strips into their rump feathers and flying back, while others simply carry the strips in their bill. Dilger crossed two species which differed in this respect. The hybrid offspring made the preliminary tucking movement and then carried the strips back to the nest in the bill. Presumably, the fixed action patterns involved in nest-building movements were inherited and were performed even though they served no function in the hybrid offspring. It appears that both behavioral and physiological characters are related to the expression of parental behavior and that both are susceptible to alteration by selective breeding.

The question of whether genes act in some direct way or by altering learning ability, responses to stimuli, or in some other way, is a problem not addressed by these experiments on behavior genetics. Nevertheless, the argument of central versus peripheral control was considered basic to the question of how behavior develops. The implication was that if behavior is centrally organized it is innate, and if it is influenced by peripheral factors it depends on experience. A number of experiments emphasized the central organization of behavior.

The occurrence of behavior at apparently inappropriate times was taken as evidence for central control. Howard (1929) showed that many birds sit in the nest before they have laid eggs. Presumably the internal or central state, and the relevant external cues were out of synchrony. As mentioned earlier, Riddle et al. (1935) showed that broody races of white leghorn respond with better incubation behavior than do nonbroody races to injections of prolactin, suggesting that the intensity of the reaction to exogenous hormones depends on innate capacities of the central nervous system. Furthermore, prolactin-injected cocks show the complete maternal behavior pattern typical of the brood hen, including brooding the chicks, leading them to food, and defending them from predators. This too implied that hormones were acting directly on central nervous mechanisms. Lorenz (1937) was the first to suggest that the central nervous system itself produced impulses which acted as specific causes of instinctive behavior patterns, and criticism was soon directed to the notion that behavior could develop centrally without any environmental influence.

Lehrman (1955) argued that hormonal modification of behavior does not necessarily imply a site of action in the nervous system. He injected prolactin into a group of doves, and then, before testing them for regurgitation-feeding of young, anesthetized the crop by injecting a local anesthetic directly into the crop wall. These birds showed a sharp reduction of regurgitation-feeding compared to a control group injected in the skin of the back. Lehrman concluded that prolactin has

15

important peripheral effects (apart from any direct effect on the central nervous system) which are relevant to the elicitation of parental behavior. These controversies once again focused attention on the need for experimental studies of development with several resulting insights.

Michel (1971) pointed out that "previous experience" need not mean previous performance of a particular response. If experienced doves are injected with progesterone for seven days and then introduced to a cage containing a nest with eggs, they incubate. Inexperienced doves do not (Lehrman and Wortis, 1960). Michel found that it was not necessary that the birds actually have the experience of a complete incubation period or of rearing young in order to respond to exogenous hormones as do experienced birds. If doves were permitted to court, nest build, and stay with their nests for only two days, the period it takes to lay the two-egg clutch, they later responded by incubating eggs after seven days of progesterone injection, even though they had never undergone a complete breeding experience.

In recent years a wide range of events have been shown to have reinforcing properties thereby allowing for learning to occur the first time that a behavior is shown. Stevenson-Hinde (1973) showed that nest building is reinforcing for canaries. This implies that adaptive, species-typical behavior, shown by the mature animal with no previous experience, can be understood in the same way as are other types of learned behavior. It is clear that the problem has shifted from attempting to dichotomize innate behavior and learned behavior, to experimental analysis aimed at understanding how behavior develops in the individual.

EVOLUTION OF BEHAVIOR

Success in each phase of parental care is achieved in widely different ways by different avian species. Ryves (1944) and Van Tyne and Berger (1971) list seven categories to which species may be assigned according to the role of the sexes in building the nest. These are: (1) both sexes build the nest; (2) the female builds but only the male provides the material; (3) the female builds without help from the male; (4) the female builds but both sexes gather the material; (5) the male builds but the female provides the material; (6) the male alone builds the nest; (7) no nest is built.

Which parent cares for the young and/or the eggs of birds also shows wide variation. The following patterns of participation are found: (1) female only; (2) male only; (3) both parents; (4) neither parent; (5) juvenile helpers; (6) unmated helpers; (7) mutual helpers of breeding birds which assist each other. Furthermore, whether or not the male or female incubates eggs does not provide information as to whether care of young after hatching will be provided by that partner.

The behavior of adults with the young is not restricted to brooding, but may include nest sanitation, or removal of feces and egg shells; cooling nestlings by providing shade in direct sunlight; feeding of fledglings or leading them to food; and protection from predators with alarm calls or injury feigning.

How did the parental behavior of birds evolve? Why is it that some birds show no parental behavior—no nest building, no incubation, no care of young. At the other extreme are species in which both parents participate in each of these phases of parental care. How did these differences in behavior among species arise and how can they be studied? The techniques used by systematists to uncover phylogenetic relationships among animals are also used to clarify the evolution of behavior. Darwin's theory of evolution by natural selection contained two arguments. First, species are related to each other by descent from common ancestors. Second, the evolution of species occurs by the process of natural selection. Evidence supporting the theory comes first from studies of phylogeny, or the evolutionary history of animals, and second from studies of function, adaptive significance, and survival value. Theoretically, the easiest way to study the evolutionary history of behavior would be to go back in time and to observe directly how behavior patterns have changed in response to selection pressures. Because this direct approach is not feasible, and because fossil records of behavior do not exist, indirect methods must be used. If the behavior of closely related animals is similar, but shows a spectrum of variation from species to species, this can provide clues to how a particular response pattern of a particular species arose. Studying adaptive radiation or divergence among species in this way yields information on the evolutionary history of behavior. The study of convergence or the analysis of similarities in behavior of distantly related forms living under similar environmental conditions provides further clues. Both of these methods are basic to evolutionary studies.

The earlier emphasis of comparative studies lay in elucidating apparently inherited patterns of response (fixed action patterns). It was assumed that if a behavior pattern is widespread in a group of related animals, then it is probably innate and similar to an ancient behavior of an ancestor of present day species. More recently evolved behaviors are likely to be less widespread. The study of living species provides insights into the evolutionary sequence. Major research efforts were directed toward discovering behavioral homologies (behaviors which are similar in origin but not necessarily in function) as clues to the evolution of response patterns.

Cullen (Paper 7) shifted the emphasis on the significance of behavioral homologies as clues to inherited behaviors in her studies of the influence of environmental factors on all aspects of social behavior. Kittiwakes (*Rissa tridactyla*), unlike other gulls, nest on narrow ledges of cliffs. Predators are rare and Kittiwakes rarely emit the alarm call. The territory around the nest is very small. Egg shell removal is not performed. Young and mate are fed directly rather than by dropping food on the ground as in other gulls. Cullen showed how adaptation to cliff nesting in the Kittiwake modified all aspects of social behavior and parental care. Her ideas were soon confirmed and extended by studies of other groups (Crook, 1964, 1965; Lack, 1968).

Hypotheses about the evolution and adaptive significance (including survival value and function) of behavior are difficult to prove. Yet they are important in integrating many apparently unrelated facts. For example, why do Black-headed gulls (*Larus ribidundus*) remove their egg shells (Tinbergen et al., Paper 27), while Kittiwake gulls do not? Black-headed gulls live on flat ground while Kittiwakes nest on narrow ledges. The selection pressures of predators is different in the two groups and egg shell removal apparently has no survival value for Kittiwakes.

Field studies of individually marked birds may provide insight into early stages of speciation (evolution of species). Norton-Griffiths (Paper 29) discovered that the techniques used by oystercatchers (*Haematopus ostralegus*) to open mussels are uniquely suited to the particular beds in which they feed. On some beds, shells are opened by hammering on the ventral surface of mussels, which are carried from the rocks to an area of suitable hard sand. On other beds oystercatchers feed primarily on mussels which are covered with sea water by stabbing downward with the bill into the gape of slightly open mussels which are still attached to rocks. By observing individual oystercatchers and their young. Norton-Griffiths (1969) was able to show that a given bird uses one or the other strategy in eating mussels. Furthermore, the young of "hammerers" feed by hammering while the young of "stabbers" feed on mussels by stabbing, and hammerers presumably mate only with hammerers and stabbers with stabbers. In time these populations could evolve other behavioral differences and become different species.

Because of their high visibility, wide distribution, the abundance of information on their life cycles and breeding habits, birds have played a prominent role in the development of model systems in the study of the evolution and adaptation of behavior. The most prominent and exciting development in this area is the recent blending of interests and ideas on ecology and population regulation, derived from the long-term studies of Lack (Papers 3 and 13; 1954, 1966, 1968, 1971) and those of Tinbergen (Paper 27; 1953, 1954, 1959, 1963, 1967) on the evolution and adaptive significance of behavior. These fields have been named "behavioral ecology" or "ecoethology" (McKinney, 1973), and are characterized by an overlap of research interests in fields which are primarily ecological with those which are primarily behavioral. Comparative studies of the diverse behaviors of birds and their adaptation to a wide variety of ecological niches continue to provide insights into the behavior of parent birds with their young.

REFERENCES

Armstrong, E. A. 1942. *Bird Display and Behaviour.* London: Lindsay Drummond Limited.

Armstrong, E. A. 1964. *The Ethology of Bird Display and Bird Behaviour.* New York: Dover Publications, Inc.

Baerends, G. P. 1957. The ethological concept "releasing mechanism" illustrated by a study of stimuli eliciting egg-retrieving in the herring gull. *Anat. Rec.* 128: 518–519.

Baldwin, P. A., and Oehlerts, D. E. 1964. Studies in biological literature and communication, No. 4: The status of ornithological literature. Philadelphia: Biological Abstracts, Inc.

Beach, F. A. 1950. The snark was a boojum. *Amer. Psychol.* 5:115–124.

Beach, F. A. 1955. The desent of instinct. *Psych. Rev.* 62:401–410.

Beer, C. G. 1962. The egg-rolling of Black-headed gulls *Larus ridibundus. Ibis* 104: 388–398.

Berthold, A. A. 1849. Transplantation der Hoden. *Arch. Anat. Physiol. Wiss. Med.* 16:42.

Bruckner, G. H. 1933. Untersuchungen zur Tierpsychologie insbesondere zur Auflösung der Familie. *Z. Psychol.* 128:1–110.

Byerly, T. C., and Burrows, W. H. 1936. Studies of prolactin in the fowl pituitary. II. Effects of genetic constitution with respect to broodiness on prolactin content. *Proc. Socl. Exp. Biol., N.Y.* 34:844–846.

Cheng, M. F., and Silver, R. 1975. Estrogen-progesterone regulation of nest building and incubation behavior of female ring doves (*Streptopelia risoria*). *J. Comp. Physiol Psychol.* 88:256–263.

Collias, N. E., and Collias, E. C. 1976. *External Construction by Animals.* Vol. 4 Benchmark Papers in Animal Behavior. Stroudsburg, Pa.: Dowden, Hutchinson & Ross, Inc.

Craig, W. 1909. The expression of emotion in the pigeons. I. The blond ring dove (*Turtur risorius*). *J. Comp. Neurol. & Psychol.* 19:29–80.

Craig, W. 1913. The stimulation and the inhibition of ovulation in birds and mammals. *J. Anim. Behav.* 3:215–221.

Craig, W. 1918. Appetites and aversions as constituents of instincts. *Biol. Bull.* 34: 91–107.

Crook, J. H. 1964. The evolution of social organization and visual communication in the Weaver birds (Ploceinae). *Behav. Suppl.* 10.

Crook, J. H. 1965. The adaptive significance of avian social organizations. *Symp. Zool. Soc. London* 14:181–218.

Cullen, E. 1957. Adaptations in the Kittiwake to cliff-nesting. *Ibis* 99:275–302.

Darwin, C. 1859. *On the Origin of Species.* London: John Murray.

Darwin, C. 1872. *The Expression of the Emotions in Man and Animals.* London: John Murray.

Dilger, W. C. 1962. *The Behavior of Lovebirds. Scientific American,* 89–98.

Dunlap, K. 1919. Are there any instincts? *J. Abnorm. Psychol.* 14:35–50.

Eisner, E. 1960. The relationship of hormones to the reproductive behaviour of birds, referring especially to parental care: a review. *Anim. Behaviour.* 8:155–179.

Goodale, H. D. 1918. Feminized male birds. *Genetics* 3:276–299.

Goodale, H. D., Sanborn, R., and White, D. 1920. Broodiness in domestic fowl. *Bull. Mass. Agric. Exp. Sta.,* No. 199:93–116.

Grosvenor, C. E., and Turner, C. W. 1958. Assay of lactogenic hormone. *Endocrinol.* 63:530–534.

Hann, H. W. 1937. Life history of the oven-bird in Southern Michigan. *Wils. Bull.* 49:145–237.

Hansen, E. W. 1966. Squab-induced crop growth in ring dove foster parents. *J. Comp. Physiol. Psych.* 62:120–122.

Harris, G. W. 1955. *Neural Control of the Pituitary Gland.* London: Edward Arnold and Company.

Hebb, D. O. 1953. Heredity and environment in mammalian behaviour. *Brit. J. Anim. Behaviour.* 1:43–47.

Herrick, F. H. 1901. *The Home Life of Wild Birds.* New York: Knickerbocker Press.

Herrick, F. H. 1911. Nest and Nest-building in Birds. *J. Anim. Behav.* 1:159–192, 244–277, 336–373.

Hinde, R. A. 1958. The nest-building behaviour of domesticated canaries. *Zool. Soc. (London) Proc.* 131:1–48.

Holst, E. von, and Saint Paul., U. von. 1963. On the functional organisation of drives. *Anim. Behaviour.* 11:1–20.

Howard, H. E. 1929. *An Introduction to the Study of Bird Behavior.* Cambridge: Cambridge University Press.

James, William. 1890. *Principles of Psychology.* New York: Henry Holt & Co.

Karten, H. S. and Hodos, W. 1967. A Stereotaxic Atlas of the Brain of the Pigeon (*Columba livia*). Baltimore, Md.: The Johns Hopkins Press.

Konishi, M. 1971. Ethology and neurobiology. *Am. Scientist* 59:56–63.

Kuo, Z. Y. 1924. A psychology without heredity. *Psychol. Rev.* 31:427–451.

Lack, D. 1954. *That Natural Regulation of Animal Numbers.* London: Oxford University Press.

Lack, D. 1966. *Population studies in Birds.* London: Oxford University Press.

Lack, D. 1968. *Ecological Adaptations for Breeding in Birds.* London: Methuen.

Lack, D. 1971. *Ecological Isolation in Birds.* Oxford: Blackwell's.

Lamb, M. E. 1975. Physiological mechanisms in the control of maternal behavior in rats: A review. *Psych. Bull.* 82:104–119.

Lashley, K. S. 1915. Notes on the nesting activities of the noddy and sooty terns. *Carnegie Instn. Publ.* 7:61–83.

Lashley, K. S. 1938. Experimental analysis of instinctive behavior. *Psychol. Rev.* 45: 445–471.

Lehrman, D. S. 1953. A critique of Konrad Lorenz's theory of instinctive behavior. *Quart. Rev. Biol.* 28:337–363.

Lehrman, D. S. 1955. The physiological basis of parental feeding behavior in the ring dove (*Streptopelia risoria*). *Behaviour* 7:241–286.

Lehrman, D. S. 1961. Hormonal regulation of parental behavior in birds and infra-human mammals. In *Sex and Internal Secretions,* ed., W. C. Young. Baltimore: Williams and Wilkins.

Lehrman, D. S., and Brody, P. 1961. Does prolactin induce incubation behavior in the ring dove? *J. Endocrinol.* 22:269–275.

Lehrman, D. S., and Brody, P. N. 1964. Effect of prolactin on established incubation behavior in the ring dove. *J. Comp. Physiol. Psychol.* 57:161–165.

Lehrman, D. S., and Wortis, R. P. 1960. Previous breeding experience and hormone-induced incubation behavior in the ring dove. *Science* 132:1667–1668.

Lofts, B., and Murton, R. K. 1973. Reproduction in Birds. In: *Avian Biology*, Vol. III. (Eds: D. S. Farner and J. R. King) New York: Academic Press.

Lorenz, K. 1935. Der Kumpan in der Umwelt des Vogels. *J. Orn., Lpz.* 83:137–213, 289–413.

Lorenz, K. 1937. The companion in the bird's world. *Auk* 54:245–273.

Lorenz, K. 1950. The comparative method of studying innate behaviour patterns. In *Soc. Exp. Biol. Symp. Physiological Mechanisms in Animal Behavior* 12, pp. 221–268. New York: Academic Press.

Lorenz, K. Z. 1974. Analogy as a source of knowledge. *Science* 185:229–234.

Lorenz, K., and Tinbergen, N. 1938. Taxis und Instinkthandlung in der Eirollbewegung der Graugans. *Z. Tierpsychol.* 2:1–29.

Maier, N. R. F., and Schnierla, T. C. 1935. *Principles of Animal Psychology.* New York: McGraw-Hill Book Company, Inc.

Marler, P. 1961. The filtering of external stimuli during instinctive behavior. In *Current Problems in Animal Behavior*, eds., W. H. Thorpe and O. L. Zangwill, pp. 150–166, New York: Cambridge University Press.

McDougall, W. 1923. *An Outline of Psychology.* London: Methuen.

McKinney, F. 1973. Ecoethological aspects of reproduction. In *Breeding Biology of Birds.* Washington, D.C.: National Academy of Sciences.

Michel, G. 1971. Experiential influences affecting progesterone induced incubation. PhD. thesis, Rutgers University.

Nice, M. M. 1937. Studies in the life history of the song sparrow I. *Trans. Linn. Soc. N.Y.* 4:i–vi, 1–247.

Nice, M. M. 1943. Studies in the life history of the song sparrow II. *Trans. Linn. Soc. N.Y.:* i–viii, 1–329.

Norton-Griffiths, M. 1969. The organization, control, and development of parental feeding in the oystercatcher (*Haematopus ostralegus*). *Behav.* 34:55–114.

Phillips, C. L. 1887. Egg-laying extraordinary in *Colaptes auratus. Auk* 4:6.

Phillips, R. E., and Youngren, O. M. 1971. Brain stimulation and species-typical behaviour: activities evoked by electrical stimulation of chickens (*Gallus gallus*). *Anim. Behav.* 19:757–779.

Riddle, O., 1935. Aspects and implications of the hormonal control of the maternal instinct. *Amer. Phil. Soc. Proc.* 75:521–525.

Riddle, O., and Bates, R. W. 1939. The preparation, assay and actions of lactogenic hormone. In *Sex and Internal Secretions*, eds, E. Allen, C. H. Danforth, and E. A. Doisy, pp. 1088–1117. London: Ballière, Tindall and Cox.

Riddle, O., Bates, R. W. and Lahr, E. L. 1935. Prolactin induces broodiness in fowl. *Am. J. Physiol.* 111:352–360.

Riddle, O., and Lahr, E. L. 1944. On broodiness of ring doves following implants of certain steroid hormones. *Endocrinol.* 35:255–260.

Ryves, B. H. 1944. Nest-construction by birds. *Brit. Birds* 37:182–188, 207–209.

Schleidt, W. 1962. Die historische Entwicklung der Begriffe in der Ethologie. *Z. Tierpsychol.* 19:697–722.

Schnierla, T. C. 1952. A consideration of some conceptual trends in comparative psychology. *Psychol. Bull.* 49:559–597.

Schnierla, T. C. 1956. Interrelationships of the "innate" and the "acquired" in instinctive behavior. In *L'instinct dans le comportement des animaux et de l'homme*, pp. 387–452. Paris: Masson.

Selous, E. 1901–1902. An observational diary of the habits, mostly domestic, of the great crested grebe and the peewit. *Zoologist* 5:161–183, 339–350, 454–462; 6:133–144.

Selous, E. 1905. *Bird-Life Glimpses*. London: G. Allen.

Selous, E. 1909–1910. An observational diary on the nuptial habits of the black-cock (*Tetrao tetrix*) in Scandanavia and England. *Zoologist* 13:400–413, 14: 23–29, 51, 176–182, 248–265.

Selous, E. 1916. On the sexual origin of the nidificatory, incubatory and courting display instincts in birds. An answer to criticism. *Zoologist* 20:401–412.

Silver, R., and Barbiere, C. 1977. Readiness to court and to incubate during the reproductive cycle of the male ring dove. *Horm. Behav.* 8:8–21.

Silver, R., and Buntin, J. 1973. Role of adrenal hormones in incubation behavior of male ring doves (*Streptopelia risoria*). *J. Comp. Physiol. Psychol.* 84(3):453–463.

Silver, R., and Feder, H. H. 1973. Role of gonadal hormones in incubation behavior of male ring doves (*Streptopelia risoria*). *J. Comp. Physiol. Psychol.* 84:464–471.

Silver, R., Reboulleau, C., Lehrman, D. S., and Feder, H. H. 1974. Radioimmuno-assay of plasma progesterone throughout the reproductive cycle in the male and female ring dove (*Streptopelia risoria*). *Endocrinol.* 94:437–444.

Skutch, A. F. 1976. *Parent Birds and Their Young*. Austin and London: University of Texas Press.

Stern, J. M. 1974. Role of testosterone in progesterone-induced incubation behavior in male ring doves (*Streptopelia risoria*). *J. Endocrinol.* 44:13–22.

Stern, J. M. 1974. Estrogen facilitation of progesterone-induced incubation behavior in castrated male ring doves. *J. Comp. Physiol. Psychol.* 87:332–337.

Stern, J. M. and Lehrman, D. S. 1969. Role of testosterone in progesterone-induced incubation behaviour in male ring doves (*Streptopelia risoria*). *J. Endocr.* 44: 13–22.

Stevenson-Hinde, J. G. 1973. Constraints on reinforcement. In *Constraints on Learning*, ed., R. A. Hinde and S. Stevenson-Hinde. New York: Academic Press.

Stokes, T. M., Leonard, L. M., and Nottebohm, F. 1974. The telencephalon, dien-cephalon and mesencephalon of the canary, *Serinus canaria*, in stereotaxic co-ordinates. *J. Comp. Neur.* 156:337–374.

Taibell, A. 1928. Riseglio artificiale di istinti tipicamenta feminili nei maschi di faluni uccelli. *Atti Soc. Nat. Mat. Modena* 59:93–102.

Tinbergen, N. 1951. *The Study of Instinct*. London: Oxford University Press.

Tinbergen, N. 1953. *The Herring Gull's World*. London: Collins.

Tinbergen, N. 1954. The origin and evolution of courtship and threat display. In *Evolution as a Process*, eds., J. S. Huxley et al., pp. 233–250. London: Allen and Unwin.

Tinbergen, N. 1959. Comparative studies of the behavior of gulls (Laridae): a progress report. *Behav.* 15:1–70.

Tinbergen, N. 1963. On adaptive radiation in gulls (Tribe Larini). *Zool. Mededelin-gen* 39:209–223.

Tinbergen, N. 1967. Adaptive features of the Black-headed gull (*Larus ridibundus* L.) In *Proc. XIV Int. Ornith. Congr.* Almquist and Wiksells, Uppsala., 43–59.

Van Tyne, J., and Berger, A. J. 1971. *Fundamentals of Ornithology*. New York: Dover Publications, Inc.

Vowles, D. M., Beazley, L., and Harwood, D. H. 1975. A Stereotaxic Atlas of the Brain of the Barbary Dove (*Streptopelia risoria*). In *Neural and Endocrine Aspects of Behavior in Birds*. Amsterdam: Elsevier Scientific Publishing Company.

Watson, J. B. 1913. Psychology as the behaviorist views it. *Psychol. Rev.* 20:158–177.

Watson, J. B. 1919. *Psychology from the Standpoint of a Behaviorist*. Philadelphia: Lippincott.

Part I

THE INITIATION OF BREEDING

Editor's Comments
on Papers 1 Through 4

The timing of all the events of the reproductive cycle, gonadal growth, migration, courtship, nest building, and incubation is influenced by the requirements of the last phase, the post-hatching period. Optimally, the young should appear when the demands on the parents in taking care of them are minimal and the probability of successful fledging is maximal. (Fledging is the interval between leaving the nest and becoming independent of all parental care.) How do the needs of the post-hatching period influence the onset of the reproductive season? What kinds of factors determine when the animals will breed?

Two sets of causes, ultimate and proximate, have been proposed to explain the selection of control systems which ensure optimal timing of breeding (Baker, 1938). Ultimate causes are those environmental factors such as food supply, predation pressure, and nesting conditions that control the efficiency or survival value of breeding through natural selection. Proximate causes provide information allowing the breeding season to start, so that later, when the eggs and young appear, conditions for their survival are best. Proximate factors include external stimuli such as temperature and day length, and endogenous cues such as circannual (endogenous annual oscillation) rhythms in sensitivity to light. In environments characterized by constant conditions, a single factor, such as the availability of food, may serve to initiate breeding and to ensure optimal conditions for the young. In environments characterized by marked seasonal changes, the sequence of events from gonadal development and migration to hatching and fledging of young takes many weeks. In this case, the organism must be informed well in advance of imminent environmental conditions. In middle and high latitudes, day length (a proximate factor) is a cue to subsequent environmental changes, and is the most reliable signal for the initiation of reproduction, though it is not directly related to the survival of young (ultimate factor).

In practice, the investigation of ultimate causes and survival value is difficult to carry out. In order to prove that breeding in spring is likelier to lead to survival of young than breeding at another time, it would be necessary to vary breeding time and estimate reproductive success and species survival. Attempts to understand the adaptiveness of behavior generally rely on investigations of the immediate effects of performance of a behavior, or immediate functions of a behavior rather than relying on direct tests of survival. The first experimental demonstration in any species that photoperiods (interval of exposure to light) exert an important influence on seasonal gonadal growth and on behavior was that of Rowan in 1925 (Paper 1). Rowan showed that the Slate-colored Junco (*Junco hyemalis*) housed in outdoor cages during a Canadian winter responds with an increase in testicular size and northward migration when exposed to long photoperiods. Almost immediately, there was a reaction against accepting too generally the hypothesis of photoperiods in the control of reproduction. Bissonnette (Paper 2) pointed to the influence of food and Lack (Paper 3) suggested that nesting conditions serve to control the timing of breeding in birds. Marshall and Disney (Paper 4) suggested that many equatorial birds have evolved a response to rainfall and its effects.

To summarize, those stimuli that are predictive of the conditions which will occur when the young appear can operate as proximate factors acting on the individual animal to signal the onset of the breeding period. The stimuli which are effective as proximate factors are characteristic of the species in its particular environment. Ultimate factors contribute to the evolution of species-typical breeding periods by selecting against those individuals bearing young in times of adverse environmental conditions. In this way environmental factors such as food supply, predators, and availability of adequate nesting conditions may determine, over time, when a species breeds.

REFERENCES

Baker, J. R. (1938). The relation between latitude and breeding seasons in birds. *Proc. Zool. Soc. Lond. 108*, Series A: 557–582.

1

Reprinted from *Nature* 115(2892):494–495 (1925)

RELATION OF LIGHT TO BIRD MIGRATION AND DEVELOPMENTAL CHANGES

William Rowan

THAT light is a factor of prime importance in the inauguration or stimulation of bird migration, has been suggested by many authors from the days of Seebohm onwards. While many of the suggestions will not bear close investigation, at least one very attractive view has been put forward by Sir E. Sharpey-Schafer. In an address delivered some years ago to the Scottish Natural History Society [1] he makes the following comments, ". . . the regularity with which migration occurs, indicates that the exciting cause must be regular. There is no yearly change, outside the equatorial zone, that occurs so regularly in point of time as the change in the duration of daylight. On this ground this may well be considered a determining factor in migration, and it has the advantage over other suggested factors that it applies to the northerly as well as to the southerly movement."

He says further " That it [migration] is a result of developmental changes in the sexual organs is improbable."

Evidently inspired by the work of the botanists Garner and Allard on what they have termed " photoperiodism," an American author [2] has lately revived this theory and has, apparently independently, come to the same conclusion as Sir Edward with regard to the absence of relation between developmental changes in the reproductive organs and migration.

On purely theoretical grounds it has always seemed to me that if the waxing and the waning of the days really in any way affect the migratory impulse, they must produce their effect through the gonads. This

[1] " On the Incidence of Daylight as a determining Factor in Bird Migration," E. A. Schäfer, NATURE, vol. 77, pp. 159-163 (December 19, 1907).
[2] " Is Photoperiodism a Factor in the Migration of Birds ? " G. Eifrig, *Auk*, vol. 41, pp. 439-444.

is not the place for theoretical discussion, and 1 merely wish to record an experiment that has just reached completion. Other corroborative work is still in progress and a critical histological examination of the experimental and normal material yet remains to be undertaken.

In September of last year I trapped a number of Juncos (*Junco hyemalis*) on their southward migration to the Middle States. These were turned into two large open-air aviaries removed from shelter of any kind. One, into which about a dozen birds were put for the experimental work, was fitted with two 50-watt electric lights. The other housed controls. Commencing on October 2, the lights were turned on at sunset (that is, while the birds were still fully active) and kept on until five minutes after dark. Each day afterwards the time was lengthened by five minutes. Taking into consideration the differences in time of sunrise, the birds thus got about three minutes longer illumination daily. On account of the fact that they went to roost at their usual time on the first day in spite of the glaring lights, and that attempts at educating them to keep awake were never wholly successful, and less so with some individuals than with others, it has proved impossible to estimate the effective light increases. For the same reason there is lack of uniformity in the results obtained.

Elimination of the warmth factor was unexpectedly successful—thanks to a severe winter—the lowest temperature to which the birds were exposed being 50° below zero (Fahrenheit).

Birds were killed at intervals of approximately two weeks, with the following results :

Dates of Killing.	Number Examined.	Size of Testes.[3]
Oct. 15	1	0·50 × 0·48
(A wild bird killed same date		·60 × ·60) [4]
Oct. 29	1	·44 × ·41
Nov. 13	1	·45 × ·53
Nov. 26	1	·60 × ·44
Dec. 11	1	·80 × ·79
Dec. 27 [5]	2	(*A*) ·90 × ?
		(Part of ribbon destroyed)
		(*B*) 1·80 × 1·54

[3] The whole series of testes was sectioned at 6μ. The first column indicates the greatest diameter of the largest section in each series in millimetres. The second is arrived at by adding the total number of sections for each series and multiplying by 0·006.

[4] The testes of this bird, still on migration, had not yet reached the winter minimum and this accounts for their large size. Diminution in size during the initial stages of the experiment is very marked.

[5] *A* was an adult bird : *B* a bird of the year.

Catastrophe overtook my control birds and I had to find substitutes. Through the kindness of the Museum of Vertebrate Zoology, University of California, I have been receiving fixed gonads of Juncos (a closely related species of approximately the same size) wintering in the Berkeley district. These are not strictly comparable with mine, therefore, but the samples include birds taken at intervals from November to early January. In spite of the California climate, the January testes are minute. My solitary female, killed also on December 27, as compared with the early January females from Berkeley, has an ovary two to two and a half times as large and with conspicuous follicles.

The two males and the female killed on December 27 were kept indoors for their last week at an average temperature of about 40°. (Their drinking water froze one night.) The marked difference in the size of the testes of the two males may probably be accounted for by their habits. *A* went to roost. in spite of the lights (the birds were together in the same cage) about an hour to an hour and a half each night before *B*. The female kept the latter company. *A* sang a good deal ; *B* incessantly. All the birds were in excellent condition when killed.

It would, therefore, appear that whatever effect daily increases of illumination may or may not have on migration, they *are* conducive to developmental changes in the sexual organs. Comparison of the normal material from Riviera-like California with the experimental product from Alberta further suggests that favourable light conditions are more potent in this respect than favourable temperatures.

2

Copyright © 1932 by Macmillan (Journals) Ltd

Reprinted from *Nature* 129(3260):612 (1932)

LIGHT AND DIET AS FACTORS IN RELATION TO SEXUAL PERIODICITY

T. H. Bissonnette

In his recent communication regarding light as a factor in causing sexual periodicity in birds and mammals, Dr. F. H. A. Marshall [1] has directed attention to several examples of the induction of sexual activity in birds by prolonged daily light periods, and to the great rarity of animals not exposed to light variations at some time. Even nocturnal animals are subject to twilight. Not all have the same degree of susceptibility to light in this respect. He also cited conditions in tropical regions, such as the Cameroons, " where environmental conditions are similar throughout the year " and " native birds have no restricted breeding season but breed at any time ".

That this is true only in a very broad sense was shown by Bates,[2] who found that some native birds breed there at any time, others only in the dry seasons, twice a year (in May, June, and July, and in November, December, and January), others only through half the year. Sex glands of both sexes often remain in breeding condition and size throughout the year. This suggests that breeding may be controlled by some factor other than light periods in some or all of these birds, even though comparatively long light periods may keep the sex glands actively producing germ cells.

Moreau [3] pointed out that, in any district, each species of bird has a rather restricted breeding season of about three months, relatively constant for any locality, but differing in neighbouring localities. He also pointed out that length of day is constant throughout the year only on the equator. It varies 40 min. at 6° N. or S. ; 70 min. at 10° ; and 3 hr. at 23·5°. In this respect, therefore, the transition from equator to poles is gradual throughout, and differences are those of degree only. Light, in its relation to sexual periodism, is a world problem.

Moreau accepts Rowan's hypothesis [4] that longer periods of wakefulness or exercise induce sexual activity in many species in addition to juncos and crows ; but finds it of doubtful value in the cases of long migrants, wintering beyond the equator, or on both sides of it. He is of the opinion that food is a controlling factor in breeding periodicity of tropical birds, the breeding seasons of which are seldom longer than three months. At Amani and Zanzibar, these seasons differ for species common to both districts. So the sexual activities of tropical birds are subject to seasonal cycles comparable with those of palæarctic birds. For him, this rules out photoperiodism for these birds, and makes climate and food supply more dominant factors in their sexual cycles. Lynes,[5] Vaughn,[6] Paget-Wilkes,[7] and others agree, and even suggest that rains and bush fires are related to the breeding cycles in Nyasaland.

Recent work by Bissonnette,[8] and Bissonnette and Wadlund,[9] on the modification of the sexual cycle of starlings in America, shows that testis activity is conditioned by daily period, intensity, and wavelength of light. The long wave red rays stimulate, and the shorter wave green rays inhibit germ cell activity. But, no matter how stimulating the light ration used, sexual regression sets in after a variable period of complete sperm formation. This period is shorter in animals brought to the climax quickly, and longer in those activated more slowly. An optimum wave-length, somewhere in the red, and an optimum intensity of light, below 185·6 foot candles of white light and nearer 29 foot candles intensity, are indicated. These results with starlings were obtained when birds were fed on a varied diet, rich in proteins, fats, and vitamins.

On the other hand, some unreported recent studies with starlings in Cambridge, using red light of a potency capable of inducing complete sperm formation in about twenty days on the rich diet used previously, indicate that, on a diet restricted to ' middlings ' grain mash, the very stimulating red light is prevented from inducing testis activity, even in 22 days. This shows that, even when the daily light ration is adequate to induce sexual activity, the type of food may be a limiting factor and, under such conditions, control the appearance of seasonal activity. This may be the case with tropical birds, as described by Bates, Moreau, and others, and may account for the apparent correlation of diet with reproductive periodicity. Then, too, light intensity varies as between rainy and dry seasons.

Variations of susceptibility, of members of the same or of different species, to changes in light rations, would cause breeding activity to occur at somewhat different dates, even in the same locality, whether species are permanent residents or migrants.

In studies of mammals in this relation, by Bissonnette,[10] Baker and Ranson,[11] Marrian and Parkes,[12] and Moore and Samuels,[13] light ration and food have been found to influence reproductive activity. When either one of these two factors remains constant and adequate for activity, the resulting activity may be varied by changes in the other factor.

It becomes more evident that, while light ration plays a great part in conditioning the periodicity of sexual cycles in many birds and mammals, it is not the only factor operating. Seasonal changes in type or quantity of food must be reckoned with, in analysing the factors effective in any particular case.

We are still far from a general law in this connexion. But all will agree with Dr. Marshall in maintaining the wide application of this principle of photoperiodism in sexual cycles.

[1] Marshall, F. H. A., NATURE, **129**, 344, March 5, 1932.
[2] Bates, G. L., *Ibis* (9), **2**, 8, 558-570 ; 1908.
[3] Moreau, E. R., *Ibis* (13), **1**, 553-570 ; 1931.
[4] Rowan, Wm., *Proc. Boston Soc. Nat. Hist.*, **39**, 152-208 ; 1929.
[5] Lynes, H., *Ibis*, **1**, 757-797 ; 1925.
[6] Vaughn, J. H., *Ibis*, 577-608 ; 1929, and 1-48 ; 1930.
[7] Paget-Wilkes, A. H., *Ibis*, 690-748 ; 1928, and 475-490 ; 1931.
[8] Bissonnette, T. H., *Am. Jour. Anat.*, **45**, 289-305 ; 1930 : *Jour. Exptl. Zool.*, **58**, 281-319 ; 1931 a : *Physiol. Zool.*, **4**, 542-574 ; 1931 b: *Science*, **74**, 18-19 ; 1932 a : *Physiol. Zool.*, **5**, 92-123 ; 1932 b.
[9] Bissonnette, T. H., and A. P. R. Wadlund, *Jour. Morph.*, **52**, 403-428 ; 1931.
[10] Bissonnette, T. H., in press.
[11] Baker, J. R., and R. M. Ranson, in press.
[12] Marrian, G. F., and A. S. Parkes, *Proc. Roy. Soc.*, B, **105**, 248-252 ; 1929.
[13] Moore, C. R., and L. T. Samuels, *Am. Jour. Physiol.*, **96**, 278-288 ; 1931.

3

Reprinted from *Proc. Zool. Soc. Lond.* 1:231–236 (1933)

NESTING CONDITIONS AS A FACTOR CONTROLLING BREEDING TIME IN BIRDS

David Lack

Zoological Laboratory, Cambridge

INTRODUCTION.

Previous work on the periodicity of the breeding season in birds has concerned the factors influencing the gonads. Rowan (1925-30) and Bissonnette (1930–32) have shown experimentally the great influence of light. Moreau (1931), Bissonnette (1932 *c*), and I. and J. R. Baker (1932) consider that at least one other factor—food—is important. It seems to have been assumed that the breeding time depends exclusively on the state of the gonads. It is the object of this paper to show that conditions at and around the nest-site may also limit breeding. The possible significance of the last factor in periodicity phenomena is then discussed.

I am deeply indebted to Mr. G. C. L. Bertram for his assistance with the following observations where they concern Bear Island (in the Barents Sea, in 74° N., 19° E.), where we spent the summer of 1932 (see 'Geographical Journal,' Jan. 1933).

THE BREEDING TIME OF ARCTIC TERNS.

Colonies of the Arctic Tern (*Sterna macrura* Naumann) were studied on islets in three lakes in the north of Bear Island. The lakes were Laksvatnet, referred to as "A," one of the Titrebekputtane, "B," and one of the Hölputtane, "C." They are shown on the map of the island published by the Norwegian Svalbard Expeditions (Oslo, 1925). These lakes were within two miles of each other, in similar open country, at about the same altitude (100–200 feet above sea-level), and close to (under 2000 yards from) the sea.

At A the breeding ground was always well above the lake water-level, at B it was submerged for a short period after the snow melt, at C for a longer period. Direct observations on this point were confirmed by the vegetation. At C it consisted of the moss found typically on ground submerged during the earlier part of the summer. At B this moss predominated, but some Phanerogams occurred, while at A it was absent, the vegetation consisting of lichens and Phanerogams typical of the herb-mat of dry ground on Bear Island.

The colonies were visited regularly, and the time of laying was noted, this being checked by subtracting twenty days, the known incubation period, from the time of hatching, also recorded. Since the colonies varied in size (each was roughly proportional to the breeding area available) it is best to use

---

* Communicated by N. B. KINNEAR, F.Z.S.
† For explanation of the Plate, see p. 237.

the date when more than half the birds had laid for comparing the breeding times.

Locality....................	A.	B.	C.
State of ground after snow melt	Dry	Waterlogged for short period.	Waterlogged for longer period.
First egg laid	20. vi.	1. vii.	13. vii.*
More than half the birds had laid by	28. vi.*	6. vii.	21. vii.*

* Possible error of a day in either direction.

The table shows that the average time of laying at C was twenty-three days later than at A, B being intermediate.

POSSIBLE FACTORS CONCERNED.

The birds at B and C were in residence before those at A had started to lay. They had not been driven away from A late in the spring. Light, temperature, and similar factors possibly influencing the gonads must have been acting equally at the three colonies, since their situations were so similar. The same applies to food, since the birds fed almost exclusively at sea, and not on their respective lakes. The marked difference in breeding time can clearly be correlated with the suitability of the ground for nesting, and with nothing else. The lakes thawed about June 19, and at A, where the ground was dry, the birds started to lay almost at once. At B and C the ground was then waterlogged. That at B dried before that at C, the birds not laying in either case until the ground was fairly dry.

It is to be presumed that the gonads of birds at the three colonies were in breeding condition at the same time, and that the nest-site factor inhibiting laying acted through the nervous system.

SIMILAR CASES.

Since the influence of nesting conditions on breeding time has been ignored in previous discussions on periodicity, some further cases are summarised. Most are taken from observations concerned primarily with other matters. Nesting conditions always control breeding time to some extent, for if a nest is destroyed before the eggs are laid a bird normally builds another before laying. Again, Chance (1922) proved that in the Cuckoo (*Cuculus c. canorus* L.) the time of laying is immediately dependent on the sight of an intended dupe building its nest, the egg being laid five days later.

Lewis (1929) showed that in the Gulf of St. Lawrence the Double-crested Cormorant (*Phalacrocorax auritus* Lesson) laid at least one week earlier at a colony in trees than at one on bare rock under a mile distant. When the birds began to lay in the trees snow still lay thick on the ground, so that suitable nesting-sites were not then available at the other colony. In this and two similar cases, where there were records for two and three years respectively, the colonies were so near that clearly food, light, etc. acted equally at each, and nesting conditions alone caused the difference in breeding time. The first colony had moved from rocks to trees only in recent years, thus putting forward their annual time of nesting.

MacGillivray (1923) observed a huge colony of Australian Pelicans (*Pelicanus conspicillatus* Temminck & Laugier) nesting on an island which was forming as floods subsided. On January 29 he noted newly hatched young in the nests on the highest part of the island, which had become uncovered some six weeks previously. But elsewhere, where the ground had not become exposed until later, the eggs were fresh or at an early stage of incubation. By February 12 the floods had receded further and the newly exposed ground was covered with fresh nests. By March 4 yet more ground was clear of the water and had been covered with nests. By this last date some of the half-grown young had left the earliest nests in the centre of the island, and fresh individuals were laying here. Hence laying was clearly limited by nesting-site conditions, fresh sites being colonised as soon as they became available. On account of this some individuals were prevented from laying until eleven weeks after the first had started.

Conditions around the nest as well as those of the actual site limited the laying time of the Pelicans, for they would nest only on islands. Hence they had to wait until the floods had subsided far enough to form islands (when they laid at once), although before this similar sites might be available on the mainland. Further (*loc. cit.* p. 173) laying ceased at one island, which, through the fall of the water, had become connected with the mainland, although at other islands on the same lake it was still continuing.

Wilkes (1928) found near Lake Nyassa that the tree-nesting *Hyliota* and *Prionus* always nest over burnt veld. If a patch of local grass is burnt before the annual bush-fires become general, the trees there are crowded with nests before the main areas have any. The bush-fires could scarcely influence the gonads, and it seems clear that breeding is immediately limited by nesting conditions.

The last two cases concerns the neighbourhood rather than the actual site of the nest. That this is important is illustrated also on Bear Island by the Northern Eider (*Somateria mollissima borealis* (Brehm)), whose breeding time at the two big colonies observed coincided with the melting of the ice round the lake islets. The melting of the ice and the breeding of the Eiders occurred some five days earlier at one islet than the other. It may be noted that in both cases the birds had arrived at least a few days before the ice broke up. The same habit was observed in the Arctic Tern colony at A (see p. 232), while Montague (1925), also Lamont, there quoted, record the same for both Eider and Tern in Spitsbergen. On Bear Island at least that nesting is delayed until the surrounding ice has melted is not correlated with food, since both species feed mainly at sea. The habit is useful, since the Arctic Fox (*Alopex lagopus*) cannot cross to the nests once the ice has melted (it was observed attempting to do so).

I am deeply indebted to the Marquess of Tavistock for data from captive birds. He writes that it is well known to aviculturists that, apart from feeding and general management, the provision or absence of a type of nest that takes the fancy of a hen bird will make all the difference to whether she will or will not lay ; even the domestic fowl in full lay will stop egg-production within twenty-four hours if moved to strange quarters. Two cases are of special interest.

From 1918 to 1922 a Barraband's Parrakeet (*Polytelis swainsonii* (Desmarest)) in his collection nested regularly in a box with a wooden bottom. The latter being unsuitable to the young, one filled with peat and decaying wood was substituted. The hen seemed not to like the new box, and did not lay, but every year she came into full breeding condition, using the breeding call and gestures, permitting the cock to pair with her, and assuming the curiously heavy "floppy" appearance of these birds when about to lay.   In 1931 the top of the same box (otherwise unchanged) was decorated inside and out with bark.   This appealed to the hen, who at once laid and reared young, as she has done in 1932.

For some years a pair of Indian Sarus Cranes (*Antigone antigone* (L.)) bred regularly in Woburn Park, until one year when an unmated female Common Crane (*Grus grus* (L.)) attached herself to them and refused to be driven away. The Sarus Cranes made their usual nest, and seemed on the point of laying, when the Common Crane laid two infertile eggs in it.   The hen Sarus, instead of laying, started to incubate these eggs, and, taking turns with her mate, sat for the full period.   Finally the eggs and the Common Crane were removed, and a few weeks later the Sarus Cranes made another nest and reared young as usual, showing that they were still capable of breeding.

Finally must be mentioned a type of behaviour different from that of all the above instances, but which may also limit breeding.   Pitt (1929) found in Norway that Fieldfares (*Turdus pilaris* L.) which had begun to build or had even laid part of their clutches suspended laying at a sudden return to cold weather.   Other species were similarly affected, a few not.   Howard (1929) states that in England breeding may be abandoned at almost any stage (with an incomplete nest, with eggs, or with young) at the onset of a cold spell, but it is essential for the change in weather to be sudden.   The birds react immediately.   These points, as Howard states, show that in such cases breeding is limited through the nervous system, not through the direct effect of starvation or low temperature on the gonads.   The case of Timmermann (1932), for which low temperature was postulated to influence egg development, seems explicable in the same way.

APPLICATION TO PERIODICITY PHENOMENA.

The foregoing cases show clearly that, though egg-laying in birds is primarily dependent on the state of the gonads, its immediate control by the nervous system may considerably modify the breeding time in nature.   Laying will normally not take place unless certain conditions concerning the site and neighbourhood of the nest are satisfied, and may also be checked by sudden cold. There may also exist other factors which can limit breeding through the nervous system.   Hence, although the periodicity of the breeding season must be primarily caused by the state of the gonads, and therefore by the factors regulating it, the factors influencing the nervous control of egg-laying cannot be ignored in interpreting the field data.

One phenomenon previously correlated with the gonads seems explicable on the latter grounds.   Many species in the Northern Hemisphere breed earlier in the southern parts of their range, as shown for the Blackbird (*Turdus m. merula* L.) by I. and J. R. Baker (1932).   When breeding is after the vernal equinox birds in the north receive more light than those in the south, and so might be expected to breed earlier, not later *.   The Bakers suggest that low temperature, perhaps indirectly through food, may partially inhibit the effect of light on the gonads.   But that Arctic Tern and Double-crested Cormorant nest earlier in the south than in the north is explicable on existing data through their breeding time being immediately determined (as shown previously) by nesting conditions dependent on the melting of snow and ice, since the last occurs earlier in the south than in the north.   In the Blackbird the phenomenon

* Rowan and Bissonnette have shown that increased light stimulates the development of the gonads.

is perhaps due to the reaction to cold weather described by Howard and Pitt. But nesting possibly depends on the appearance of spring foliage, which comes later in the north than in the south ; the case of Barraband's Parrokeet shows how particular a bird may be as to conditions at the nest-site. These reactions, first to cold weather and secondly to particular nesting conditions, may well account between them for all species which breed earlier in the south than in the north. Only if it were shown that the gonads mature later in the north (of which there is not evidence at present) need the phenomenon be considered due to low temperature or food scarcity partially inhibiting the effect of light on the gonads.

That birds breed late in a late summer seems due to the same factor as that causing them to nest later in the north, where each year the summer comes later than in the south, since arguments parallel to the above are applicable. Further, in the Arctic but not in Temperate regions there is widespread non-breeding in a late summer. Thus on Bear Island in 1932 the season was late, birds began to breed late, and many—60 to 80 per cent. in most species— did not breed (Bertram and Lack (1933)). It is difficult to believe that any factor activating the gonads is absent throughout a late summer in the Arctic (where, on Bear Island at least, the summer seems normal when it does come) but yet is absent only temporarily in a late summer in Temperate regions. But if, as suggested, late breeding is due to the late appearance of suitable nesting conditions the facts are explicable as follows.

The light ration is presumably the same in a late as in a normal season; hence the birds may be expected to be in breeding condition at the normal time. Bissonnette and Wadlund (1932) have shown that, even with optimum light conditions, the gonads regress after a certain period of full development. In the Arctic suitable nesting conditions appear late in the summer even in a normal season, and in a late season much later. Non-breeding in the latter case may well be because birds have passed out of breeding condition before nesting conditions have become suitable. In Temperate regions nesting conditions are not delayed to nearly the same extent, hence there a late season does not produce non-breeding. It may also be noted that birds matured most quickly are the first to regress (Bissonnette and Wadlund, *loc. cit.*), which implies that birds regress sooner in the Arctic, where the daylight is longer, than further south.

Manniche (1911) attributed the non-breeding of swimming birds in Greenland in a late season to lack of food, because ice stayed all the summer in the inlet ; but dropped eggs of the Arctic Tern were found, suggesting that the birds of this species were in breeding condition, and breeding may have been pre- vented by nesting conditions, since ice stayed round the nesting islets (see p. 233). Three other species (Long-tailed Duck, King Eider, and Red-throated Diver) fed not in the inlet but on freshwater lakes, which were clear of ice. Manniche may have been correct for the Northern Eider, whose feeding ground was blocked by ice and whose ovaries (*loc. cit.* p. 25) may not have attained breeding con- dition ; but on Bear Island in 1932 the sea-ice disappeared in early June, hence the non-breeding of the Eider was not due to this cause here. Therefore though Bissonnette (1932 c) states that the gonads are not activated by light if food is reduced, Manniche's conclusion must be treated with caution. The main phenomena of a late season are more simply correlated with the delayed appearance of suitable nesting conditions.

There are probably other ways in which nesting conditions influence breeding periodicity. This may especially apply to the Tropics, where the incidence of suitable nesting conditions is often determined by the periodic bush-fires or rains. Further, photoperiodicity seems negligible, and some other factor, perhaps food, must cause the periodicity of the gonads (Moreau (1931), Bissonnette (1932 c)). Many workers (Moreau, *loc. cit.* p. 567, also Lynes (1930)) correlate breeding time with nesting conditions, though the only proved case seems that of Wilkes, cited earlier. Further data, very kindly supplied me by Admiral Lynes, were extremely suggestive. It is unfortunate that so little is known as yet of the breeding season in the Tropics.

SUMMARY.

·1. The differences in the time of laying at three colonies of the Arctic Tern (*Sterna macrura* Naumann), which amounted to twenty-three days between the extremes, were correlated with nesting conditions, not with the state of the gonads.

2. This and similar cases show that, though laying is primarily dependent on the state of the gonads, it is immediately controlled by the nervous system, through which nesting conditions, sudden cold, and perhaps other factors can limit breeding. These latter factors on occasion considerably modify the breeding time in nature, and may hence be important with regard to the phenomena of breeding periodicity.

3. That in the Northern Hemisphere birds breed earlier in the southern parts of their range seems explicable through the later appearance of suitable nesting conditions and later disappearance of cold weather in the north.

4. The last two facts may also explain why birds breed late in a late summer. It is further suggested that the widespread non-breeding in a late season in the Arctic, which is not paralleled in Temperate regions, is caused through the birds' gonads having regressed before nesting conditions become suitable, though the adverse effects of food scarcity on the gonads may explain some cases.

REFERENCES.

BAKER, I. and J. R. 1932. Proc. Zool. Soc. Lond. pp. 661–667.

BERTRAM, G. C. L., & LACK, D. 1933. Ibis, ser. 13, vol. iii. pp. 283–300.

BISSONNETTE, T. H. 1930 a. Amer. J. Anat. xlv. pp. 289–305.

BISSONNETTE, T. H. 1930 b. *Ibid.* xlvi. pp. 477–497.

BISSONNETTE, T. H. 1931 a. J. Exp. Zool. lviii. pp. 281–319.

BISSONNETTE, T. H. 1931 b. Physiol. Zool. iv. (4), pp. 542–574.

BISSONNETTE, T. H. 1932 a. ' Science,' lxxiv. pp. 18–19.

BISSONNETTE, T. H. 1932 b. Physiol. Zool. v. (1) pp. 92–123.

BISSONNETTE, T. H. 1932 c. ' Nature,' cxxix. p. 612.

BISSONNETTE, T. H., & WADLUND, A. P. R. 1932. J. Exp. Biol. ix. pp. 339–350.

CHANCE, E. 1922. The Cuckoo's Secret.

HOWARD, H. E. 1929. Introduction to the Study of Bird Behaviour, pp. 28–31, 62.

LEWIS, H. F. 1929. The Natural History of the Double Crested Cormorant, pp. 38–40. Ottawa.

LYNES, H. 1930. Ibis, ser. 12, vol. vi. Cisticola Suppl. p. 41.

MacGILLIVRAY, W. 1923. ' Emu,' vol. xxii. pp. 162–174.

MANNICHE, A. L. V. 1911. Danmark-Ekspeditionen til Grønlands Nordøstkyst 1906–1908, Bind v. nr. 1.

MONTAGUE, F. A. 1925. British Birds, vol. xix. pp. 139–140.

MOREAU, R. E. 1931. Ibis, ser. 13, vol. i. pp. 553–570.

PITT, F. 1929. *Ibid.* ser. 12, vol. v. pp. 53–71.

ROWAN, W. 1925. ' Nature,' cxv. pp. 494–495.

4

Reprinted from *Nature* 180(4587):647–649 (1957)

EXPERIMENTAL INDUCTION OF THE BREEDING SEASON IN A XEROPHILOUS BIRD

By Dr. A. J. MARSHALL

Department of Zoology and Comparative Anatomy, St. Bartholomew's Medical College, University of London

AND

H. J. de S. DISNEY

Department of Agriculture, Dodoma, Tanganyika Territory

ALTHOUGH there have been suggestions to the contrary, it is unlikely that the breeding seasons of truly equatorial vertebrates are controlled by photoperiodicity, although at 16° N. (Senegal)[1], and even at 3·22° S. (Tanganyika)[2], the sexual cycle of *Quelea quelea* (the red-billed weaver finch) has been advanced by artificial illumination. Rather, many equatorial and other species have evolved a reproductive response to rainfall, or its effects. There has been adduced, however, little evidence concerning the role of rainfall *per se*, as distinct from the various environmental effects (for example, the sudden appearance of green grass, green seeds, animal food) of which it is an essential cause[3].

Q. quelea is a sexually dimorphic, highly nomadic weaver-finch that occurs in flocks of up to one million strong over great areas of Africa below the 30-in. isohyet[4,5]. Such flocks cross the equator and may breed on either side of it irrespective of minor photofluctuations. Between March and June, 1956, five hundred young *Quelea* were trapped soon after leaving the nest in the Dodoma area (lat. 6° 11′), Tanganyika. The juvenile age-groups were distinguished by ringing, and one hundred birds were placed in each of four cages measuring 18 ft. × 6 ft. × 6 ft. These were furnished as follows :

Cage I : Nesting bush (*Acacia mellifera*) of traditional kind, dry seeds of white millet (*Panicum miliaceum*) and bulrush millet (*Pennisetum typhoideum*), trough-water, and long dry grass of the genus (*Cynodon*) the bird uses for building material.

Cage II : As Cage I, but equipped also with a sprinkler simulating natural rainfall for two hours a day (=approx. 0·5 in.).

Cage III : As Cage I, but additionally with fresh *green* grass-seeds, long *green* nesting grass and trays of just-germinated green millet.

Cage IV : As Cage III, but additionally with a sprinkler for two hours per day (0·5 in. of 'rain') and quantities of harvester termites (*Hodotermes mossambicus*), grasshoppers, and dipterous larvae.

The difficulty of maintaining supplies of green grass, green grass-seeds and live larvæ during the dry season was overcome by growing appropriate grasses under irrigation and establishing an insect hatchery. This plantation grass was augmented by fresh grass (*Echinochloa stagnina*) collected by Africans from a distant swamp and dispatched by goods train three times a week. Grasshoppers and termites were collected daily. The cages were 10–30 yards apart, and optically isolated by hessian screens and sorghum stalks. Food, grass and water were replenished twice a day.

The experiment was begun on June 30. In all groups the birds almost immediately tried to weave nests. Building was done only by the males, and in synchronized surges within groups. Frequently, too, there were synchronized demolitions followed again by concerted building. Weekly counts were made and after two months current nesting attempts were recorded as follows :

Date	Cage I	Cage II	Cage III	Cage IV
September 1	1	3	21	13

Thus, indications were obtained that for these birds (which were still physiologically incapable of reproduction) green grass is an important factor in nest building. It appeared to be physically impossible for long, brittle, dry grass to be used, and wild populations never attempt to do so.

The experimental procedure was now varied as follows : On September 10 all nests were demolished in Cages I and III. To Cage I, where building under dry conditions had been negligible, green nesting grass was added. From Cage III green seed and trays of germinated millet were eliminated so that both these cages now had dry food and green nesting grass. The following results were obtained :

Date	Cage I	Cage III
September 9	1	22
,, 10	—	—
,, 11	9	7
October 1	17	20

There is, then, good evidence of the importance of long green grass as a nesting stimulus. However, the above figures refer to nesting *attempts* in which imperfect structures were built. It was found that complete nests were made only when the grass had produced long stems (for the frame of the nest), long blades (for tying the frame together), and seed-heads.

Although rain did not raise the number of building attempts it did appear slightly to accelerate the assumption of breeding coloration in the female. In Cages II and IV the females began to change colour before the others, but in all cages the visible changes took three weeks. (It was considered inexpedient to handle the birds so it is possible that initial moult stages may have been unobserved.) In all experimental cages the females changed colour before the wild population. In a reserve cage provided with dry millet, resting bush, trough-water and no building material whatever, the colour changes were delayed and remained approximately in phase with those of the wild population. Therefore, it seems possible that close association with males engaged in the manipulation of nesting material may have exerted a stimulus that operated neurohumorally and accelerated colour change. In the experimental cages full nuptial colour was complete in both sexes before November 20 so that, even without rainfall, green grass or animal food, the birds were physically prepared for reproduction before the 'rainy season' expected in December or January.

The first eggs were laid (in Cage III) on December 24. One egg was found on the ground. The other was laid in a nest : it was incubated, developed an embryo, but was destroyed by intruder birds just before hatching. Before the initial ovulation only 4·01 in. of natural rain fell and for more than a month (Nov. 28–Dec. 31) there fell only 5·09 in.

This natural precipitation was negligible in quantity compared with the artificial rain provided in Cages II and IV and did not lead to reproduction in Cages I, II or IV. The manufacture of complete nests was slightly retarded in early January when our plantation grass failed to produce seeding heads. By January 28, however, stems bearing seeds were again available, and many nests were completed. More young appeared in Cage III on March 1 and 5, and on the latter date the first eggs hatched in Cage IV. The incubation period is twelve days and there had been no natural rain between Feb. 12 and March 1.

In Cage II (for months supplied with a daily 0·5-in. of artificial rain, dry food and dry nesting grass), a few birds at length managed to weave sodden dry grass into nests. One egg was found on the ground on April 12 and an egg was laid in a nest on May 14. This disappeared the next day, probably destroyed by intruders.

Altogether, thirty young were hatched in Cage III. As successive groups of young appeared, a temporary supply of insect food was provided but only five chicks were successfully reared. In Cage IV five young were hatched, of which none survived. Deaths were generally caused by the attacks of adults. The begging posture of the chicks somewhat resembles that adopted by soliciting females and this was often responsible for severe injury by neighbouring males. The 1956–57 rainfall was very unevenly distributed, with periods of no effective precipitation for as long as six weeks. The only reproduction that occurred in *Q. quelea* over an area of many hundred square miles was in our Cages III and IV.

The present initial experiment was hampered by the periodic failure of our plantation grass to mature sufficiently, by the collapse of the nesting bushes due to the ravages of weevils and, perhaps, by the presence in the cages of too many birds. However, from the above results it appears that :

38

1. There exists in *Q. quelea* an innate urge to build nests which, in the confines of a cage well supplied with food, will find expression within about three months of leaving the nest, and before individuals have assumed nuptial plumage.

2. Successful nest-building cannot take place until the environment provides green grass with stems and blades sufficiently long to be used by the males for weaving. By this time (and when the females will accept the nests) the green grass is seeding. It is not yet known whether the seed-heads exert a stimulus on the females. It has been previously shown[4] that the young are fed principally on green seeds from the age of five days, but that during the first five days of life the parents bring them insects (providing essential amino-acids). Yet this kind of protein food does not appear to constitute a stimulus to reproduction. Inevitably, of course, animal food becomes available at the same period as the seed-heads. Thus, if the nomadic flocks of *Quelea* find enough suitable green nesting grass, the amino-acids essential to the early development of the young will be there as well.

3. Experimentally, rainfall in itself lowers, rather than increases, the number of ovulations (but not the number of *complete* nests).

4. An internal rhythm of reproduction exists, and this controls in both sexes the assumption of sexual coloration of beak, soft parts, plumage and, in the male, the onset of sexual display, performed on or near the incomplete nest, that leads to copulation. This rhythm is modifiable by external conditions, including possibly rainfall and other external effects (for example, social stimulation). Many birds of both sexes came into breeding dress when only $5\frac{1}{2}$ months old, and reproduction was shown to be possible at the age of nine months. Thus the breeding which occurs in this highly nomadic pest in southern Kenya in December and January could be by birds themselves hatched as recently as the previous March or April in Tanganyika.

Fuller observations, including particularly those concerning comparative gonad development in relation to external conditions and display in caged and wild populations, will be published elsewhere.

We are indebted to the director (Mr. J. R. Soper) and the assistant director (Mr. N. R. Fuggles-Couchman) of the Department of Agriculture, Tanganyika Territory, whose kind co-operation made our work possible. One of us (A. J. M.) was enabled to work in East Africa through the courtesy of the East African Fisheries Research Organization and its director (Mr. R. S. A. Beauchamp).

[1] Morel, G., and Bourliere, F., *Alandra*, **24**, 119 (1956).

[2] Marshall, A. J., and Disney, H. J. de S., *Nature*, **177**, 143 (1956).

[3] Marshall, A. J., *Mem. Soc. Endocrinol.*, **4**, 75 (1955).

[4] Disney, H. J. de S., and Marshall, A. J., *Proc. Zool. Soc.*, **127**, 379 (1956).

[5] Morel, G., and Bourlière, F., *Bull. Inst. Franc. Afrique Noire*, **17**, 618 (1955).

Part II

PAIR FORMATION AND NEST BUILDING

Editor's Comments
on Papers 5 Through 8

The two basic facets of reproductive behavior are the sexual and the parental aspect. The strictly sexual component involves the coming together of males and females for mating. Mating, involving crouching by the female, mounting by the male, and cloacal contact of the pair, lasts a few seconds. Given that 90 percent of avian species are monogamous (Lack, 1968), the copulatory response forms a small part of the association of the mated pair. What other functions do the pre-copulatory and post-copulatory behavior serve?

The most obvious function of courtship display is the stimulation of gonadal development in both males and females. Lehrman (1959) showed that the courtship behavior of the male Ring Dove stimulates ovarian development in the female. Burger (1953) found that male starlings (*Sturnus vulgaris*) housed in heterosexual groups showed a more rapid rate of testicular growth in the spring than males housed in homosexual groups. Another function of courtship displays may be primarily related to pair formation and mate retention rather than gonadal development, and is thus basically behavioral rather than physiological. For successful reproduction coordination must be achieved between behavioral and physiological events. For example, nest building (behavioral response) must preceed egg laying (physiological response) and the two must be closely coordinated.

It is convenient to describe courtship, pair formation, nest building, etc., as though they occur in a linear temporal sequence. However, parental behavior has its beginnings in birds with the formation of a pair bond, which can occur weeks or even months before the appearance of the eggs. Sometimes a distinction is lost between sexual and nonsexual phases of development. Among some of the lovebirds (*Agapornis spp.*), heterosexual pairs form when the birds are about two months old and still have their juvenile plumage. In some species the events during a post-hatching critical period when imprinting occurs influence mate selection in adulthood (see Hess, 1977). The blurring of a distinction between sexual and nest-building behavior is seen in the display of bowerbirds, who decorate their nests with items such as flowers, bones, buttons, and bits of egg shells. Marshall (1952) argued that this display at the nest serves to attract the female and is under the same hormonal control as copulation and nest building. Selous long ago concluded that the origin of nest building lies in movements resulting from sexual frenzy during copulation. (See Armstrong, 1965: 30–42, for a further discussion of this point.)

Turning to the nests themselves, it is clear that these take a wide variety of forms, from the shallow "scrape" of the Killdear Plover (*Charadrius voliferous*), to the elaborately woven hanging baskets of some of the weaver finches (Ploceidae) and orioles (Icteridae). Despite the variability of nests to be seen even within a single family of birds, the members of species generally show remarkable uniformity in the type of site selected, the materials used, and the manner of construction. Similarity or *convergence* in nest form and structure between unrelated species often furnishes significant clues to the nature of the ecological forces operating in the evolution of the "characteristic nests." The paper by von Haartman (Paper 5) illustrates the use of the convergence approach in the study of adaptions to hole nesting in different groups of birds. Adaptive *radiation* in a group of closely related species is shown in an important paper by Dilger (Paper 6). Dilger described in great detail the behavior of eight species of parrots (*Agapornis*). In a subsequent experiment Dilger (1962) crossed two species of parrots which differed in their nest-building behavior and studied the building responses of the hybrid offspring.

The value of comparative analysis is again made clear in the detailed studies of the behavior of many species of gulls by Tinbergen and his colleagues. Most gulls nest on the ground. The Kittiwake, an exception, breeds on tiny ledges of steep cliffs. The classic paper by Cullen (Paper 7) clearly demonstrates that the shift to cliff nesting in Kittiwakes modified all aspects of social and parental behavior. At the time Cullen reported her work, the behavior of many other gull species had already been described. Thus, Cullen had at hand rich sources for the development of ideas about the adaptive significance of behavior.

Quite a different aspect of nest site selection in Kittiwakes is seen in the paper by Coulson (Paper 8). Gulls which nest in the center versus the edge of a colony fledge more young and have a lower "divorce rate." Centrally located pairs of gulls which successfully rear young one season are more likely to mate with each other the next season than are pairs which nest in the periphery and fail to rear young. The pair bond and its stability is of great consequence for cooperative sharing of the work of parental care, and breeding success in turn has consequences for the maintenance of the pair bond.

REFERENCES

Armstrong, E. A. (1965). *The Ethology of Bird Display and Bird Behavior: An Introduction to the Study of Bird Behavior.* Dover Publications, Inc.: New York.

Burger, J. (1953). The effect of photic and psychic stimuli on the reproductive cycle of the male starling (*Sturnus vulgaris*). *J. Exp. Zool., 124,* 227–239.

Dilger, W. C. (1962). The behavior of lovebirds. *Sci. Am.,* 89–98.

Hess, E. H., and S. B. Petrovich (1977). *Imprinting.* Benchmark Papers in Animal Behavior. Vol. 5. Dowden, Hutchinson & Ross, Inc.: Stroudsburg, Pa.

Lack, D. (1968). *Ecological Adaptations for Breeding in Birds.* Methuen: London.

Lehrman, D. S. (1959). Hormonal responses to external stimuli in birds. *Ibis, 101,* 478–496.

Marshall, A. J. (1952). Display and the sexual cycle in the Spotted Bowerbird (*Chamydera maculata, Gould*). *Proc. Zool. Soc. Lond., 122,* 239–252.

5

Reprinted from *Evolution* 11:339–347 (1957)

ADAPTATION IN HOLE–NESTING BIRDS

Lars von Haartman

University of Helsinki, Helsinki, Finland

Received February 10, 1957

A considerable number of birds nest in holes of some kind. If we restrict ourselves to Passerine species only, we arrive at the following figures for the percentages of hole-nesters (table 1). The essential thing is not so much the number of species as the number of pairs. As there are fairly good estimations of the numbers of land birds nesting in Finland (Merikallio), I have included the percentages in question.

Both in Palearctic and Nearctic woods the numbers of hole-nesters are quite large, if we take into consideration that suitable nesting-holes are far from abundant. The premium on nesting in holes was first studied by Mrs. Nice in her classical work on the Song-Sparrow. In a number of hole-nesting species she found about 65% of the eggs resulting in fledglings, whereas the corresponding value for open nesters was only 43%. This difference has since been confirmed by large bodies of data.

The drawback is the severe inter- and intraspecific competition for nesting-holes. It is easy to demonstrate that there are too few natural tree-holes for the demands of the hole-nesting birds.

(1) In the beginning of my Pied Flycatcher investigation (v. Haartman, 1949) after catching the males I used to bring them home to ring them. This repeatedly allowed other males to take over the territory during the short absence of its former owner. Thus, at one nest-box six males succeeded each other within a few days.

(2) By putting up nest-boxes one is usually able to increase enormously the number of hole-nesting birds. This has been proved in many study-areas in Finland, Germany, Holland, and England (cf.

for instance v. Haartman, 1956; Creutz, Campbell). Before nest-boxes were put up in my study area in S.W. Finland there were hardly 10 pairs of Pied Flycatchers. Now there are about 70, and I am sure the population can still be increased. An even more drastic increase was brought about by Pfeifer in a small protected area in Germany. A corresponding population increase was found in the Tits.

TABLE 1. *Number of species*

Nest	Eastern North America (223)	Palearctic (279)	No. of pairs in Finland
Open	72%	61%	54%
Domed	6%	11%	23%
Niches	4%	4%	7%
Holes	17%	24%	16%

Such population statistics raise the question of whether the classical theory about food as the ultimate cause limiting population density is not overdone. In the case of the hole-nesters it seems obvious that the number of holes, and not the amount of food, mostly acts as an ecological limiting factor, determining the maximum number of nesting pairs. Unfortunately, we know very little about matters in woodlands uninfluenced by man.

(3) The interspecific competition is very strong, too. In Northern Europe the Great Tit and the Pied Flycatcher are the main competitors for the nest-boxes. The struggles not infrequently end in the death of the weaker Flycatcher. In my study area more Flycatchers were probably killed during nesting time by Great Tits than by birds of prey. A very unfair fighting method is adopted by the Wry-

neck (*Iynx torquilla*). If a hole is already occupied by another bird, the Wryneck simply empties out the nesting material and eggs. One pair can in this way destroy a considerable number of Pied Flycatcher's nests, without ultimately choosing any of them as a nesting site. Next to man, the Wryneck is the worst destroyer of Pied Flycatcher's nests (v. Haartman, 1951).

TERRITORIAL BEHAVIOUR

Both these facts, the safety of the nests and the severe competion for the nesting sites, have caused the evolution of a number of adaptations in the behaviour and biology of the hole-nesting birds.

Competition for nest-holes instead of food leads to certain characteristics in their territorial behavior. In most open nesters the male seems to defend an area, the encounters taking place along its borders. In the Pied Flycatcher (v. Haartman, 1956) the fights are concentrated at the very center of the territory, around the nesting-hole. The Pied Flycatcher's home is his castle. The same may be said of many other hole-nesters, e.g. the Starling (Schüz, and others), the Jackdaw, *Corvus monedula* (Lorenz, 1931), and the English Sparrow (Daanje).

In the Chaffinch (*Fringilla coelebs*), which finds a suitable nesting place in every tree, the male first occupies a territory, the nesting-site being chosen afterwards by the female. In the Pied Flycatcher, the choice of territory occurs in quite a different way. Only if the male finds a satisfactory hole, does he begin to sing and display other territorial behavior. In choosing a territory, the terrain in which the hole is situated is of relatively slight importance. A suitable hole seems to be the most important key stimulus inducing the male Pied Flycatcher or Great Tit to choose a territory. This is easy to understand: nothing guarantees that a certain forest area, which offers plenty of food, also offers a nesting-hole, whereas the reverse is more likely to hold true.

An analogous reaction is found in some territorial fishes. Species with a very specialized nesting site (e.g. *Gobius microps*) choose the nesting-site first and then show territorial behavior (Nyman), whereas less specialized species (e.g. *Gasterosteus aculeatus*) choose the territory first and then the nesting-site (Tinbergen).

COURTSHIP

The courtship of the Pied Flycatcher (v. Haartman and Löhrl, v. Haartman, 1956, etc.) is in many respects typical of the hole-nesting Passerines. If a female approaches the nest of an unpaired male, he flies to the nest entrance, singing an excited song. If the female follows him, he jumps in. The female usually stays outside. Only when the performance has been repeated often enough, does she dare to follow him into the inside of the nesthole. This behavior evidently functions as an advertisement: good-looking bachelor with own apartment wants a mate. This type of courtship necessitates that the male has only a small territory, as he must always be at hand near to the nesthole, not only to drive other males off, but also to demonstrate it to the female.

In the Collared Flycatcher (*Muscicapa albicollis*), a very close relative of the Pied Flycatcher, the male has a similar nest-demonstration (Löhrl). Even the excited song is the same, in spite of the normal territorial song being quite different. Also, the male Red-breasted Flycatcher, *Muscicapa parva*, seems to have a comparatively similar display (Michel). The behavior is evidently homologous in these species.

But even in a number of other hole-nesters which are not related to the Flycatchers, the male demonstrates the nesting-hole to the female in a rather similar way. This is a fairly good example of analogy or convergence, though from a field where it is less well known than in morphology or physiology.

The male Redstart (*Phoenicurus phoenicurus*) demonstrates the nest-hole, showing alternately the red color of his back and the conspicuous coloring of his head and throat. The white spot on the male's front somewhat resembles that of the Pied and Collared Flycatchers. Its function may be the same, to enhance the visual effect of the male's sitting in the nest-entrance with the head turned towards the female (Buxton).

The male Great Tit demonstrates the nest-hole by sitting in the entrance, and picking at its lower border (Hinde). In the Starling (Wallraff), the English Sparrow (Daanje), and even in owls (Kuhk), nest-demonstrations are known. The male Starling sings a special excited song when showing the nesting-hole. So, also, does the male Wren (*Troglodytes troglodytes*), which builds a domed nest before the arrival of the female, and shows it by creeping into it (Kluijver, Armstrong).

The Spotted Flycatcher (*Muscicapa striata*) nests in niches. Evidently there is no surplus of these, either. The male demonstrates ownership of a niche to the female by sitting in the niche and moving his tail up and down, raising his crest, and singing a very unmusical song (v. Haartman, 1949). However, not only niches, which might do as presumptive nurseries, are demonstrated, but also quite unsuitable places, such as narrow branches etc. Nest-demonstration has in these cases lost its original significance of showing a nesting-place. It has become a symbolic action, only serving to ensure pair-formation. The same may, to a certain extent, be said of the nest-demonstration in the Jackdaw (Lorenz, 1931).

Again, we find a number of analogies among fishes. The males of *Gasterosteus aculeatus* and *pungitius* (Tinbergen, Morris) on the one hand, and the male *Gobius microps* (Nyman) on the other, have a highly analogous way of guiding the female to the nest-entrance and of stimulating her to enter the nest.

The nest-demonstration in many of these hole-nesting species has probably developed phylogenetically from displacement nest-building, released in the male by the female.

POLYGAMY, HISSING NOTES, RELEASING OF THE GAPING REACTION IN THE YOUNG

Male polygamy has been observed in numerous birds. In European Passerines, polygamy has been reported in at least 32 species (v. Haartman 1954, etc.). More than half of these, i.e. 18, nest in holes or niches, or have domed nests. Kluijver has pointed out that safe nests are an important prerequisite for the evolution of polygamous behavior, as the females have to care for the broods more or less unaided. This is plausible, though it should be noted that the biology of open-nesting species may be, on the whole, less well known.

Sibley has pointed out that the young and the incubating adults of many different species of hole-nesters give a hissing, snakelike note when disturbed. Examples are the Wryneck, many Tit species, and the young of *Phylloscopus sibilatrix*.

The gaping reaction of the hungry young in a number of hole-nesters is released by darkening of the nest (cf. e.g. Lorenz, 1935, and v. Haartman, 1953). Experiments have shown (v. Haartman, unpubl.) that this reaction is innate in the Pied Flycatcher.

EGG COLOR

Morphological characters which are not used tend to disappear during evolution. This is probably due to the fact that most mutations have a negative effect, causing the disappearance of something (Huxley), and secondly to the fact that characters which are not used tend to be harmful.

A well known example is afforded by the egg color. In open-nesting species it is mostly cryptic, in hole-nesters [1] often

---

[1] In the following, the term "hole-nesters" refers to all birds whose nests are not open, including species building domed nests or using niches of some kind.

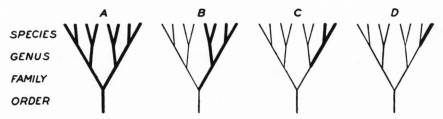

FIG. 1. Hypothetical family-trees of hole-nesters of order (A), family (B), generic (C), and specific (D) rank. Cf. text. —— = hole-nesters, —— = open nesters.

unspotted, white or slightly bluish. It is not certain whether the white color has to be looked upon merely as a disappearance of unnecessary color, or if it has a positive selection value, enabling the incubating bird to see its eggs more easily.

This rule of egg color has, however, a great many exceptions. Probably only oologists have realized how many. These exceptions are, in my opinion, mostly related with the taxonomic position of the species concerned. The habit of breeding in holes has not, of course, evolved at the same time in all hole-nesters, but is in some of them a relatively old, in others a relatively new acquisition. It is hardly possible to estimate exactly how old it may be in the different species. Instead, we can use their taxonomic position as a measure.

Firstly, there are species belonging to an order consisting principally of birds nesting in holes (including niches and domed nests). The family-tree of such an order may be represented as in figure 1 A. In the following, I will refer to these species as hole-nesters of order rank, or primary hole-nesters.

TABLE 2. *Method of classification including cases of uncertain nesting habit*

Family	Genus	Species	Classification as hole-nester of:
holes	holes	holes	Family rank
open	holes	holes	Generic rank
unknown	holes	holes	Generic rank
open	open	holes	Specific rank
open	unknown	holes	Specific rank
unknown	open	holes	Specific rank

Secondly, the order to which the species belongs may principally consist of open-nesters, its family, however, being hole-nesters. A schematic family-tree of such hole-nesting species of family rank is seen in figure 1 B.

Thirdly, the hole-nesting species may belong to a genus, consisting principally of hole-nesters, this genus belonging to a family of open-nesters. The family-tree of such a hole-nesting species of generic rank is shown in figure 1 C.

The fourth main possibility is that a hole-nesting species occurs in a genus of principally open-nesters. Such a secondary hole-nester, of specific rank, is likely to have a family-tree such as is shown in figure 1 D.

Besides these alternatives, there are a number of difficult cases, occurring for instance when a genus is represented only by two species, one of which nests in holes, the other not. Table 2 shows how the classification was undertaken in ambiguous cases.

Now it is quite clear that the primary hole-nesters have evolved their nesting habit much earlier than the secondary hole-nesters. Even the hole-nesters of generic rank must, on average, be older than hole-nesters of specific rank. On the whole, the taxonomic position of the hole-nesting species must be in some way proportional to the time that has elapsed since their ancestors acquired this nesting habit. I do not need to stress that this is a statistical rule which cannot be applied to individual cases.

As regards the birds, I have not man-

TABLE 3. *Egg color in North American[2] and Palearctic Passerine birds*

Nest	Rank	Eggs unspotted	Clutch size	Incubation[3] period in days	Nestling period in days
Holes, niches, domed	Order	100% (1103)[4]			
	Family	36% (76)	9.2 (40)	13.8 (43)	19.0 (34)
	Genus	35% (72)	5.0 (51)	13.8 (21)	14.2 (18)
	Species	20% (23)	4.5 (13)	12.6 (5)	12.3 (4)
Open	All ranks	10% (331)	4.1 (279)	12.9 (114)	12.8 (108)

[2] References.—Palearctic Passerines: Lack, Dementiev, and, for a few species, Hartert.—North American Passerines: Bent, and for the families *Ploceidae, Icteridae, Thraupidae,* and *Fringillidae:* Pough, who, however, treats only Eastern and Central North American species, and does not give any information as to incubation and nestling times, and numbers of subspecies.

[3] In calculating clutch size, incubation period, and nestling period, the Corvidae have been excluded.

[4] Number of species in brackets.

aged to obtain data regarding the absolute age of the taxonomic categories. For three other Vertebrate classes (*Mammalia, Reptilia,* and *Pisces*), Rensch (1954) gives the following age estimates:

orders 65–270 million years
families 25–80 million years
genera 15–50 million years
species 0.1—a few million years

Thus, orders are much older than families, and genera much older than species, the difference between families and genera being relatively less.

Table 3 demonstrates the egg color in the North American and Palearctic Passerine birds. Moreover, it includes the 4 orders (*Psittaci, Trogones, Coraciidae, Pici*) which consist of hole-nesters.

The values in table 3 convincingly show that systematic status is as important a factor in determining the egg color as way of life. Secondary hole-nesters mostly have spotted eggs, primary ones always unspotted eggs. The hole-nesters of family and generic rank have an intermediate position, being closer to secondary hole-nesters.

Two facts may help to explain this rule. Firstly, the hole-nesters of lower rank may be satisfied with less safe nest-sites. Relatively large numbers of them use domed nests, niches, or holes with a wide entrance. It is impossible exactly to measure the influence of this factor, as the safety of the nests of the different species is known only in very few cases.

Another possibility is that the evolution of egg color requires a certain time. The advantage (or the mutation pressure) of the eggs being unspotted may be so small that reduction of egg color takes a very long time. Certainly, this factor is of importance in a great many cases. The Tits, for example, are very well adapted hole-nesters, using only fairly safe nesting-holes. In spite of this, their eggs are slightly spotted. I think these spots have to be looked upon as an insignificant rudiment, comparable with the vermiform appendix in man.

The opposite case, secondarily open-nesting species with white eggs, also occurs. The order of *Psittaci* consists of 316 species, of which the vast majority are hole-nesters and have white eggs. The few species secondarily nesting in the open (*Myiopsitta monachus, Geopsittacus occidentalis,* and the genus *Pezoporus,* cf. Makatsch) have maintained this white egg color.[5]

[5] A similar case has been described by Poulsen. Birds secondarily nesting on the ground have deep nests (similar to those of tree-nesting birds) and do not retrieve eggs into the nest from outside it, as primary ground-nesting birds do.

The occurrence of white eggs in hole-nesters of different rank shows a certain correlation with the absolute age of these categories. There are marked steps between hole-nesters of order and family rank, and, again, between those of generic and specific rank, but no significant difference between those of family and generic rank.

CLUTCH SIZE

The clutch size in the Passerine birds concerned is shown in table 3. As with egg color, we note that adaptation does not take place immediately, but evolves slowly. Hole-nesters of specific rank hardly have an increased clutch size at all.

It sounds paradoxical that hole-nesters should have larger clutches. D. Lack supposes that safe nests enable the young to develop slower. Hence, the parents should be able to rear a larger brood. In species with open nests, it would be more important to get the young fledged as soon as possible than to rear many of them.

Another alternative may be that non-migratory birds are proportionately often hole-nesters (v. Haartman, 1954). This holds true at least with the North European Passerines. The winter mortality in non-migratory species is, of course, much heavier than in migratory ones. If clutch size is adapted to compensate for mortality (which, on the whole, seems uncertain), the non-migratory species, and hence also the hole-nesters, must have larger clutches.

As one pair, on average, gets only two descendants reaching maturity, the mortality must be the heavier the larger the clutch. This means a more rigorous selection, unless we assume that the differential mortality is restricted only to the very young individuals, being wholly unselective. Theoretically, a larger number of offspring should cause a more rapid evolution. So far, no positive evidence of the influence of number of offspring on the speed of evolution has been presented.

TABLE 4. *Correlation between clutch size and average number of subspecies per species*

| Clutch size | Numbers of subspecies per species | |
	Palearctic species[6]	North American species[7]
2.5– 3.49	5.6 (5)	1.5 (19)
3.5– 4.49	4.7 (14)	2.1 (69)
4.5– 5.49	4.5 (85)	2.0 (47)
5.5– 6.49	6.4 (41)	3.0 (14)
6.5– 7.49	10.7 (9)	4.8 (13).
7.5– 8.49	13 (5)	
8.5–11.49	16 (9)	

[6] The numbers of subspecies according to Dementiev.

[7] According to Bent; the families *Icteridae*, *Thraupidae*, and *Fringillidae* are not included, cf. table 3.

On the contrary, reference has been made to the rapid evolution of elephants and other big animals with small litters. However, these selected examples can only prove the very trivial fact that the number of offspring is not the only major factor determining the speed of evolution. It seems necessary to consider all species belonging to a larger group, such as, for instance, the Passerines. On doing this, we find the correlation between clutch size and average numbers of subspecies per species shown in table 4.

As hole-nesters have large clutches, this means that they also have more subspecies.

Now, I do not wish to claim that the question of the influence of clutch size on numbers of subspecies is finally answered by these statistics. The problem is more complicated. The species with large clutches are comparatively often non-migratory, and this status may act, in itself, as a factor causing rapid subspeciation owing to increased site tenacity (Rensch, 1933) and, even more, larger climatic differences in the wintering areas in the non-migratory species (Salomonsen).

INCUBATING AND NESTLING PERIODS

The incubating and nestling periods are on the average longer in hole-nesting birds

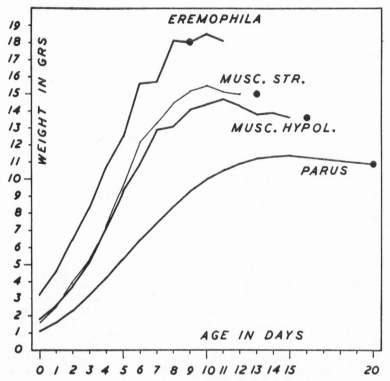

Fig. 2. Weight increase in young of *Eremophila alpestris* (cf. Pickwell), a typical open nesting species, *Parus caeruleus* (cf. Gibb), a typical hole-nesting species, *Muscicapa striata* (Sommardahl, unpublished), and *Ficedula* (*Muscicapa*) *hypoleuca* (v. Haartman, 1954). ● = day of leaving the nest.

(table 3). Again, we note the importance of taxonomic position, besides breeding biology. Neither incubating nor nestling period are prolonged in secondary hole-nesters.

The period of weight increase in the young of hole-nesters is only moderately prolonged (v. Haartman, 1954). The prolongation of nestling time is due above all to the time elapsing between the young's reaching full weight and leaving the nest. The young of hole-nesters are mostly well fledged on leaving the nest, the young of open-nesters often leave the nest on foot many days before they are able to fly.

The young of related species seem to follow the same course of development, irrespective of their nesting site. The Pied and Spotted Flycatchers [8] are good examples. One would expect a very slow development in the Pied Flycatcher and a very rapid one in the Spotted Flycatcher, as the former is very successful in its nesting, and the latter very unsuccessful. In the last two years I have not seen any Spotted Flycatcher's brood fledged (the Jay, *Garrulus glandarius*, seems to be a very severe predator), whereas nearly every Pied Flycatcher's brood was successful. Nonetheless, the weight increase in the young of the two species is almost identical (cf. fig. 2). Young Pied Flycatchers stay a few days longer in the nest. Thus, the period of remaining in the nest after attaining full weight is more

[8] These species are not very nearly related. Vaurie places them in different genera, *Ficedula* and *Muscicapa*.

adaptable than the weight increase. In *Motacilla alba, cinerea,* and *flava,* the clutch size, incubation period, and nestling period seem to be fairly similar (the nestling period is somewhat prolonged in *M. alba*), in spite of the fact that the two first-mentioned species nest in fairly safe niches, whereas the last-mentioned uses open nests.

SUMMARY

A number of analogous (convergent) adaptations occur in hole-nesting birds. Their main causes are the safe nesting sites and the keen competition (both intra- and interspecific) for them. Competition has caused certain characteristics in behavior, such as development of territorial behavior only after finding a suitable hole, fighting for a nesting-hole instead of for a territorial area, and the male's demonstration of the nesting-hole in courtship-display.

The safety of the nesting site has caused other adaptations, such as frequent polygamy, hissing notes in incubating adults and in the young, lack of cryptic color in the eggs, large clutch size, and slow development of eggs and young. A long period of evolution is needed, however, before a species is able to take full advantage of the safety of the nesting site. In primary hole-nesters the adaptations are much more clearly developed than in secondary hole-nesters. Tits, for instance, have maintained spotted eggs, open-nesting Parrots white eggs. The Pied and Spotted Flycatchers have nearly identical growth curves in the young, in spite of the one having very safe and the other very unsafe nesting sites, and so on.

The number of subspecies was found to show a positive correlation to clutch size (hence, hole-nesting birds tend to have larger number of subspecies). This may be due to stronger selection in species with more offspring, and/or the fact that species with large clutches are relatively often non-migratory.

LITERATURE CITED

ARMSTRONG, E. A. 1955. The Wren. London.

BENT, A. C. 1942. Life histories of North American Flycatchers, Larks, Swallows, and their allies. U. S. Nat. Mus. Bull. 179.

———. 1946. Life histories of North American Jays, Crows, and Titmice. Ibid. 191.

———. 1948. Life histories of North American Nuthatchers, Wrens, Thrashers, and their allies. Ibid. 195.

———. 1949. Life histories of North American Thrushes, Kinglets, and their allies. Ibid. 196.

———. 1950. Life histories of North American Wagtails, Shrikes, Vireos, and their allies. Ibid. 197.

———. 1953. Life histories of North American Wood Warblers. Ibid. 203.

BUXTON, J. 1950. The Redstart. London.

CAMPBELL, B. 1955. A population of the Pied Flycatcher (*Muscicapa hypoleuca*). Acta XI. Congr. Intern. Ornith., p. 428–434.

CREUTZ, G. 1955. Der Trauerschnäpper (*Muscicapa hypoleuca* (Pallas)). Eine Populationsstudie. Journ. Ornith., **96**: 241–326.

DAANJE, A. 1941. Über das Verhalten des Haussperlings (*Passer d. domesticus* (L.)). Ardea, **30**: 1–42.

DEMENTIEV, G. P., and N. A. GLADKOV. 1951–54. Ptizi Sovjetskogo Sojuza. I–VI. Moskva.

GIBB, J. 1950. The breeding biology of the Great and Blue Titmice. Ibis, **92**: 507–539.

V. HAARTMAN, L. 1949. Der Trauerfliegenschnäpper. I. Ortstreue und Rassenbildung. Acta Zool. Fenn. 56, 104 pp.

———. 1951. Der Trauerfliegenschnäpper. II. Populationsprobleme. Ibid. 67, 60 pp.

———. 1953. Was reizt den Trauerfliegenschnäpper (*Muscicapa hypoleuca*) zu füttern? Vogelwarte, **16**: 157–164.

———. 1954. Der Trauerfliegenschnäpper. III. Die Nahrungsbiologie. Acta Zool. Fenn. 83, 96 pp.

———. 1956. Territory in the Pied Flycatcher, *Muscicapa hypoleuca.* Ibis, **98**: 460–475.

———, and H. LÖHRL. 1950. Die Lautäusserungen des Trauer- und Halsbandfliegenschnäppers, *Muscicapa h. hypoleuca* (Pall.) und *M. a. albicollis* Temminck. Ornis Fenn., **17**: 85–97.

HARTERT, E. 1910. Die Vögel der paläarktischen Fauna. Berlin.

———. 1932. Die Vögel der paläarktischen Fauna. Ergänzungsband. Berlin.

HINDE, R. A. 1952. The behaviour of the Great Tit (*Parus major*) and some other related species. Behaviour, Suppl. II, 201 pp.

HUXLEY, J. 1948. Evolution, the Modern Synthesis. London.

KLUIJVER, H. N., J. LIGTVOET, C. VAN DEN OUWELANT, and F. ZEGWAARD. 1940. De levensvijze van den Winterkonig, *Troglodytes tr. troglodytes* (L.). Limosa, **13**: 1–51.

KUIIK, R. 1949. Aus der Fortpflanzungsbiologie des Rauhfusskauzes, *Aegolius funereus* (L.). Ornithologie als biologische Wissenschaft, p. 171–182.

LACK, D. 1947. The significance of clutch size. I–III. Ibis, **84**: 302–352; **90**: 25–45.

LÖHRL, H. 1951. Balz und Paarbildung beim Halsbandfliegenschnäpper. Journ. Ornith., **93**: 41–60.

LORENZ, K. 1931. Beiträge zur Ethologie sozialer Corviden. Ibid., **79**: 67–120.

——. 1935. Der Kumpan in der Umwelt des Vogels. Ibid., **83**: 137–213, 287–413.

MAKATSCH, W. 1954. Die Vögel der Erde. Berlin.

MERIKALLIO, E. 1954. Über die Anwendung der quantitativen Untersuchungsmethode zur Ermittlung der regionalen Verbreitung und der Zahl der Vögel in Finnland. Acta XI. Congr. Intern. Ornith. Basel, p. 485–494.

MICHEL, J. 1907. Meine Beobachtungen über den Zwergfliegenfänger. Ornith. Jahrb., **18**: 1–18.

MORRIS, D. 1952. Homosexuality in the Ten-spined Stickleback (*Pygosteus pungitius* L.). Behaviour, **4**: 233–261.

NICE, MARGARET M. 1937. Studies in the life history of the Song Sparrow. I. Trans. Linn. Soc. New York 4, 247 pp.

NYMAN, K. J. 1953. Observations on the behaviour of *Gobius microps*. Acta Soc. Fauna Flora Fenn., **69**: 5, 11 pp.

PFEIFER, S., AND K. RUPPERT. 1953. Versuche zur Steigerung der Siedlungsdichte höhlen- und buschbrütender Vogelarten. Biol. Abhandl. 6, 28 pp.

PICKWELL, G. B. 1931. The Prairie Horned Lark. Trans. Acad. of Science of St. Louis 27, 153 pp.

POUGH, R. H. 1949. Audubon bird guide. Small land birds. New York.

POULSEN, H. 1953. A study of incubation responses and some other behaviour patterns in birds. Vidensk. Meddel. Dansk Naturhist. For. 115, 139 pp.

RENSCH, B. 1933. Zoologisches Systematik und Artbildungsproblem. Verh. deutsch, Zool. Ges. 1933, p. 19–83.

——. 1954. Neuere Probleme der Abstammungslehre. Berlin.

SALOMONSEN, F. 1955. The evolutionary significance of bird-migration. Kongel. Dansk Vidensk. Selsk. Biol. Medd., **22**: 6, 62 pp.

SCHÜZ, E. 1942. Biologische Beobachtungen an Staren in Rossitten. Vogelzug, **13**: 99–132.

SIBLEY, CH. G. 1955. Behavioral mimicry in the Titmice (Paridae) and certain other birds. Wilson Bull., **67**: 128–132.

TINBERGEN, N. 1951. The Study of Instinct. Oxford.

VAURIE, CH. 1953. A generic revision of Flycatchers of the tribe *Muscicapini*. Bull. Amer. Mus. Nat. Hist., **100**: 453–538.

WALLRAFF, H. G. 1953. Beobachtungen zur Brutbiologie des Stares (*Sturnus v. vulgaris* L.) in Nürnberg. Journ. Ornith., **94**: 36–67.

6

Reprinted from *Z. Tierpsychologie* 17:649–653, 667–683, 684–685 (1960)

The Comparative Ethology
of the African Parrot Genus *AGAPORNIS*

By William C. Dilger

Received June 21th, 1960

INTRODUCTION

The genus *Agapornis* is composed of several closely related and allopatric species which readily breed and hybridize in captivity. Moreover, they demonstrate a fairly consistent trend from a relatively "primitive" form *(cana)* to the most "highly evolved" one *(p. nigrigenis)*. In addition, many behavior patterns show both quantitative and qualitative differences, greatly facilitating studies of the evolution of species-typical behavior.

All of the described species are being studied, with the exception of *swinderniana* which has not yet been obtained. The minimum number of individuals studied was about fifteen *(p. lilianae)* and the maximum about forty (both *roseicollis* and hybrids between *roseicollis* and *p. fischeri*). These birds were observed both in indoor breeding cages in which one pair at a time was observed and in large outdoor flight cages in which several pairs at a time were observed. Several small, specially designed, cages were also used to obtain photographic records of certain behavior patterns (see Dilger, 1958, for a complete description of these).

All of the behavior of each species was first carefully described and then evaluations were made of the function, survival value, causation, and evolution of each behavior pattern. The methods used were those commonly employed by ethologists and described, for instance, by Tinbergen (1959).

At first, observations were recorded directly in notebooks but later by speaking into a small tape recorder and then transcribing these notes into notebooks. This enabled uninterrupted observations to be made and was a vastly superior method. Hundreds of feet of motion picture film were taken with the aid of a Bell and Howell time and motion camera. This technique permitted the study of behavior exactly as it occurred in time and space, being especially valuable in studying the exact interactions of two or more individuals.

All species were fed a diet consisting of French's Parakeet Seed (a mixture of red and white millet and canary seed), French's Conditioning Food (a diet supplement containing a variety of small seeds, dried milk products, and other ingredients including several vitamins), cuttle bone, quartz gravel, and fresh water. All but the small-billed *cana* and *pullaria* relish sunflower seeds as well.

GENERAL BIOLOGY

The genus *Agapornis* belongs to the subfamily Psittacinae of the parrot family, Psittacidae. This subfamily has a wide distribution in Africa, Asia, Australia, and the Americas. The next most closely related genus is *Loriculus* (the Hanging Parrakeets) of Asia.

Peters (1937) recognizes nine species of *Agapornis: cana* (Madagascar or Grey-headed Lovebird), *taranta* (Abyssinian or Black-winged Lovebird), *pullaria* (Red-faced Lovebird), *swinderniana*, (Swindern's or Black-collared

Lovebird), *roseicollis* (Peach-faced Lovebird), *personata* (Black-masked Love-bird), *lilianae* (Lilian's or Nyassaland Lovebird), and *nigrigenis* (Black-cheeked Lovebird).

Both NEUNZIG (1926) and HAMPE (1957) express the opinion that *fischeri*, *personata*, *lilianae*, and *nigrigenis* are subspecies of one species. Our obser-vations, augmenting those of NEUNZIG's and HAMPE's support this conclusion. It is interesting to note that those who were most familiar with these birds in life all agree as to the relationships of the four forms. Although MOREAU (1948) recognized their close relationship, he did not recommend combining them into one species. He did, however, recognize the fact that *lilianae* and *nigrigenis* are particularly closely related and that *fischeri* and *personata* are somewhat less closely related. Our observations support this conclusion. These four forms agree in having a rather wide area of naked white skin surrounding the eyes and will be referred to as the "white eye-ringed forms".

The relationships among the remaining species are not as close as among the white eye-ringed forms. However, *taranta* and *pullaria* seem more closely related to each other than to any of the others. The relationships of *swinder-niana*, a species poorly known in the wild and almost unknown in captivity, are obscure although we feel that it is probably most closely related to *rosei-collis*. *Agapornis cana* has long been isolated on the island of Madagascar and has retained many primitive characters. Its closest living relatives seem to be *taranta* and *pullaria*. The white eye-ringed forms are most closely related to *roseicollis* and were probably derived from a *roseicollis*-like ancestor.

Cana
Pullaria
Taranta
Swinderniana
Roseicollis
Personata Fischeri — A
P. Personata — B
P. Lilianae — C
P. Nigrigenis — D

Fig. 1. Map showing the ranges of the species of *Agapornis* (after MOREAU, 1948)

The range map (Fig. 1) modified from MO-REAU (1948), summarizes the geographic distribu-tion of *Agapornis;* notice that these species have an essentially allopatric dis-tribution which may par-tially explain why they have not developed, as a result of selective pressu-res stemming from the re-sults of the formation of mixed pairs, effective spe-cies isolating mechanisms. Also, the differences among them that do exist are not very effective as isolating mechanisms and are probably the result of either other selective pres-sures or random fixation occurring in present or past small populations. This si-tuation is of considerable advantage to us, permitting productive interspecific pairs to form in the laboratory.

MOREAU (1948) provides the best and most recent summary of present knowledge of the life history of these birds in the wild. Except for *swinder-niana* (inhabiting tropical rain forest) all inhabit areas of wooded grasslands.

The little information available from the wild and from our own observations in the laboratory indicate that *cana*, *taranta*, and *pullaria* are largely solitary nesters. However, *roseicollis* and especially the white eye-ringed forms seem to be rather strongly colonial. Colonialism is coincident with the lack of sexual dichromatism. This may be an advantage because each individual would be constantly signalling the same information from its plumage pattern and color (even the juvenals in the white eye-ringed forms), thus facilitating the coordination of group activity.

All lay from three to eight plain white eggs, one every other day until the clutch is complete. Occasionally two eggs will be laid on successive days and rarely two days will elapse between layings. We have found them to be determinate layers; that is, they can not be encouraged to lay a larger clutch by removing eggs as they are laid. Laparotomies performed on birds which have just completed their clutches revealed the presence of a number of atretic ovarian follicles graded in size. This indicates that some physiological condition "shuts off" the gonadotropic hormone complex to the active ovary and not that only the proper number of follicles become active in the first place.

Some incubation begins with the first egg laid; consequently, the young hatch at different times, causing a considerable differential in the sizes of individuals in the same brood. Incubation takes about 23 days. We calculated the exact incubation periods for a number of marked eggs and the results are shown in Table 1. Only the female incubates but the male may sit in the nest with her, close against her side and facing in the same direction, but not actually on the eggs.

Table 1. Showing the mean adult weights, incubation periods, and fledging times for *Agapornis*

	Mean Weights of Adults		Mean Incubation Periods		Mean Fledging Time	
	grams	no.	days	no.	days	no.
cana	28.2 (♂♂)	8	23.1	8	43	4
	31.4 (♀♀)	8				
taranta	43.9 (♂♂)	8	25.5	8	49.3	4
	52.8 (♀♀)	6				
pullaria	37.4 (♂♂)	12				
	38.0 (♀♀)	6				
roseicollis	55.4 (♂♂)	18	23.3	21	42.8	18
	56.0 (♀♀)	20				
personata fischeri	48.7 (♂♂)	12	22.7	10	38.4	9
	53.1 (♀♀)	11				
p. personata	49.5 (♂♂)	8	23.0	6	44.3	3
	56.4 (♀♀)	9				
p. lilianae	38.0 (♂♂)	4	22.2	9	44.2	5
	43.0 (♀♀)	6				
p. nigrigenis	37.9 (♂♂)	7	23.6	10	40.6	10
	43.0 (♀♀)	6				

Upon hatching, the young are covered with down; rather dense and reddish in *roseicollis* and the white eye-ringed forms and sparser and white in *cana*, *taranta*, and *pullaria*. The parents feed the young by regurgitation. There is no evidence that the crops of the parents contribute anything to the diet as is the case with pigeons and doves.

The young grow rapidly and the natal down is soon obscured by a second down coat (of varying shades of grey, depending upon the species) coming in amidst the natal down and immediately preceding the juvenal plumage. The young of *cana* and *taranta* are conspicuously slower in their development as compared with *roseicollis*, and *roseicollis* is considerably slower in this development than are the white eye-ringed forms (Fig. 2). Hybrids between *roseicollis* and *p. fischeri* are intermediate in this regard.

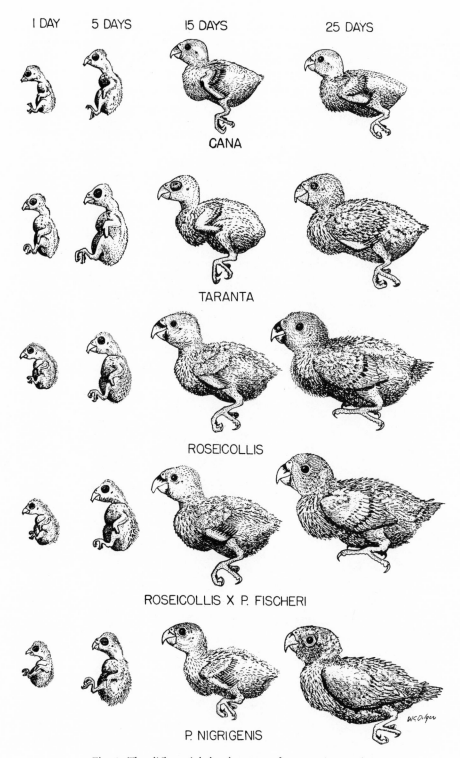

Fig. 2. The differential development of young *Agapornis*

The young fledge in about 43 days but there seem to be slight species differences in this regard (Table 1). The young are fully feathered and capable of strong flight immediately upon leaving the nest. They sleep in the nest cavity at night as long as the parents permit them. They are decreasingly dependent upon their parents for food for about two weeks after fledging, at which time they become independent, although they begin trying to husk seeds for themselves immediately after leaving the nest.

The young may become sexually mature at about 80 days, which would be about 37 days after fledging (one young male *roseicollis*). However, most birds begin to behave sexually at a considerably later date; usually after the post-juvenal molt is complete at about four months.

Although the evidence is somewhat contradictory, these birds seem to be seasonal nesters in the wild. Our laboratory birds (many of which were wild-caught), however, nest more or less continuously. They seem to be relatively independent of varying day lengths and may depend in the wild upon such environmental features as the onset of rains, food availability, or some combination for their initial stimulation for breeding. In captivity they have constant access to a surplus of food, nesting material, nest sites, and so forth. In many instances, eggs are laid before the young of the previous brood have fledged. In such cases, both parents demonstrate over-lapping sexual and parental care behavior.

[*Editor's Note:* Material has been omitted at this point.]

5) *Reproductive Behavior:* Pair Formation. — Little precise information is available on pair formation. We know that these birds all form pair bonds very quickly and ordinarily remain paired until one of the partners dies. Pair bonds may be broken by removing the birds from one another's sight and hearing for several weeks. We also know that birds still in juvenal plumage will form pair bonds which are usually heterosexual but are sometimes homosexual. These latter bonds seem to break up spontaneously in time. Adults, if given a choice, will ordinarily only form heterosexual pair bonds. Adult *roseicollis* and the white eye-ringed forms (all sexually non-dichromatic) easily form homosexual bonds if potential mates of the opposite sex are not available. The sexually dichromatic species have never been observed to form such homosexual attachments. However, two male *pullaria* were once seen to form a temporary and rather loose bond but this species is not as strikingly dichromatic as are *cana* and *taranta* and this, in some way, may have contributed to the formation of the bond. On the other hand, the homosexual pairs of *roseicollis* and the white eye-ringed forms seem to be as difficult to disrupt as are normal heterosexual pairs. As might be expected, one member of these latter homosexual pairs (regardless of the sex involved) always assumes the male sexual role and the other the female sexual role. This, once established, remains absolutely consistent. A homosexual male pair will not, however, inspect a nest site nor demonstrate any interest in nest material. A homosexual female pair will occupy the same nest cavity and both birds will collect, carry, and utilize nest material and both will simultaneously lay eggs and incubate.

The exact mechanisms of pair formation are obscure. If a group of unpaired adults are liberated into a large flight cage they will form a rather well-knit flock (especially the more social forms). They tend to perform their various maintenance activities in concert. Each bird seems to attempt to interact socially with a number of others, however. This consists of attempts to preen others, to sit closely with others and so forth. Soon it becomes apparent that certain of these combinations seem to be more fortunate than others and the two birds tend to seek each other's company in preference to other birds. This bond becomes progressively stronger until in just a few hours the pair bonds seem to have become firmly established. It is not clear which sex takes the initiative. CINAT-THOMPSON (1926) found that female Budgerigars (*Melopsittacus undulatus*) choose males to which they are most attracted on the basis of color pattern and voice. It may be that in *Agapornis* the males "test" a variety of females by attempting to preen them, sit with them, and so forth until they find one which accepts their presence most easily. Aside from these activities, the most important "pair cementing" activity seems to be courtship feeding (see p. 679 for description and discussion).

Nest-building. — Interest is shown in possible nest sites soon after adults form their pair bonds or soon after juvenal pairs attain their adult plumage. The female in particular investigates entrance holes and soon chooses a cavity within which to built her nest. The male remains close to the female throughout these activities but his interest seems to be in her and not particularly in the nest site.

The nesting of *pullaria* is somewhat different. As far as is known this species chooses hard, earthy, arboreal ant nests as nest sites. These are burrowed into; the tunnel usually turning to one side before the chamber is excavated. Sometimes the tunnel turns upward so that the chamber lies over the tunnel (PRESTWICH, 1957; PERRY, 1959; SERLE, 1957). Probably the female does the excavating (HAMPE, 1937). However, the evidence is somewhat contradictory. Pitman in Moreau (1948) says that both sexes excavate and Prestwich (1957) says, "the task of excavating falls mainly to the female. The males show great enthusiasm but are more of a hindrance then any real help." Perry (1959) says on page 120 that "... the pair succeeded in making a tunnel right through the ant mound!" On page 121, however, Perry says, ... "the female was busy excavating into the cork slab. This went on for several days until a deep tunnel had been made and the hen eventually disappeared altogether." Based on the evidence for *pullaria* and on what we know of male activity in the other species, it is probably safe to say that the male, although closely attentive to the female, does little or no excavating.

Agapornis cana, taranta, and *pullaria* are rather solitary nesters in the wild. This is borne out by our observations on our captive birds which show little social cohesiveness when several pairs are attempting to nest in the same outdoor flight cage; indeed, these birds engage in much inter-pair aggression under such conditions. On the other hand, *roseicollis* and the white eye-ringed

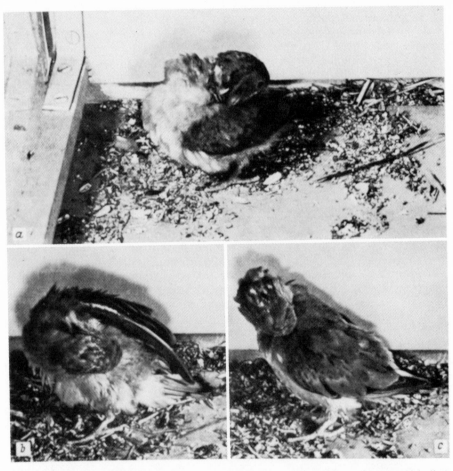

Fig. 15. An adult female *Agapornis taranta* tucking nest material (sunflower seed hulls) in her feathers; a) in her rump, b) in her flank, and c) in the side of her neck. Note that she has many missing feathers which have been used to augment the nest material, and the extensive ruffling of the entire plumage. This method of tucking is typical for *cana* females as well

Fig. 16. Nest of *Agapornis cana* (a) and nest of *A. taranta* (b). Note the small fragments of material which make a soft pad under the eggs and young

forms tend to be quite colonial in their nesting. The nest-box tends to be defended by the resident pair in all species but often strangers are allowed to sit on it. The entrance hole itself, however, is always defended. Most of the defence is carried out by the male; apparently because the female is simply not available much of the time as a result of incubation and brooding. Both sexes will threaten or attack intruders regardless of sex in *roseicollis* and the white eye-ringed forms. On the other hand, the sexually dichromatic *cana* and *taranta* have a tendency for males to defend against other males and females against females. The females of these species are particularly aggressive toward strange females and vigorous fights sometimes occur, especially during the time nest sites are being selected. Males of these species are typically far more tolerant of intrusion.

Normally, only females of all species prepare and carry nesting material, which consists of pieces of bark, leaves, plant stems, wood fibers, paper, etc. In addition to such materials, females of the white eye-ringed forms also utilize twigs, chewing them from larger branches or picking them up from the ground. Females of *cana, taranta,* and *pullaria* bite off very small pieces of nesting material and carry them, many pieces at a time, thrust amidst the feathers of the entire body. The entire plumage is ruffled as if to receive such material every time a piece is being placed in the feathers. The feathers are compressed again during each bout of cutting the next piece. Fig. 15a, b, and c shows a female *taranta* tucking small pieces of material (sunflower seed hulls) and also shows the characteristic appearance of the plumage of *taranta* females while nest-building. *A. taranta* has the unique habit of also utilizing its own feathers as nest material. These are obtained while the bird is in the nest and it is not clear whether she actually plucks herself or whether the feathers fall out naturally. If they are plucked, the male may help as her head also ordinarily shows much feather loss. We have never seen males of *cana, taranta,* or *pullaria* cut or carry nesting material. However, PERRY (1959) observed a male *pullaria* carrying nest material in the feathers of his back. These species do not carry very much material and its very nature precludes the possibility of building any sort of structure. Instead, the material serves as a soft pad lining the shallow saucer-shaped scrape in the detritus at the bottom of the cavity (Fig. 16a and b). In the wild, *pullaria*, at least, may sometimes carry no nesting material (SERLE, 1957).

Agapornis roseicollis females also tuck material amidst the feathers but only those of the lower back and rump are utilized, and only these are erected

during bouts of tucking. Moreover, the pieces of material cut by *roseicollis* are much larger and take the form of strips having quite uniform characteristics relative to length, width, and straightness. Several pieces are tucked before the bird flies off to the nest. Fig. 17a shows a female *roseicollis* about to tuck a strip of nest material. The nature of this material (long strips) permits a structure of sorts to be built within the cavity. They typically fashion a rather deep cup in which the eggs are deposited (Fig. 17b).

Fig. 17. a) An adult female *Agapornis roseicollis* about to tuck a strip of nest material in her rump feathers. Note that only the rump feathers are ruffled to receive this material, and the uniformity of the strips cut, judging from where she has been cutting. b) Nest of *Agapornis roseicollis*. Note the rather deep cup made possible by the long strips of nest material

The white eye-ringed forms carry nest material only in their bills. HAMPE (1936), however, had a young male *p. fischeri* which often stuck nesting material in the feathers of its rump. These forms, as mentioned above, carry twigs as well as strips of paper, bark, or leaves which they have cut. This latter material seems more variable in its characteristics than it is in *roseicollis* and is probably a reflection of the fact that it no longer has to be transported in the feathers. The ability to carry twigs as well as other material allows these birds to build rather elaborate nests consisting of a completely roofed chamber in which the eggs are deposited and a narrow tunnel leading up to the cavity entrance (Fig. 18). MOREAU (1948) states that such a nest is a structure, "... that could hardly exist unless it were supported below and on all sides." This is definitely not the case. The material is carefully positioned and inter-

Fig. 18. Nest of *Agapornis p. fischeri*. Note the completely covered chamber and the twigs used. The entrance tunnel is not visible but extends upward to the cavity entrance

woven in such a way that the entire nest can be removed from the nesting cavity without its losing its shape or identity in any way. The highly developed motor patterns employed by these birds, which result in such an elaborate structure, are being studied at the present time in our laboratory by Mrs. William KEETON, and this information along with facts concerning the evolution of nest building in the genus will be the subject of a forthcoming paper.

We do agree with MOREAU (1948), however, that the carrying of material thrust admidst the feathers represents a primitive condition for the genus. We also agree with him in doubting the explanation put forward by others that this method of carrying serves to free the bill for climbing into awkwardly situated nest entrances. White eye-ringed birds all climb well with the aid of the bill even though they may be carrying nest material at the same time.

Males of *roseicollis* and the white eye-ringed forms have been seen to occasionally cut nesting material and carry it about. It is exceedingly rare, however, for them to carry material into the nest. Interestingly enough, male nest material cutting and carrying is only done in connection with heterosexual pairs. Homosexual male pairs show no interest in nest material. This observation, coupled with observations on normal pairs, leads us to conclude that such activity in males is largely, if not entirely, the result of mimesis afforded by the nest material preparation by the female. We know that mimetic behavior between members of a pair is particularly strong. Homosexual pairs of males are not stimulated to respond to nest material if adjacent females are active in nest building although such activity is clearly "contagious" (both visually and auditorily) among adjacent females.

The preparation and utilization of nesting material in all of these species does not seem to be as dependent upon a specific phase of the reproductive cycles as it is in many birds (e. g., BENOIT, 1956:234). Paired females of all species (unknown for *pullaria*) prepare and carry nesting material throughout the year in the presence of nesting material and a nest site. This activity occurs regardless of the stage of reproduction (or non-reproduction). However, the amount of such activity, as reflected in the amount of nest material utilized, varies markedly with the stage of the cycle in the white eye-ringed forms. Details of this for *roseicollis, p. fischeri, p. lilianae,* and *p. nigrigenis* are given in Fig. 19. It may be seen that two peaks of activity occur; one

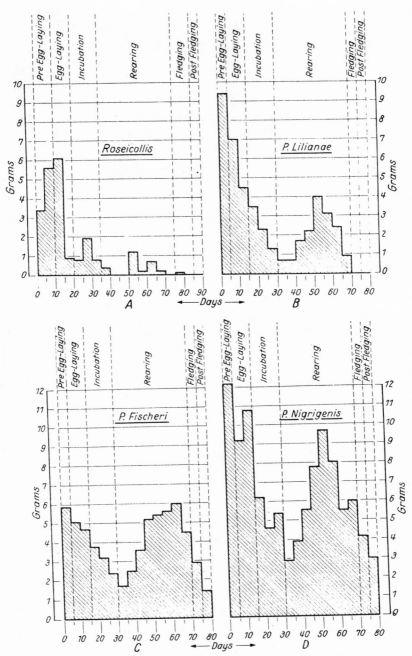

Fig. 19. Graph showing the nest material utilization (in grams) by a) *Agapornis roseicollis:* 2 females, 2 cycles; b) *A. p. lilianae:* 1 female, 1 cycle; c) *A. p. fischeri:* 2 females, 11 cycles; and d) *A. p. nigrigenis:* 1 female, 7 cycles. Successive five-day means were plotted

just before and during egg-laying and the other during the period the young are developing in the nest. The least activity occurs during incubation and during the period between nestings. The tendency to carry nest material is strong during incubation but the female is mostly unavailable to engage in such activities. When she does make an occasional brief appearance, in order to defecate or to eat or drink, she ordinarily performs a strong bout of comfort movements and then takes a few moments to engage in a brief but intense bout of nest material cutting and makes at least one trip into the nest with material. However, during the period between nestings, the female, who now has plenty of time, cuts and carries steadily but in a rather desultory fashion. It is apparent that the drive for nest building is high throughout the reproductive period and that the peaks of activity are the result of superimposed conflicting activities which inhibit nest-building. Nest-building activity does not seem to be so markedly divided into peaks in *cana* and appears to be somewhat intermediate in *roseicollis*.

The placement of material is governed by a complex of stimuli which have not been worked out thoroughly in every case. This will be a part of the forthcoming paper by KEETON. In brief, it may be mentioned that (aside from constructing the nest itself) cracks of light; splinters of wood, twigs, etc. behind which softer material can be wedged; and fresh feces all prompt the deposition of nesting material. The response to fresh feces functions during the nestling stage as a nest sanitation measure. However, experimentation determined that this response is present during egglaying and incubation. During the nestling stage the amount of nest material deposited varies directly with the amount of feces deposited. However, a few days before the young fledge, their feces gradually take on the drier characteristics of adult feces and here the nest material deposition drops markedly. We have added and substracted to the amount of feces deposited in nests and have always obtained a corresponding adjustment in the amount of nest material carried. Foreign objects in the nest, including dead young, are not carried out but are covered with nesting material.

Older young of the white eye-ringed forms do not tend to remain in the nest chamber, but eventually spend most of their time clustered near the entrance hole.

Sexual Behavior. — Males of all species come into sexual condition before the females. We do not know what stimulates the males but they seem at least relatively independent of varying day lengths, judging from the breeding success we have observed under varying light conditions. Males perform a variety of courtship activities ("courtship" being used in the sense of MORRIS, 1956) and at first the females respond overtly with "indifference", active avoidance, or aggression. As time goes on, however, females clearly gradually begin to have a stronger and stronger tendency to behave sexually; until finally the time arrives when they will permit copulation. One function, therefore, of male sexual displays is to stimulate females until their physiological condition will permit them to exhibit sexual behavior (long term arousal). This is in contrast to the similar male activities which serve as stimuli for short term arousal (permitting copulations to occur) after females are in sexual condition. Long term arousal may take anywhere from a few days to two or three weeks. Short term arousals usually take place in a few seconds to several minutes. The speed of the long term arousal seems to be dependent upon the persistence of the male's sexual activities and upon other environmental conditions such as availability of nest sites, nesting material, interference from other birds and so on. Short term arousal seems to be dependent not only upon the sexual activity of the male but largely upon conflicting tendencies in the female (feeding, resting, comfort activities, etc.) and, primarily, upon nest-building which occupies a large part of her time. Male sexual behavior remains strong throughout nest-building, egg-laying, and into incubation, but wanes markedly during late incubation and remains weak until the next reproductive cycle (which may overlap with rearing young). Males of all species have essentially the same activities comprising their sexual behavior. In order to fully understand these activities it must be realized that the two strongest tendencies ordinarily present in a "courting" male are to avoid the female (fear) and to approach her (sex or "mating" in Morris's, 1956, terms). Thus, a male *Agapornis*, particularly of *roseicollis* and the white eye-ringed forms, has the typical conflict of FaM according to Morris's system where "f" refers to flee, "a" to attack, and "m" to mate. The female, on the other hand, has the typical conflict that can be expressed as fAm (in other words, little tendency to flee, considerable tendency to attack, and little tendency to behave sexually). Of course, for copulation to occur both must eventually achieve the situation symbolized as faM, and the sexual activities of the male function to achieve this state. The following is a description of these sexual activities with comments on their differences, similarities, functions, causations, and evolution.

Switch-sidling. — This activity is clearly the result of conflicting tendencies to approach and not to approach (or even flee from) the female. The male sidles along the perch toward or away from the female, repeatedly turning around as he does so. When sidling toward the female, the male typically turns away from her. Causally, this seems to be very much like the "pivoting" described for various small birds (e. g., the Zebra Finch, *Poephila guttata*, Morris, 1954). Weaker tendencies to Switch-sidle cause the omission of the turning component, or they are expressed merely as intention movements to turn while sidling. The speed, direction, and duration of Switch-sidling depends primarily upon signals from the female. If she gives the slightest sign of becoming less aggressive and/or more apt to behave sexually (e. g., slight fluffing of the plumage, particularly of the face and/or a slight leaning forward), the male will sidle toward her more rapidly. If the female signals that she is more aggressive (e. g., any plumage sleeking, especially the face; and/or becoming more erect on the perch) the male will sidle away more rapidly. How far the male travels away from her depends largely upon when she signals a more amenable attitude. If she doesn't do so the male will not continue sidling away indefinitely but will usually fly over her and land on her other side. The female may sometimes respond more strongly by attacking or, more rarely, fleeing.

Carpals Held. — Only *pullaria* males utilize this display and it consists of the bird, with sleeked plumage, raising the carpals out from the body and then, standing rather erect, moving about stiffly, and orienting frontally to the female. This is strangely reminiscent of the Aggressive Upright used by gulls and is probably homologous to the Carpal Flash of *taranta* males. The tail is also spread during high intensity bouts and this display seems to be caused by fear conflicting with sexual or aggressive motivation.

Head-bobbing. — This has been derived from the head-bobbing associated with courtship feeding (see below). The movements consist of repeated bobbing of the head and vary somewhat from species to species, tending to be more numerous and faster in *cana*, *taranta*, and *pullaria* and fewer and slower (more exaggerated) in the white eye-ringed forms. They are intermediate in these respects in *roseicollis*. In *roseicollis* and the white eye-ringed forms, the males, during high intensity courtship bouts (characterized by extreme persistence despite repeated thwartings and rapid transition from one display to another) perform many Head-bobbings but will not feed the female even if she should proffer her bill. This is why we believe that this movement is at least partially ritualized. This particular use of Head-bobbing has not been seen in *cana, taranta,* or *pullaria.* Head-Bobbing may occur during Switch-sidling and seems independent of the distance from the female. Fig. 20/21 shows the differences in form of this movement in *cana, roseicollis,* and the white eye-ringed forms.

Fig. 20. The forms of the head-bobbing associated with courtship feeding in a) *Agapornis cana,* b) in *A. roseicollis,* and c) in *A. p. fischeri.* Note the various amounts of feather ruffling and amplitudes of the movements; also note the differences in body position

Displacement Scratching. — This is a particularly interesting display because it nicely demonstrates the evolution of a display from its non-signal precursor (headscratching). All species will, upon being thwarted, scratch the head with a foot. In *cana, taranta,* and *pullaria,* this movement is indistinguishable from ordinary headscratching. In *roseicollis,* the scratching move- ments are faster and often directed at the bill, rather than at the head plu- mage; they also occur much more frequently than in *cana, taranta,* and *pullaria.* In the white eye-ringed forms, the movement is still faster and most often directed at the bill; here it occurs most frequently of all. A scratching bout may consist of several scratchings in both *roseicollis* and the white eye- ringed forms. In addition to these changes throughout the evolutionary series, the foot employed shows differences from species to species. Originally, the leg nearest the female is the one employed; probably because this leg is already lifted as an intention movement to mount and may, therefore, be an example of what LIND (1959) has termed an activation of an instinct caused by transitional action. The leg nearest the female seems to always be employed by *cana, taranta,* and *pullaria.* However, in *roseicollis* the ratio of employ- ment of the near foot to the far foot is 3:2. In *p. fischeri* and *p. personata* this ratio is 7:6, and in *lilianae* and *nigrigenis* this ratio is very close to 1:1. Dis- placement Scratching in these latter five forms seems to have approached neurophysiological emancipation. We do not yet have precise figures on the frequency of occurrence of scratchings directed at the bill as opposed to those directed at he cheek. However, frequently when a bird scratches its cheek it will immediately afterward flick its head; this is apparently a movement ha- ving the function of rearranging disturbed feathers. Consequently, those scrat- chings directed at the cheek should be more often followed by head flicks than those directed at the bill. The percentages of the time Displacement Scratchings are followed by head flicks for various species are as follows: *roseicollis* (8%), *p. fischeri* (2%), *p. personata* (3%), and even lower in *p. nigrigenis* and *p. lilianae.* Our data for *cana, taranta,* and *pullaria* are too meager for confident evaluation but subjective observation makes it quite clear that the percentages would be much higher than for *roseicollis.* Fre- quently only intention movements for Displacement Scratching are given.

These frequently occur when the external situation, responsible for the thwarting, resolves before the scratching can be completed, or if the male apparently has a rather low sexual motivation when thwarted.

Tail-wagging. — This comfort movement, described above, is frequently employed as a displacement activity by *roseicollis* and the white eye-ringed forms. However, it is not seen when the male is exhibiting low intensity courtship behavior as is Displacement Scratching. It is as if Tail-wagging, being unritualized, is more effective as a displacement activity than is the highly ritualized Displacement Scratching, especially during very strong thwartings.

Displacement Preening. — Apparently, only *cana* normally demonstrates highly ritualized Displacement Preening. Immediately upon being thwarted, during a bout of precopulatory activity, the male makes rapid, small amplitude, sideways jerks of the head. The true nature of these movements may be seen when the thwarting is especially strong. Here, the sideways movements have greater amplitude and the bill then frequently briefly probes the feathers of the sides of the upper breast. Interestingly, one male *roseicollis* was seen to do this two or three times, although this display is apparently extinct in all but *cana*.

"Pseudofemale" Soliciting. — The performance of this activity described on page 677 is so common during the precopulatory bouts of *roseicollis* and the white eye-ringed forms that it must be regarded as a normal activity. It may be given merely as an intention movement, or the complete act may be performed. It is not usually maintained for more than a second or two at the most and always occurs in situations characterized by the thwarting of a strong tendency to behave sexually (mounting the female). However, it seems to have little or no value in alleviating the "uncomfortable" conditions of thwarting as it is ordinarily sandwiched between intense Displacement Scratchings and/or Tail-waggings. The female, under such circumstances, usually makes at least intention movements to mount the male and sometimes does so briefly. Such male soliciting is always facilitated by his back being toward her.

It is, of course, well known that most, if not all, the motor patterns associated with sexual behavior in vertebrates are available for performance in both sexes. The dependence of the sexual motor patterns, exhibited by *roseicollis* and the white eye-ringed forms, on what the partner is doing becomes especially apparent when the situation is somewhat aberrant. For example, sometimes a bout of precopulatory activity will take place on the top of a nest box, a platform seven and a half inches square. Under this circumstance, as the male sidles after the female and she sidles away, a circular chase around the box top results; both birds then alternately adopt the sexual motor patterns of the other (attempting to mount and soliciting). It is as if the very tight, circular course taken by the birds confuses them as to who is following whom!

Such "pseudomale" and "pseudofemale" behavior has never been seen in the sexually dichromatic species.

Squeak-twittering. — This, the only vocalization associated with male precopulatory behavior, has an interesting history. Basically, it is composed of a long series of repetitive notes. In *cana*, this vocalization is made up of a great variety of notes, relatively low in pitch and not at all musical or rhythmical. It is uttered most often after the female has gone into the nest box or has otherwise become absent during a bout of precopulatory behavior. In this context it seems to be associated with a certain amount of thwarting.

The same is true for *taranta;* although here the vocalizations are somewhat harsher and more rhythmical.

In *roseicollis,* Squeak-twittering is higher in pitch, less variable in frequency, and distinctly rhythmical. These conditions are even further pronounced in the white eye-ringed forms. In the former as well as the latter, this vocalization is nearly always associated with Displacement Scratching, which suggests that Squeak-twittering may also be associated with thwarting.

Supplanting. — Supplanting attacks on nearby birds increase markedly as the intensity of male precopulatory behavior rises. I would have expected males to be increasingly "oblivious" to other outside influences as their preoccupation with precopulatory behavior increased. It seems, however, that in the face of repeated thwartings during precopulatory behavior, the males become increasingly aggressive but are unable to direct this to their females and redirect, instead, to nearby birds. Our data for *roseicollis* and the white eye-ringed forms indicate that such supplanting attacks are most frequently followed by mountings or attempts to mount.

Copulation. — This is the least variable and terminal appetitive activity of a successful bout of "precopulatory" behavior. The male mounts the female by stepping up on her back; not by flying up and landing on her as is done by many birds. The female may solicit copulation by crouching with the body held nearly horizontally, wings raised (but not much extended), and the head and tail raised. The feet also assume a wide stance on the perch and the feathers around the cloaca are parted. In very strongly soliciting females the head and tail may even be raised beyond the vertical. In these cases the plumage is also fluffed more than usual. A weak tendency to solicit is characterized by a slight wing raising. As the tendency becomes stronger the wings are raised even more (they are never quivered), and finally the head and tail become involved along with the crouch. No vocalizations are associated with female solicitation. The female will often solicit in response to a varying amount of male precopulatory behavior, but more often she does not actually solicit until the male makes an intention movement to mount. Rarely, a female will solicit without any immediate prior activity on the part of the male. In these cases the male simply mounts without any preliminary display.

As soon as the male mounts, he grasps the female's flank feathers in his feet, lowers his tail to one side of hers, and makes cloacal contact with a twisting, thrusting motion. Copulation consists of a series of these movements repeated with rhythmical regularity (about two per second). This activity may last from a few seconds to as long as over six minutes. After every dozen or so thrustings the male changes sides (places his tail on the other side of the female's) very quickly; so quickly that the normal rhythm is scarcely broken. Dismountings before sperm emmission are usually initiated by the female. She may threaten the male with Gaping over her shoulder or simply leave, either sidling or flying off (simply sidling away ordinarily does not dislodge the male unless she also threatens). In the few cases where the male has been seen to dismount on his own accord, it has been to supplant another bird which has approached too closely. In these cases the female usually remains in the copulatory posture, and the male immediately returns and resumes. After a prolonged copulation, the female remains in the copulatory posture for a few seconds after the male dismounts. Her cloaca then appears swollen, wet, and reddened if there has been sperm emmission; if not, the cloaca retains its normal appearance.

Males do not hold on with the bill during copulation but will occasionally use the closed bill as a crutch for support during mounting or once in

69

a while during copulation. Males often flap the wings as if to aid in their balance. Females may also do this, but not so commonly.

Females of the sexually non-dichromatic species hold the head tilted sideways with the cheeks and throat ruffled during copulation. The feathers of the wrists are commonly ruffled also. These are the features most visible to the male during copulation (Fig. 21 a). It is as if males of these species have to be constantly reminded as to the sex of their partners and the females have thus evolved a compensatory "diethism" in the absence of sexual dichromatism. If such females abandon the cheek ruffling and head tilt during copulation, the male usually dismounts at once and resumes precopulatory activities.

Fig. 21. a) Copulation in *Agapornis roseicollis*. Note the ruffled wrists and tilted head with ruffled cheeks and throat. This is typical of *personata* females as well. b) Copulation in *Agapornis taranta*. Note the sleeked wrists and untilted head devoid of any ruffling. This is typical of *cana* and *pullaria* as well. The open mouth (squeaking) is species-typical for *taranta* females only

Females of *cana*, *taranta*, and *pullaria* do not hold the head tilted to one side nor are the cheeks, throat, and wrists ruffled (Fig. 21b). A monotonous series of rapid squeaks is uttered constantly by *taranta* females during copulation; these are the only females which vocalize in this context. This vocalization sounds somewhat like Squeak-twittering in males but is more regular in frequency and may have an appeasement function. The males of the dichromatic species are not typically as afraid of their mates as are the others. There seems to be less conflict between attack and or escape with sex.

If sperm emission occurs, there are no more precopulatory activities for some time (ordinarily a half hour to an hour or more elapses before the next such bout). Ordinarily, many incomplete copulations occur throughout the day. However, no copulations under about two and a half minutes in length seem to result in sperm emission.

There are no postcopulatory displays by either sex in any species of *Agapornis*. The females do not even engage in bouts of plumage ruffling and shaking, preening, and so forth. Ordinarily, both birds remain quite still for some time after the male dismounts.

Changes in Precopulatory Activities in Time. — As a pair (particularly among the white eye-ringed forms and *roseicollis*) has progressively more breeding experience, the precopulatory displays change markedly. All of the components remain in their species-typical forms but their frequencies undergo alteration; Switch-sidling becomes less "intense" and Squeak-twittering eventually almost disappears in some cases. Displacement Scratching becomes less frequent in response to typically fewer thwartings. It is as if the birds gradually get to "know" each other better and their interactions become smoother because of increased sensitivity to each other's individual response variations. For reliable quantitative comparisons between species it is therefore necessary to compare birds having similar amounts of breeding experience.

Courtship-feeding. — This is a particularly interesting activity since, like nest material preparation, transportation, and utilization, it apparently is influenced by successive physiological causal factors, as well as having several functions. First of all, courtship feeding occurs throughout the year. Hence, it seems to be another one of the many activities between mated pairs and probably serves, along with these, to constantly reinforce the pair bond and to ameliorate any inhibitions there may be for close bodily contact between the birds.

Birds may often be seen to engage in other activities during courtship feeding (for example, various stretching movements of the wings and legs). Once, a pair of Budgerigars were seen to engage in courtship feeding during copulation!

Courtship feeding, however, becomes much more frequent in occurence with the onset of reproductive activities, and may actually be one of the activities which tend to stimulate the female reproductively. This behavior continues throughout the period of sexual activity and gradually grades into behavior characterized by the male feeding the female while she is incubating and brooding young. Feeding the female while she is incubating probably provides an important proportion of her nourishment as she typically briefly emerges from the nest cavity only three or four times a day. Later, when the young are being raised, this behavior not only helps to nourish the female but enables the female to provide the very young nestlings with well processed food since it has been partly digested by both parents. When the young become older, both parents feed them directly and now they receive essentially undigested, whole hulled seeds. At this time courtship feeding is relatively infrequent and the male continues to feed the young until they are several days out of the nest. The frequency of courtship feeding again increases. The head-bobbing component of courtship feeding differs from species to species (Fig. 20), but these differences are most exaggerated in *roseicollis* and the white eye-ringed forms during the sexual phase of the cycle and may, therefore, reflect the presence of sexual motivation superimposed on courtship feeding. More likely, however, these special feather adjustments may reflect the typical fear of these males toward their females and indicate a degree of submissiveness during sexual activities.

One of the reasons I believe that courtship-feeding is not prompted by sexual motivation is the fact that it frequently immediately follows a successful copulation but clearly sexual activities never do so.

All of the activities thought of as precopulatory may sometimes be associated with courtship feeding during the sexual phase of the cycle. It may be that courtship feeding has its own unique responsible internal state which can be somehow linked to other drives which may lower or raise the threshold of its performance.

Only the male feeds the female in *roseicollis* and the white eye-ringed forms (Fig. 22a). Much plumage fluffing and ruffling is done by these females, indicating their submissiveness. In *cana* and *taranta* the female frequently

Fig. 22. Courtship feeding a) in *Agapornis roseicollis* Note the ruffling and fluffing in the female. This is representative of *personata* females as well. b) male *taranta* feeding female (note lack of fluffing or ruffling in female)

feeds the male; more often in *cana* (Fig. 23). No plumage ruffling occurs here. It appears that, during the evolution of this genus, there has been an increasing tendency to relegate courtship feeding to a strictly unilateral activity with the male doing the feeding. This may have been possible because of the increasing stimulatory value, through the series (toward colonialism) of other birds outside

Fig. 23. Courtship feeding in *Agapornis cana*. Note the lack of fluffing or ruffling. a) male *cana* feeding female, b) female *cana* feeding male

the pair which would perhaps compensate for any possible reductions of stimulation associated with a lack of reciprocity in courtship feeding.

Recognition of Young. — After many attempts to achieve the rearing of young by foster parents by swapping egg clutches, we found that none of the species normally having red-downed young would feed newly hatched white-downed young and *vice versa*. However, the pairs among which we made the swappings had all raised many prior broods of their own. We then attempted to get *roseicollis* (red-downed young) to raise white- downed young by giving such eggs to a pair which had never bred before. In this we were completely successful. Moreover, when we gave this pair red-downed young of the same age as the white-downed young they were rearing, they refused to feed the former. Young of different species which are well-feathered appear much more unlike each other than they do at hatching, yet these birds may be placed in nests of other species and raised with no difficulty.

At the present time we are concentrating on greater comparative quantification of the behavior of these birds and have begun to investigate the effects of early experience and genetics on the development of species-typical behavior.

Summary

The comparative ethology of the parrot genus *Agapornis* is described and discussed (development of the young, maintenance activities, agonistic behavior, and reproductive behavior). It is apparent that there is a fairly uniform evolutionary trend in the genus represented by the living forms from *cana* through to *personata nigrigenis*. This trend involves the gradual loss and acquisition of various behavioral and morphological characters. Certain of these trends sometimes tend to reverse themselves. Certain of the more important trends are listed below:

1) loss of sexual dichromatism, *pullaria* being intermediate in this regard.
2) development of increasing colonialism, *roseicollis* probably representing an intermediate condition.
3) early loss of both nest defence displays and ritualized displacement preening.
4) increasing ritualization of displacement scratching and, finally, the appearance of unritualized tail-wagging as a displacement activity.
5) gradual loss of carrying nest material in the feathers and the acquisition of carrying such material in the bill.
6) increase in size of strips of nest material and, finally, the utilization of twigs.
7) increase in complexity of nests from a simple pad (*cana, taranta,* and *pullaria*) to a well-formed cup (*roseicollis*) to an elaborate substantial structure featuring a roofed nest chamber and tunnel connecting it to the outside entrance (*personata*).
8) loss of reciprocal courtship feeding and the ritualization of the head-bobbing component as a precopulatory display.
9) blue rumps develop (*pullaria*), become strongly developed (*roseicollis* and *swinderniana*), begin to be lost (*p. fischeri* and *p. personata*), and are lost (*p. lilianae* and *p. nigrigenis*).
10) development of increasing exaggeration of head color and pattern.
11) gradual development of more predictable sexually-specific behavior relative to attack, escape, and sex (coincident with gradual loss of sexual dichromatism).

Some features are species-specific and must have developed after the ancestral population had severed its genic connections with the rest of the genus. Some of these are: 1) pale bills in *cana* and again in *roseicollis*, 2) Carpal Flashing in *taranta* males, 3) squeaking in copulating *taranta* females, 4) burrowing into tree ant nests by *pullaria*, 5) developing black bill in *swinderniana*, 6) use of feathers as nest material by *taranta*, 7) development of white, naked skin around eyes in *personata*, and 8) carrying nest material in bill by *personata*.

Sleeping upside down must have been ancestral to both *Agapornis* and *Loriculus*. It is still retained by *Loriculus* but spottily by *Agapornis* (strong in *pullaria*, and weak in *taranta*).

In the less social species, particularly in *cana*, agonistic behavior has become associated with rather elaborate threat and appeasement displays but little inhibition in actual fighting if the displays fail in their function. The results of such fighting, though rare, are disastrous. On the other hand, the more social species have little ritualization associated with threat but the fighting itself has become highly ritualized (Bill-fencing). Here, the ritualized fighting and inhibition associated with biting serves to prevent damage to contestants; a distinct advantage among birds having many contacts with others of their kind.

The development of highly ritualized displays from non-signal precursors is readily apparent in this genus. It is also apparent that there are not a large number of displays associated with sexual or agonistic behavior. This suggests that display extinction must be as normal as display evolution and that there must be very strong competition among existing displays from the standpoint of their value as signals prompting certain responses in the recipient. The selective pressure molding precopulatory displays, for instance, probably initially stem from the inherent response capabilities of the recipient. Those activities capable of achieving quicker and/or more profound effects are probably selected for over others, all else being equal. The loss of highly ritualized displacement preening in all but *cana*, the development of highly ritualized displacement scratching in *roseicollis* through to *personata*, and the appearance of displacement tail-wagging in these latter birds probably represent such a struggle for display survival. It is interesting, in connection with this, that an "extinct" display will sometimes appear rarely in an individual of a species normally no longer exhibiting this behavior. For instance, the Displacement Preening of *cana* has been seen in its highly ritualized form once or twice in a male *roseicollis*.

The head has obviously become exceptionally important in the behavioral interactions of these birds and much activity and special color patterns are associated with this (courtship-feeding, Bill-fencing, reciprocal preening, strongly contrasting bill colors, and the development of conspicuous eye-rings).

Preliminary investigations on hybrids (the subject of a forthcoming paper) and on birds variously deprived of certain early experiences lead to the conclusion that many, if not all, species-typical patterns are dependent upon an intricate complex of both learned and not-learned methods of acquiring such behavior. This subject is being explored further at present. The individual development of Bill-fencing, nest building, and recognition of newly hatched young are cases in point.

Acknowledgements

I wish to express my appreciation to the R. T. French Co. and to the National Science Foundation (N. S. F. G-5518) for financial aid. Many people were of assistance during the course of the study and in criticising the manuscript but I would like to particularly thank Mr. and Mrs. Alan BROCKWAY, Mr. and Mrs. Robert FICKEN, Dr. Robert GOODWIN, Miss Helen HAYS, Mrs. William KEETON, and Miss Jane SELIGSON. Drs. Daniel LEHRMAN, Konrad LORENZ, and Niko TINBERGEN helped me with many useful and stimulating discussions. Mr. David ALLEN was especially helpful with photographic matters. I wish also to thank Prof. O. KOEHLER for preparing the German Summary. I hasten to add that none of these people are responsible for any errors of omission or of commission which may have occurred.

[*Editor's Note:* The foreign language summary has been omitted at this point.]

Literature cited

ADLERSPARRE, A. (1938): Dimorphismus des Jugendkleides und Nestbau bei *Agapornis*. J. Ornith. 86, 248. • BENOIT, J. (1956): Etats physiologiques et instincts de reproduction chez les oiseaux . *in:* L'Instinct dans le Comportement des animaux et de l'homme. Paris. • CINAT-THOMPSON, H. (1926): Die geschlechtliche Zuchtwahl beim Wellensittich *(Melopsittacus undulatus* Shaw). Biol. Zentralbl. 46, 543. • DILGER, W. (1958): Studies in *Agapornis*. Avic. Mag. 64, 91. • HAMPE, H. (1937): The nesting habits of *Agapornis pullaria*. Avic. Mag. 2, 148. • HAMPE, H. (1957): Die Unzertrennlichen. Pfungstadt. • LIND, H. (1959): The activation of an instinct caused by a "transitional action". Behaviour 14, 123. • MOREAU, R. (1948): Aspects of evolution in the parrot genus *Agapornis*. Ibis 90, 206, 449. • MORRIS, D. (1954): The reproductive behaviour of the Zebra Finch *(Poephila guttata)*, with special reference to pseudofemale behaviour and displacement activities. Behaviour 6, 271. • MORRIS, D. (1956): The function and causation of courtship ceremonies. *in:* L'Instinct dans le comportement des animaux et de l'homme. Paris. • NEUNZIG, R. (1926): Zur Systematik und Biologie der Gattung *Agapornis*. Verh. Ornith. Ges. Bayern 17, 112. • PERRY, J. (1959): Breeding the Red-faced Lovebird *(Agapornis pullaria)* in South Africa. Avic. Mag. 65, 119. • PETERS, J. (1937): Check-list of Birds of the world, Vol. 3 Cambridge. • PRESTWICH, A. (1957): Breeding the Red-faced Lovebird *Agapornis pullaria*. Avic. Mag. 63, 1. • SERLE, W. (1957): A contribution to the ornithology of the eastern region of Nigeria. Ibis 99, 371. • TINBERGEN, N. (1959): Comparative studies of the behaviour of gulls (Laridae): a progress report. Behaviour 15, 1.

7

Reprinted from *Ibis* 99:275-302 (1957)

ADAPTATIONS IN THE KITTIWAKE TO CLIFF-NESTING.

By Esther Cullen.

(*Department of Zoology and Comparative Anatomy, University of Oxford.*)

Received on 8 August 1956.

INTRODUCTION.

In 1952 I undertook a study of the breeding behaviour of the Kittiwake *Rissa tridactyla* in order to extend the scope of studies of various gulls, initiated by Dr. N. Tinbergen at Oxford. At that time accounts of the behaviour of this species were fragmentary (Selous 1905, Bent 1921, Perry 1940, Fisher & Lockley 1954), though recently the situation has been altered by a paper by Paludan (1955).

My observations were made while living on one of the Farne Islands off the Northumberland coast, where I was able to watch the birds through three seasons, 1953–55, and more briefly in 1952 and 1956. Although my observations were all made in these colonies there seems no reason to doubt that the behaviour described is typical for the species.

It is well known that the Kittiwake, together with its close relative *Rissa brevirostris*, differs from most gulls in two ways : outside the breeding season they live in the open sea rather than in the neighbourhood of land, and when they do visit the land to breed they nest on tiny cliff-ledges. Other gulls occasionally nest on ledges but these are far larger than the little shelves, sometimes only four inches wide, on which a Kittiwake can stick its nest. The present paper sets out some of the peculiarities of the Kittiwake's behaviour compared with the ground-nesting gulls and attempts to show that these peculiarities can be related to the cliff-nesting habit. A more detailed account of other aspects of the behaviour of the species is in preparation.

Since the great majority of gulls nest on the ground one may presume that this was also the ancestral breeding habitat of the Kittiwake. This view is supported by two facts : (1) Kittiwakes' eggs retain to some extent the cryptic pattern of blotching although this can be of little value, as every nest is marked conspicuously by a flag of white droppings (p. 288); (2) the young Kittiwakes are able to run under suitable conditions, though not quite as well as young ground-nesting gulls. This is an unusual feature for a species nesting in such precarious sites.

The advantage of cliff-nesting is certainly that it reduces predation. The nests seem fairly safe not only from ground-predators but also from such aerial ones as large gulls. I never saw a single Kittiwake chick or egg taken from the nest by the Herring Gulls *Larus argentatus* or Lesser Blackbacks *Larus fuscus*, which frequently preyed on the eggs and young of the Eider Ducks *Somateria mollissima* and terns nesting on the ground nearby; Nicholson (1930) remarks on a similar immunity even in the Kittiwakes nesting within a foot or so of the nests of Iceland Gulls *Larus leucopterus*. Perry (1940) observed two pairs of Herring Gulls repeatedly robbing young Kittiwakes from their nests but such predation does not seem to be the rule. It is reported however that in some places in the Shetlands Great Skuas *Catharacta skua* are a greater danger to Kittiwakes' nests. Lockie (1952) on Hermaness found them feeding largely on Kittiwakes' eggs and young and Venables & Venables (1955) also report nest-robbing by them in Foula. But heavy predation of nests seems restricted to few places (L. S. V. Venables, personal communication) and in the colony in Noss for instance the Skuas seem to specialise more on catching the Kittiwakes in the air (Perry 1948).

The Kittiwakes' security from large gulls probably results from the difficulty which they would often have in landing on the small perches, and this is enhanced by the complicated wind-eddies in such places. In strong winds this turbulence is sometimes so awkward that even the agile Kittiwake has to circle again and again before it can alight where it wants to.

In addition to this protection from other species of gull the Kittiwake does not suffer from the depredations of its own species as do the ground-nesting gulls. Unlike the other gulls, the Kittiwake does not take eggs or young of other birds but eats mainly fish and plankton. This difference in diet is a subsidiary reason why the Kittiwake's eggs are safer than the ground-nesters'. But even if Kittiwakes did prey on their own species they would suffer less than the other gulls because their nest-sites are much less accessible to ground-predators who might cause the parents to leave their nests exposed to be robbed by neighbours.

Throughout this paper I have contrasted the behaviour of the Kittiwake* with that of the ground-nesting gulls as a group. Only two species, the Black-headed Gull *Larus ridibundus* and Herring Gull have been at all fully studied, but quite a lot is known about the other gulls in a more fragmentary way. Although more studies are desirable, on the existing information there seems a good indication that in many respects the ground-nesters behave alike. For information about the behaviour of the ground-nesters I have used published information (recent references summarised in Moynihan 1955 and Tinbergen 1956) and drawn extensively on personal communications from N. Tinbergen, M. Moynihan, R. & U. Weidmann. I have also had the opportunity of seeing something myself of the behaviour of Herring and Lesser Black-backed Gulls on the Farne Islands.

* The behaviour of *Rissa brevirostris* is very little known but appears to be similar to that of *Rissa tridactyla* (Bent 1921).

FIGHTING METHODS AND RELATED BEHAVIOUR.

The ground-nesting gulls have various ways of attacking, the Kittiwake has specialized on one and this seems to have had a number of repercussions on the Kittiwake's behaviour.

According to Tinbergen (1953) and Moynihan (1955) an attacking Black-headed or Herring Gull tries to get above the enemy and peck down, or it may try to grasp any part of the other and then pull. If it happens to grasp the beak it sometimes tries to twist the other's head from side to side and thus may upset its balance. But this twisting seems rather rare : it is reported from the Common Gull *Larus canus* (Weidmann 1955), Tinbergen has seen it occasionally in the Herring Gull (personal communication) and I have observed it twice in the same species.

The Kittiwake on the other hand attacks by darting its head forward horizontally and always tries to grasp the opponent's beak. If it succeeds it twists the head from side to side. As most fights start on the ledges, which comprise the birds' whole territories, this method is very effective and often the opponent is twisted off the sill. It may however hang on to the other's beak and both tumble into the sea together, where they go on twisting with beaks still locked, ducking each other's heads under the water.

It would be difficult for the Kittiwake to use the other fighting methods of the ground-nesters; on their small perches they obviously cannot pull backwards and the structure of the cliffs often does not allow a bird to get above the other to peck downwards. On the water the ground-nesters' methods would probably be equally inefficient and it may be no mere coincidence that both times I saw the beak-twisting method used by Herring Gulls they were fighting on the water.

In connection with its specialized fighting method the Kittiwake seems to have lost a particular threat posture, which has been found in all the other 16 species of gull observed. In this posture, the " aggressive upright " (Moynihan 1955) or " upright threat " (Tinbergen 1953), the neck is stretched upwards and often a little forward and head and bill are pointed down. Tinbergen (1953) interprets the stretched neck and the attitude of head and beak as an intention-movement of pecking down at the opponent from above and later authors essentially agree with this interpretation in other species (Moynihan 1955, Weidmann 1955). As the Kittiwake does not fight in this way it is not surprising that it does not have the aggressive upright posture.

With the Kittiwake's specialization on one way of fighting the beak seems to have become an important stimulus in aggressive encounters. The fact that an attacking bird always goes for the beak of its opponent shows that the beak is important in directing the attacks. But it also appears to be a strong releasing stimulus. This can be concluded from observations in three situations : (1) when a female joins an aggressive male and tries to stay beside him and to prevent him attacking her, (2) when two females compete for one male, each trying to drive the other away, and (3) when one bird is trying to conquer a nest from another bird or from another pair. In each of these situations the bird who is attacked, whom I shall speak of as the visitor or intruder, may fight back, but often it seems too frightened to attack and turns its beak away hiding it in the breast-feathers, at the same time erecting the feathers of the neck. The attacker tries to get hold of the opponent's beak from either side, twisting and turning its neck over the other's; but if the beak is sufficiently hidden it does not succeed and usually after a while the attempts to attack stop, though the aggressor could easily peck other parts of the intruder's body. Should the intruder's posture relax so that it shows its beak again the owner of the nest starts to attack once more and tries to grab the beak and throw the other off the ledge.

It should be added that the beak-hiding does not always succeed in pacifying an opponent and a very aggressive bird may peck the neck or body of an intruder whose beak is out of reach. There is no doubt however that by tucking its beak out of sight a visitor can stay on a ledge beside the owner for much longer than it otherwise could. On one occasion I saw an intruder crouching on a ledge in this way beside the owner pair for a whole hour. A pair may ignore the presence of a stranger on their ledge to such an extent that they perform various friendly displays together such as mutual head-tossing, sometimes even standing on the intruder as if unaware of it, whereas a stranger with its beak visible would never be tolerated.

To test the importance of the beak experimentally I used models of Kittiwake heads, which were cast of a plastic material (" Welvic " paste) from the same mould and then painted. I placed them on Kittiwakes' nests and counted the number of pecks delivered at them during a standard time after the bird had alighted on the nest. In some preliminary tests to determine the attack-releasing effect of the beak I presented two models in succession, one with a yellow beak like the adult's and one without a beak. Too few birds responded successively for a definite conclusion, but the results at least suggest that the head with the beak releases more attacks than the one without.

The experiments show more convincingly that the beak directs attacks. There were seven individuals who attacked the model with the beak, making 286 attacks in all. Of these attacks 98% (all but seven) were directed at the beak itself.

From general observations of fights it appears that the beak is not such an important stimulus in the ground-nesting gulls as it is in the Kittiwake. In particular the ground-nesters do not as a rule aim their attacks at the beak of the opponent (p. 277). Huxley & Fisher (1940) and R. Weidmann (personal communication) made some model-experiments with Black-headed Gulls which give information on the attack-directing effect of the beak in this species. Huxley and Fisher used stuffed birds and a corpse which they decapitated after a few experiments, presenting head and body separately. They found that the whole specimens were attacked and so was the head alone, but only a few pecks were aimed at the beak of the latter and none at all at the beaks of the whole birds. Weidmann repeated the experiments and also found that a far smaller percentage of the attacks were aimed at the beak of the head-model in the Black-headed Gull than I found in the Kittiwake though exact figures are not available. From general observations it seems probable that the beak directs attacks equally little in the other ground-nesters.

In releasing attacks the importance of the beak in the ground-nesters is more uncertain. I have said earlier that in the Kittiwake the effect of the beak-hiding demonstrated the attack-releasing effect of the beak. The ground-nesting gulls have a head-turning movement rather similar to that of the Kittiwake, the so-called " head-flagging " (named by Noble & Wurm 1943), but its effect is less conspicuous than in the Kittiwake and no experiments have been made to investigate its function. It has however been previously inferred without actual experiments that the beak had a certain attack-releasing effect in the Black-headed Gull (Tinbergen & Moynihan 1952).

None of the ground-nesting species shows such an elaborate hiding of the beak as the Kittiwake, they merely turn or jerk the face away from the opponent in the head-flagging. There cannot be any doubt that the movements are homologous, especially as the Kittiwake at a lower intensity also merely turns the face away, but one can understand why the Kittiwake has the elaborate beak-hiding : firstly because the beak may have a stronger attack-releasing value and secondly because the situation in which the Kittiwake does the elaborate beak-hiding does not occur in the ground-nesting gulls. There, a frightened bird which is attacked but wants to stay near the opponent or on the territory can always take a few steps away from the attacker to a place where it is not so frightened and does not provoke the other so much. But the Kittiwake has only the choice between staying very close to the other bird or leaving the territory altogether. Thus the intruding bird's greater fear and the attacker's increased aggressiveness might have led to the Kittiwake's exaggerated beak-hiding.

When young Kittiwakes are threatened they show the same head-turning movement as the adults and, as in the adults, it " appeases " the opponent, i.e. it tends to stop him from attacking. Young ground-nesters apparently do not head-flag at all. Adult gulls, Kittiwakes as well as ground-nesters, head-flag (or hide the beak) when they are pecked or threatened but want to stay. This situation hardly ever arises for young ground-nesters, for while they may be attacked by strangers, unlike young Kittiwakes, they can usually run away. This must be the reason that the Kittiwake is the only species known where the young shows the head-turning movement.

A young Kittiwake is not often pecked by a stranger. In the first weeks after hatching the parents guard the nest so well that a strange bird is seldom able to land there. But later the young are more often left alone and an adult who is searching for territory may land on their nest. When this happens the adult may attack the young. On the other hand fights between nest-mates over food are common and some kind of appeasement seems necessary then. There are usually two young in a nest, one a day or two older than the other. They may sit peacefully together for hours, but hostility flares up when the parent comes to feed them. As soon as the younger starts to beg for food or tries to get the food, the older gives it a peck or prepares to do so and the smaller at once turns its head away. This stops the older from attacking and while the smaller looks away the older often can get the food for itself. Only when the older chick is satisfied is the younger one able to get food.

In the young ground-nesters fighting between brood-mates does not seem to occur in connection with food or for any other reason (personal communications, R. & U. Weidmann, N. Tinbergen). The need for appeasement does not therefore normally arise.

I created this need artificially for two Herring Gulls and a Black-headed Gull which were placed in Kittiwakes' nests. They were fiercely attacked by their Kittiwake foster-brothers and wire-netting fastened round the rim of the nest prevented them from escaping. Even in these circumstances the young ground-nesters showed no trace of the head-turning. As all three chicks had been hatched in Kittiwakes' nests this difference in behaviour must be innate.

The young ground-nesters were attacked more fiercely by the Kittiwake foster-brothers than young Kittiwakes would be in the same situation. Even two Kittiwake chicks which from hatching were always in the company of a young Herring Gull and Black-headed Gull respectively remained extremely hostile towards their companions whereas two young Kittiwakes settle down more quickly. This continued hostility might well be due to the lack of head-turning in the ground-nester chicks.

It seems that the appeasement effect of the beak-hiding in the adult Kittiwake is mainly due to the removal of the beak from view. But when a young Kittiwake turns its head away it not only conceals its beak, but it also shows off the black band across the nape, which is present only in this plumage (Fig. 1). As the band is erected and displayed only in this situation one may infer that it is a special appeasement structure supporting the effect of hiding the beak. Of all the 44 species of gull which Dwight (1925) described, only the young Kittiwake and the young of *Rissa brevirostris* have such a band. This supports the idea of its function, since only the young of these two species must stay where they are however severely they may be attacked.

A slight difficulty remains. I have suggested that the head-turning of the young has been evolved in connection with fighting between the nest-mates, but this fighting is most fierce in the first days after hatching, before the neck-band develops. In these early fights the tiny chicks will hide their beaks just like the adults, squatting low in the nest-cup and tucking the beak under the breast. This gesture tends to appease the opponent although the displaying bird presents nothing but the pink skin of the neck shining through the white down. These early fights establish a kind of peck-order between the chicks and later when the black band has developed actual fights between the young are rarer. But the head-turning remains extremely common, especially at feeding times. There is however reason to think

FIGURE 1.   Young Kittiwake beak-hiding.

that it is still important in preventing fights and an additional structure, the black band, may indeed be more necessary than at first, because the young are now much stronger.

We may conclude therefore that the young Kittiwake who has both the head-turning movement and the black neck-band has developed these characters as a consequence of cliff-nesting.

The head-turning in the Kittiwake seems to work because it conceals the weapon from the opponent and Tinbergen & Moynihan (1952) have suggested basically the same explanation for the head-flagging in the Black-headed Gull. But Lorenz has put forward another interpretation of similar movements, based mainly on his observations on dogs and wolves. He holds that the movements have appeasement effect because the inferior animal presents the most vulnerable part of its body to the enemy, who has a specific inhibition to attack the spot. This explanation cannot hold for the Kittiwake and probably not for the gulls in general, since in exceptional circumstances the neck may be pecked severely and the victim does not seem to take serious harm from it. It even appears uncertain at present whether Lorenz's explanation can hold for wolves and dogs, since, according to R. Schenkel (personal communication), Lorenz has misinterpreted the appeasement ceremony in these animals.

ADVERTISING DISPLAY OF THE MALE.

The Kittiwake differs from the ground-nesters in the way in which the new pairs form at the beginning of the breeding season.

In the Kittiwake the males occupy their territories, i.e. their nesting ledges, as soon as they arrive at their nesting cliffs from the winter quarters. They advertise themselves there with a special display which attracts females to them and it is on the nesting ledges that the pairs form. Unmated ground-nesting gulls usually do not seem to go straight to their breeding territories after they have arrived in the nesting area, but may spend the first few days either on neutral ground, the so-called " clubs " (Tinbergen 1953) or " pre- " or " pairing-territories " (Tinbergen & Moynihan 1952, Moynihan 1955). Here as a rule the pairs form and only later do they seem to occupy the actual breeding territories. In some species pairs may form outside the breeding season (Drost 1952) or perhaps even away from the breeding grounds (Moynihan, personal communication).

There may be several reasons why the Kittiwakes go straight to their nesting places while other species apparently do not. Firstly the Kittiwakes appear to be frightened of the land : they avoid sitting on the top of the nesting cliffs in the first few days after the occupation of the colony and it seems that the nesting ledges are the only places on the cliffs where the birds dare to land at this time. Secondly the Kittiwakes may have to stake their claims early because the number of suitable ledges is restricted and the competition for nest-sites is severe. On the other hand for the ground-nesters there is probably little shortage of nest-sites and they can therefore afford to wander about in clubs and pairing-territories.

The Kittiwake males advertise themselves to the females by a particular display, the " choking " (named by Noble & Wurm (1943) in the Laughing Gull *Larus atricilla*). Figure 2 shows the posture of the body in a choking bird and in this position head and neck are rhythmically jerked up and down with a nodding movement of a frequency of about 3 nods per second. Un-mated females are attracted by this display and land beside advertising males.

Choking occurs in other species of gulls, but the Kittiwake is the only species known regularly to use it when advertising for a mate. Unmated Black-headed Gulls use another display, the " oblique " posture with the " long call " (Moynihan 1955) and from recent observations it appears that unmated Herring Gulls also may use the long call as advertisement display (Tinbergen 1956).

On the other hand there are indications that both in the Herring Gull (Goethe 1937) and Black-headed Gull (R. & U. Weidmann, personal communication) an unmated male may occasionally choke in order to attract a mate. In these species and also in the Common Gull one can see much more often the attraction exerted by the choking display between a mated pair (Tinbergen 1953, Weidmann 1955). One bird, usually the male, walks away from its mate towards the nest or a prospective nest-site and chokes there; and this at once causes the female to approach and join in.

From these reports of the choking of the ground-nesters two points emerge. Firstly that choking attracts a mate and even though it may not be the usual display by which the unmated male first allures a female, it is not surprising that the Kittiwake has developed it for this purpose. Secondly it appears that the ground-nesting gulls tend to go to a nesting place in order to choke

FIGURE 2.   Kittiwake " choking ".

and this may be the reason why the Kittiwake males, whose pairing territory is the nesting ledge, have specialized more than any other gull in advertising themselves by choking.

The link between choking and the nest-site can be partly understood when the origin of the display is explained. The origin of choking has been discussed by several authors, most recently Moynihan (1955). Two different explanations have been put forward : one is that most if not all elements of the display are derived from nest-building movements; the other is that the pattern is derived from feeding the chicks. For various reasons, which will be published in a later paper, I believe the first explanation to be true and that choking behaviour is mainly derived from a displacement nest-building activity. It has been suggested (Armstrong 1950, Tinbergen 1952, Lorenz 1953 and more recent authors) that in situations where displacement activities might be expected external stimuli may play a part in determining which particular activity is chosen and it might well be that for this reason choking so often occurs at the nest-site.

It should however be stressed that this influence of the nest-site is not direct, because when a Kittiwake male advertises on the water or on the cliff-top, as happens occasionally, he still chokes as he would when advertising on the nest. The influence of the nest-site must be supposed rather to have acted phylogenetically in selecting for this particular type of advertisement display.

COPULATION.

Almost all the Kittiwake copulations take place on the nesting ledge. Only exceptionally are pairs seen to copulate elsewhere, for instance on the cliff-tops, and these are birds who do not own a ledge.

During copulation the female sits down; and Paludan (1955) has pointed out that this is different from the behaviour of a female ground-nester, who remains standing during the whole performance, often moving a little to and fro, apparently to keep her balance. Paludan has also pointed out that this method of copulating would be most unsuitable in the Kittiwake, as the female has only the tiny ledge to move on.

That this adaptation is not immediately dependent on the external situation is shown by the fact that the female sits down even during copulations on flatter places, where she would have plenty of room to move about.

NEST-BUILDING.

Nobody who has seen the loosely built nests of one of the ground-nesting gulls will expect that such a construction could be stuck onto the small shelves of the Kittiwake cliffs. And indeed the Kittiwake has a nest-building technique which is much more elaborate than the ground-nesters'.

To begin the nest the Kittiwakes mainly bring mud or soil, often mixed with roots or grass-tufts. Such trips for mud sometimes alternate with trips for grass, seaweed or other fibrous material. One bird usually deposits its material sideways over its shoulder with a rhythmic downward jerking movement of the head, performing about three to four jerks per second; at each downstroke the beak is opened and in this way the material, which often sticks to the beak, is jerked down piece by piece onto the nesting platform. Then the bird starts to trample on the ledge as if it were marking time. A bird trampling intensely does about six to seven steps a second and may go on trampling for up to half an hour with only short intervals. This behaviour is poorly directed as the bird often tramples a long time beside the nest-material on the bare rock, apparently without noticing the mistake. But as the pairs go on building in that way for several hours they succeed in the end in stamping the mud and the fibrous material to a firm platform, which sticks to the narrow ledges and enlarges them and produces horizontal shelves on amazingly steep slopes. In the later stages of building the birds collect mainly material such a dry grasses, which they deal with in the same way as the mud : jerking it down and trampling on it, so as to form the nest itself on the reinforced mud-foundation.

In other gulls nest-building is simpler; they collect no mud but only fibrous material and they hardly trample on it at all. The nests produced are piles of material loosely interwoven and this is all that is necessary on the ground.

Published descriptions (e.g. Moynihan 1953) of the depositing movements in ground-nesting gulls do not mention the downward jerking movement which I have described. However the downward jerk sometimes occurs in the Black-headed Gull (R. & U. Weidmann, personal communication) and Herring Gull (own observation). These species however do not seem to jerk as vigorously as the Kittiwake, usually performing only one movement, which is therefore easily overlooked, and only occasionally a few jerks in succession. They apparently do not need to jerk as vigorously and as long as the Kittiwake, because the material they use does not stick to the beak while the Kittiwakes often have great difficulty in getting the sticky mud or humus onto the ledge. I counted the number of jerks per depositing bout in eight individuals, which all brought material of both kinds (sticky, containing mud or soil and non-sticky, containing only grasses or seaweed), and found that 45 bouts with sticky material averaged 12·5 jerks, while 24 bouts with non-sticky material averaged only 2·6.

It is not known at present whether the differences found between species in the number of jerks are merely due to differences in the nest-material normally used or whether they are more intrinsic. But there are strong indications for other nest-building activities that the specific differences depend not only on the immediate external stimuli : for instance once the Kittiwake has covered the mud with grass its nest is similar to that of a ground-nester, yet it still tramples intensely on the dry stuff, while the ground-nesters show only traces of the movement. But this still does not decide whether the difference is innate or whether experience plays a part. On the other hand the fact that there is plenty of mud available to the ground-nesters and yet they do not collect it suggests that at least this difference between the Kittiwake and the other species is innate.

Another outcome of the different nest-building techniques is the different shape of the nest-cup; the Kittiwake's is relatively deeper than the ground-nester's, thus holding the eggs more safely. This is necessary because if the Kittiwake's are accidentally kicked out, they almost always fall and are broken, whereas the ground-nesting gulls retrieve theirs in such an event (Kirkman 1937, Tinbergen 1953).

The nest-building behaviour of the Kittiwake has another peculiar feature : the birds collect nest-material in groups. On the Inner Farne the collecting grounds were on the grassy top of the island and one would often see a single bird flying around hesitating to alight anywhere until it saw other birds alighting or already on the ground.   In this way parties of twenty birds or more would quickly assemble at one place.   A few birds after a while might detach themselves from the group and fly away with what they had gathered, while others joined the remaining body, so that there was a busy coming and going for a time, until suddenly the whole party would leave the place together.   Late-comers who either had no time to collect material or were just going to land flew off with the others empty-handed.   One could easily see that the birds were frightened on the flat ground even when they were in a flock.   One could not approach them within forty yards or so, whereas on the nest one could almost touch them.

This social collecting of nest-material is known only in the Kittiwake. The ground-nesters set off alone; they find their material either in the breeding colony itself or in places similar to where they nest, whereas the Kittiwake has to land on exposed flat ground and one can therefore understand its greater reluctance to collect alone.   As in many birds the flock seems to offer protection from predators and only the sense of security given by the flock seems to overcome the birds' reluctance to land on the ground.

There are other indications of the birds' fear of the land : on the Inner Farne they never flew over the island except when collecting nest-material and this avoidance of the land has struck other observers (Naumann 1820, Bent 1921).   Further, early in the season they will not assemble on the preening places on top of the island and away from the colony as they do later.   They restrict their visits to the ledges and rocks just above and below the nesting cliffs and only start to land on the preening places away from the colony when the gathering of nest-material has begun.   It seems thus that only the birds' strong urge to collect nest-material can overcome their fear of the unfamiliar land.   On the preening places, as on the collecting grounds, the birds always remain shy.

The birds do not gather nest-material all through the nest-building season. They have spells of building, during which collecting birds move to and fro between the colony and the collecting grounds and these active phases are interrupted by long spells when none or hardly any of the birds are building.   This synchronization is very remarkable and unusual for a gull-colony and might have the function of insuring that a bird finds companions on the collecting grounds.   It might be brought about by a strong mutual stimulation to build or by a certain stimulus to which the birds respond simultaneously, or by a combination of both.   While I have no evidence for or against the first possibility there is some evidence supporting the second. From my observations in the first two seasons I found that most of the big building outbreaks occurred during or after rain.   In 1955 I noted weather and building-activities in a colony of about fifty nests during five days, each divided into four roughly equal periods, at the height of the nest-building phase.   In ten of the thirteen periods without rain there was little or no building (i.e. building at not more than three nests) and in three there was a lot (i.e. building at four nests and mostly more), whereas in all seven periods with rain a lot of building was recorded.

Although these figures are small they show that rain is correlated with the building activity in a Kittiwake colony $(P < 1\%)$.   Apart from the function of the rain-stimulus in synchronizing the birds' building, collecting during and after rain might be advantageous to the Kittiwake, since during that time it probably is easier to pick up mud and humus from the softened ground.

The Kittiwakes are very ready to steal nest-material from unguarded nests, to such an extent that half- or almost finished nests may be dismantled completely.   Most pairs however do not leave their nests undefended once they have started to build seriously; the mates take turns in guarding the nest, in contrast to their behaviour in earlier days when they frequently were both absent from the colony, sometimes for several days together.

Ground-nesting gulls often leave their nest alone until the first egg is laid and they steal material much more rarely (personal communication, N. Tinbergen, R. Weidmann).   This difference again seems to be connected with cliff-nesting; it is obviously more convenient for a Kittiwake to collect from a nearby source on the cliff than from collecting grounds of which it is more frightened.

We can thus summarize the adaptations connected with nest-building : the Kittiwakes have specialized in collecting their material in groups as a consequence of their fear of open ground, which is itself apparently the result of cliff-nesting.   They collect mud and soil in addition to fibrous material and work the material further with jerking and trampling.   In this way they build a fairly safe nest on their small perches, and they guard it well from robbery by their neighbours.

CONCEALMENT AND DEFENCE OF BROOD.

I have mentioned earlier that the Kittiwakes' nests are much better protected from predators than the ground-nesters'.   Because of this security the Kittiwake has been able to give up a number of behaviour patterns and morphological features which protect the eggs and young of the ground-nesting gulls.

When a predator approaches a colony of ground-nesting gulls the birds which can see it will start to give the alarm call and fly up when it is still at a distance; other birds join them even when they cannot see the enemy. (As Tinbergen (1953) has pointed out, this wariness at the nest must be correlated to the fact that the adult birds are very conspicuous whereas the eggs and young, being cryptically coloured and widely spaced, cannot be found easily.) Disturbed ground-nesters will moreover pursue an intruder and, particularly near hatching time and when they have young, they will swoop at it and may even strike it. This probably deters or at least distracts predators from searching diligently for the cryptic eggs or young (Tinbergen 1953).

In the Kittiwake it is more difficult to evoke these anti-predator reactions than in the ground-nesters. The birds allow an intruder to approach them far more closely before they react and very often they fly up without giving the alarm call at all. Some individuals can actually be lifted off the nest in broad daylight, even in the exceptional Kittiwake colony in Denmark where the birds nest on the ground (Paludan 1955). It is very striking how the alarm call is heard much less often in a Kittiwake colony than in a colony of ground-nesters, apparently because the Kittiwakes are safer and do not need to warn others so much.

The Kittiwakes also rarely attack predators at all. When I climbed among the nests I was hardly ever swooped at and never struck; and this applies also to the ground-nesting Kittiwakes studied by Paludan (1955). In one year on the Inner Farne a Herring Gull was seen several times to catch a young Kittiwake in the air. Even when it did so only a couple of feet from a nesting cliff the adult birds left their nests merely to hover in a completely silent cloud over the scene without interfering, while the chick was screaming and trying to defend itself against the powerful beak of its attacker. Such an assault in a ground-nesters' colony undoubtedly would have provoked violent attacks by the adult birds nesting nearby.

Although eggs and young ground-nesting gulls are cryptic they would be easily detected if the nest itself was not cryptic also. During incubation the ground-nesting gulls leave the nest in order to defaecate and when the young hatch the parents carry away the egg-shells. Apart from having any hygienic function, these actions are probably important in concealing the nest, as a collection of the white droppings or the white inside of an empty shell might easily betray the nest to a predator hunting by eye. A comparison of this behaviour in some species of terns points to the same conclusion (J. M. Cullen, personal communication).

On the other hand a Kittiwake while standing on the nest simply lets its droppings fall over the rim of the nest and a brooding bird merely gets up for defaecation. In this way soon a most conspicuous white flag forms below each nest. The egg-shells too are just left lying on the nest until they are accidentally knocked off. This usually happens soon after hatching, but may take several weeks.

Young ground-nesters are not only cryptically coloured but all the species which have been watched also show cryptic behaviour. When a predator approaches they run to hide under cover, and crouch. By contrast the young

Kittiwakes are most conspicuously coloured : in down they are white with a light grey back and in the juvenile plumage they are like the adults with the addition of some black bars.   In typical nest-sites it is impossible for them to run when danger approaches and Salomonsen (1941) reports that even in a Kittiwake colony on the ground the half-grown young remained on their nests when he came near (while the young Black-headed Gulls from the nests around all ran away).   Yet Kittiwake chicks will run when taken from the nest, put on the ground and frightened.

On the other hand two young Herring Gulls and one Black-headed Gull which were hatched and reared in Kittiwakes' cliff-nests would all have run over the edge of the nest had I not prevented them by fastening wire-netting round the nest-rim.   From this and Salomonsen's observations we can conclude that the difference in running between the young of the different species must be innate.

CLUTCH-SIZE.

Lack (1954 and previously) has suggested that in many birds the size of the clutch is adapted to correspond to the maximum number of young which the parents can, on the average, supply with food.

To test this hypothesis for the Kittiwake a number of nests were checked daily in a more or less random sample in the Kittiwake colony on the Inner Farne in 1953, 1954 and 1955 and the data were used to calculate clutch-size, nesting success etc.   In addition other nests were checked less often and might be expected to yield rather different results since eggs or young might disappear between two checks.   Clutch-size, hatching success and survival-rate were not significantly different for the two groups (P <5%) and the data have been therefore combined in Tables 1 and 2.   A few young were not able to fly by the time I left the island, but in view of their age I have regarded them as surviving in compiling Tables 1 and 2.

The clutch-size distribution in 138 nests checked daily was 21 (15 %) with one egg, 104 (75%) with two and 13 (9%) with three.

Table 1 *a* shows that the hatching success increases with clutch-size. In order to discount this effect and to make comparison possible with the group of experimental nests mentioned below the nesting success of the Kittiwake was calculated on the basis of the number of young hatched rather than on the number of eggs laid (Table 1 *b*).   Since the numbers of families of one and three were small I adjusted the family size in some additional nests artificially, where necessary adding a young of a suitable age.   Table 1 *c* combines the results from these nests with those whose clutch-sizes were undisturbed and shows that families of three have a lower survival rate than families of one and two. (Survival of families of two is significantly higher than that of families of three (P <3%).)   The results obtained so far from this single colony fit with one requirement of Lack's hypothesis but it should be noted that the number of young produced per nest is nevertheless greatest for the families of three.

Lack has emphasized the importance of food-shortage in determining the number of young which can be raised and there is some circumstantial evidence that food-shortage may account for mortality of young Kittiwakes. As noted above, young nest-mates regularly establish a peck-order which is important mainly at feeding times and when the parents have only a little food the superior chick may be the only one to be fed. This food-fighting suggests that food must at times be short. Furthermore the Kittiwakes' eggs hatch asynchronously, a device in birds which is thought to lessen the bad consequences of food-shortage (Lack 1954).

TABLE 1.   *Survival of Kittiwake eggs and young in relation to (a) clutch-size and (b) and (c) number of young hatched per nest. Only first clutches included.*

Clutch-size	No. of pairs	Young hatched		Young surviving	
		No.	Per cent hatching		
1	26	13	50	12	
2	130	182	70	164	(a)
3	18	40	74	32	

Size of family	No. of pairs	Young surviving		Average no. young per pair	
		No.	Per cent		
Undisturbed nests					
1 young	26	24	92	0·9	
2 young	91	164	90	1·8	(b)
3 young	9	20	74	2·2	
Both undisturbed and experimental nests					
1 young	31	28	90	0·9	
2 young	91	164	90	1·8	(c)
3 young	19	45	79	2·4	

In spite of this indirect evidence of food-shortage I was not able to see any direct signs of starvation except at a few nests and in some of these the parents seemed at fault, for nearby nests with the same number of young were being supplied adequately. Food-shortage might, however, be more evident in other colonies and under different conditions.

On the other hand food-shortage is probably not the only reason for a greater mortality in larger families and of the young whose deaths are recorded in Table 1 I judged that no more than half can have been due to starvation.

91

Chicks sometimes fall off their nests, or part of the nest ; or even the whole of it, may collapse, precipitating the young into the abyss.   One must suppose that the more young there are on a nest the more likely they are to fall, both because they have less room to stand and because their combined weight is greater.   I recovered some of the young which had fallen and found them perfectly healthy and they were in a number of cases successfully reared by foster-parents.   I seldom saw a chick actually fall from its nest, but when a chick was missing, one could get some idea of the likelihood that such an event had taken place from the condition of the nest.   From observations in a group of about twenty nests which I studied most intensely over three years, it would seem that the disappearance of young is just as often due to falling as to starvation.

It appears therefore that both shortage of food and the collapse of nests might account for the greater mortality in larger families in the Kittiwake.

The Kittiwake usually lays two eggs (p. 289) and so does its close relative *R. brevirostris* (Bent 1921), but the usual clutch-size of the gulls is three, except the Ivory Gull *Pagophila eburnea* which usually lays two (Buturlin 1906, Stark & Sclater 1906, Bent 1921, Baker 1935, Witherby *et al.* 1944).

Many young ground-nesting gulls are killed by predators, including members of the same species (Goethe 1937, Kirkman 1937, Darling 1938, Paynter 1949, Paludan 1951).   The Kittiwake young have a much lower mortality (Table 2) presumably thanks to their different nesting habit which protects them from predation and the disturbances which go with it.   The mortality of the young Kittiwakes, at least on the Inner Farne, is so low that in spite of their small clutch the birds produce on the average more young per nest than is in many cases reported of the other gulls with their larger clutches (Table 2).

Comparing my own data for the Kittiwake with Paynter's (1949) for the Herring Gull and Paludan's (1951) for the Herring and Lesser Black-backed Gull it appears that so many young ground-nesters die soon after hatching that within a week (Paludan) or from about three weeks (Paynter) there are fewer young per pair in these species than in the Kittiwake.   Other observers also remark on this high early mortality in the Herring Gull, Lesser Black-backed Gull and Black-headed Gull without giving quantitative evidence (Goethe 1937, Kirkman 1937, Darling 1938, Lockley 1947, Tinbergen 1953, R. & U. Weidmann, personal communication).

The available data thus suggest that through much of the nestling period the Kittiwake parents have as many or more young to feed than the ground-nesters.

Lack has repeatedly stressed that clutch-size in a species cannot be adapted to adult mortality, but he points out (1954) that mortality at an early age might be a subsidiary factor affecting the clutch-size.   Even if the number of young which could be raised by the Kittiwake and the ground-

TABLE 2. *Nesting success in different species of gull. Different methods of sampling may account for some of the differences in the results.*

		No. of		No. of young		Per cent hatched young surviving	Average young per pair surviving	Reference
		Pairs	Eggs	Hatching	Surviving			
Kittiwake	1953–55	179	348	240	211	88	1·18	this paper
Herring Gull	1936	59	126	108	46	43	0·78	Darling (1938)
	1937	68	198	189	77	41	1·13	ditto
	1943	90	270	243	less than 131	<54	<1·46	Paludan (1951)
	1944	87	371	206	41	20	0·47	ditto
	1947	100				c. 51	c. 0·51	Paynter (1949)
Lesser Black-backed Gull	1936	43	93	86	c. 50	c. 58	c. 1·16	Darling (1938)
	1937	71	206	196	c. 104	c. 53	c. 1·47	ditto
	1943	120	354	216	less than 32	<15	<0·27	Paludan (1951)
	1944	112	362	220	12	5	0·11	ditto

nesting gulls were the same, one would expect that with relatively little early mortality, a consequence of cliff-nesting, the Kittiwake would evolve a smaller clutch-size than the other species.

FEEDING YOUNG.

The Kittiwake feeds its young in a different way from the ground-nesting gulls and this can be correlated with the different nesting habitats.

Young ground-nesters leave the nest a few days after hatching and may be fed at any place in the territory. Young Kittiwakes are confined to the nest until they can fly, that is, when they are about six weeks old, and have to be fed there. All gulls feed their young by regurgitation and in the ground-nesting species the parents very often drop the food to the ground and the young pick it up. But a young Kittiwake takes its food from the throat of the parent and food is rarely dropped. Gull's food is often half-digested and therefore cannot be picked up completely and it is very likely that soon a little heap of rotting food would collect in a Kittiwake's nest if the birds fed in the ground-nesters' way. This would not only smear the young, but it might also develop into a source of disease. For birds restricted to the nest it obviously is advantageous to have a more hygienic feeding method.

The Kittiwake keeps its nest clean in another way. It immediately picks up remains of food and other strange objects which may have fallen into the nest and either swallows them or, more frequently, flings them away with a vigorous head-shake. This habit has not been observed in the ground-nesters (N. Tinbergen, R. & U. Weidmann, personal communication), who do not seem to need it.

There are other behaviour patterns connected with the feeding of the young which seem to have changed in the Kittiwake. When a ground-nesting gull comes with food, its young are often hidden under cover, away from the place where the adult alights and if they are hungry they will run towards it with a characteristic food-begging movement : from a horizontal posture the so called " hunched posture " (Moynihan 1955) or " attitude of inferiority" (Tinbergen 1953) they quickly stretch their neck up vertically and withdraw it again, repeating these movements in succession so that head and neck perform a kind of pumping movement (Tinbergen 1953, Moynihan 1955, R. Weidmann personal communication). In the stretched phase they open the beak widely and give the food-begging call as they withdraw the neck again. One might guess that the function of this alternation and particularly the stretching of the neck is to make the young conspicuous to the parent who may alight some yards away. This explanation is confirmed by the behaviour of a young Black-headed Gull which I reared, for the pumping movement was much more pronounced when some distance away than when close to its " parent ".

By contrast the young Kittiwake lacks both the pumping movement and the need to make itself conspicuous for it is always on the nest.

It may be noted that in the food-begging of adult ground-nesting gulls, before courtship-feeding and copulation, a time when the pair is standing close together on the ground, the pumping is absent.

A related difference is that the Kittiwake lacks a special feeding call such as that which attracts the young ground-nester to the parent.

FLYING MOVEMENTS IN THE YOUNG.

Both the Kittiwakes and the ground-nesting gulls perform incipient flight movements when they are small, but this behaviour develops differently later in life.

From a very early age, before the wing-feathers have grown, the downy chicks of the ground-nesters flap their wings and at the same time jump up into the air. As the young grow these jumps get higher, the wing-flapping more vigorous and gradually it develops into proper flying.

The young Kittiwakes on the other hand, although they start wing-flapping from about the same age as the ground-nesting gulls (from the third day, Kittiwake, own observation ; from the fifth day, Black-headed Gull, personal communication R. Weidmann; from the seventh to ninth day, Herring Gull, Goethe 1955), do not perform the movement as vigorously as the other species and they do not start to jump into the air as early as the young ground-nesters. When eventually they start to do so, a few days before they fly, they never lift their feet for more than an inch or so from the nest, whereas much smaller ground-nesting chicks can be seen jumping a foot or two in the air.

This lack of vigorous flying movements in the young Kittiwake is certainly an adaptation connected with cliff-nesting for wing-flapping as vigorous as ground-nesters' would endanger their lives.

The orientation of the chicks when wing-flapping also differs with the species. The ground-nesters face into the wind, but the young Kittiwakes always face the wall. Kittiwake chicks spend much of their time orientated in this way, even when they are resting or sleeping and in the first days after hatching they hardly ever face outwards. This probably is a habit which prevents them from falling down. It is therefore not surprising that during a vigorous movement like the wing-flapping the young face the wall; moreover they often would not have room to spread their wings if they were facing the other way. When the young Kittiwakes prepare for their first real flight, however, they have to face outward. Thus their wing-flapping and jumping does not grade imperceptibly into proper flying as in the ground-nesters but their first flight is much more of a new achievement.

RECOGNITION OF YOUNG.

Several authors have claimed that ground-nesting gulls and terns are able to distinguish their own young from strangers (Watson & Lashley 1915, Culemann 1928, Dircksen 1932, Goethe 1937, Kirkman 1937, Palmer 1941, Tinbergen 1953, Pfeffer 1955). The experiments on which these claims are based are often unsatisfactory and in many cases do not distinguish between personal recognition by the parent of the young and other factors, such as the recognition by the young of the parent and possible differences in behaviour which this may lead to, which themselves may partly be the means by which the parent " recognizes " its young. Most of the experiments however show that if young more than a few days old are exchanged from different nests the parents drive them away. Tinbergen's experiments with Herring Gulls show that the parents begin to distinguish between their own young and strangers when their own young are about five days old and Pfeffer-Hülsemann (1955) claims the same of the Common Gull. Tinbergen stresses that this ability to recognize the young is very different from the ability to recognize the eggs, which depends almost entirely on location, and that this difference corresponds to the behaviour of eggs and young, as the eggs remain in the nest while the young soon start to move about. Furthermore it appears that the parents start to recognize their own young at the time when the young begin to move about actively. Thus

TABLE 3. *Exchange of young Kittiwakes from different nests. In each test, except the last, the whole brood was replaced by strangers, of the same number and age as the original young.*

Age in days	Age-group	No. of tests	Results
0–6	1	2	accepted
	1–2	4	,,
7–13	2	8	,,
	2–3	2	,,
14–20	3	12	,,
21–27	4	5	,,
28–34	5	1	pecked a little (see text)
Only one of two young exchanged	4	1	accepted

one may infer that the recognition enables the parents to attend to their own chicks by the time they are old enough to wander about. According to this explanation one would expect that the Kittiwakes, whose young, like the eggs, do not move from the nest, would also lack the ability to distinguish between their own young and those of strangers.

To examine this possibility I exchanged a number of young from different nests for five minutes (Tables 3, 4, 5) or longer (Table 6). Chicks of one nest were sometimes presented to several strange parents, and chicks which had been exchanged before were sometimes used again in an older age-group. In the first series I exchanged broods of the same age and the same number of chicks (Table 3). In another series I substituted young of different ages either for eggs or for young without altering the number of young (Table 4) and in some further experiments the number of individuals in the nest was altered as well as the nature of the contents (Table 5). In all the tests except

TABLE 4. *Young Kittiwakes of different age-groups exchanged. The whole brood was replaced by the same number of strangers. All accepted.*

Age-group of own brood	Age-group of replacing brood	Number of tests
2nd clutch ca. 1 week before hatch	3	1
near hatch	2	2
1	2	2
1–2	4	2
2	1	1
4	1–2	3
4	5	1
5	4	1
6	4	1

TABLE 5. *Number of individuals and nature of contents of nest changed in Kittiwakes' nests. All accepted.*

Owner's brood	Change made
1 young (age-group 1)	added 1 Kittiwake (age-group 1).
2 young (age-group 1)	added 1 Kittiwake (age-group 1).
1 young (age-group 4)	added 1 Kittiwake (age-group 4).
1 egg (pipped)	added 1 Shag (newly hatched).
1 young (2–4 days)	added 1 Shag (newly hatched).
2 young (1–3 days)	older replaced by a Herring Gull (newly hatched).
1 young (6 days)	added a Herring Gull (1 day).

the one in Table 3 with a young bird of the fifth age-group the parents showed the same behaviour to the strange young as they usually do to their own, which varies with the age of the chicks : small young are brooded, larger ones just guarded. Sometimes a young one was preened a little or even fed. The young Kittiwakes behaved towards the strange parents very much as they did towards their own; they usually stood beside them and might preen or peck at the adult's beak, begging for food, etc. It appears therefore that under the experimental conditions the parents do not recognize their offspring up to an age of at least four weeks either by number or the state of development, let alone by more subtle individual characteristics.

On the other hand Paludan (personal communication) found that the parents may attack their own young outside the nest, which confirms that they " recognize " their brood by location.

It is difficult to exchange young more than about four weeks old because from this age they are liable to jump off their ledges when approached. Of the three tests I made with young of this class the stranger was accepted in two. Once however the foreign chick was slightly pecked although its behaviour appeared not to differ from that of the parents' own young. These experiments are supported by other observations in which young who could

TABLE 6.   *Strange young left in experimental nests to be reared after tests listed in Tables 3 and 4. (The whole brood was replaced without altering the number of young.)*

Number of exchanged young per nest	Age-group of own young	Age-group of strange young	Number of tests	Result
1	egg (ca. 1 week before hatch)	3	1	reared
1	1	1	2	ditto
2	1–2	1–2	2	ditto
2	3	3	2	3 reared, 1 died prematurely
1	6	4	1	reared

fly (and were therefore at least 5½ weeks old) landed on strange ledges. Of twelve cases where a young bird landed on a nest with young, the stranger was pecked by the parent in six in a clearly hostile, although mostly inhibited way. In six cases the young was not molested and once it was even fed. In four of these intrusions the parent saw the chick landing and in three of the four it attacked; in four cases the young was on the nest first and in three of the four the parent accepted it. This and further observations suggest that a juvenile approaching the nest provokes attack more strongly than a juvenile already on the nest and this may apply even to the birds' own offspring.   There are however signs that the parents can really distinguish their young from strangers at this age : an adult may threaten one of its own young who lands but does not actually attack it and I have seen a parent responding to the calls of its own young which landed on a strange nest. I suspect therefore that also in the cases when the strange fledglings were treated like their own young the birds might be aware of the change.

Summarizing these facts, it can be said that there is no evidence of personal recognition between parent and young in the Kittiwake up to four or five weeks after the young have hatched, but that the parents seem to discriminate later. Thus the behaviour of the Kittiwake agrees with what was expected and confirms the idea about the function of recognition in the ground-nesters.

So far I have discussed mainly the recognition between adults and young, but the reaction of the chicks themselves to strange nest-companions is also interesting. Nest-mates know each other and fight strange young fiercely. Because of this hostility, it is difficult to foster a strange chick into a family with young, if at least one of the young is more than a few days old. Small strange young usually get used to each other after some fighting, but in larger ones the squabbling goes on and even after the loser has adopted the appeasement posture the other sometimes goes on pecking him fiercely for a long time.

This hostility of the young towards strangers is particularly surprising in view of the fact that the parents apparently do not distinguish their own young from strangers before the young fly, but the most probable explanation is that the young birds' hostility towards a stranger has another function from that of the parents'. In the ground-nesting gulls the parents' aggression is presumably a means of spacing out the families and ensuring that a stranger is not fed. Of this the Kittiwakes have no need. On the other hand the the fighting between the nest-mates is probably concerned with establishing the peck-order, already mentioned, whose function it is to determine the order in which the young are fed. In the normal course of events the peck-order is settled at an early age, when both the young are small and weak and easily tired, but if two strange young encounter one another on a nest (as happened during experiments) this may lead to a longer battle before one is subdued. It may be remarked that the outcome of these battles between strangers was a foregone conclusion as it always happens, in my experience, that the larger chick wins, whether or not it is the rightful owner of the nest.

To summarize this section we may say that the apparent inability of parents and young to recognize one another in the Kittiwake is another effect of cliff-nesting and the consequent isolation of the nests of different pairs from one another.

CONCLUSION AND SUMMARY.

The Kittiwake is probably derived from a ground-nesting gull and the change to cliff-nesting was presumably an anti-predator device. Because predation is less, the species seems to have lost a number of behaviour patterns and morphological features possessed by other gulls. On the other hand it has had to acquire a number of adaptations to suit its new life. Signal and non-signal movements have been altered : some have been modified, like the beak-hiding or the nest-building movements, others have been lost, like several anti-predator devices or the aggressive upright posture. There also seem to be a few new acquisitions like the black neck-band of the young and the collecting of mud.

Adaptations to fit an animal for a new kind of life are of two kinds. Some are inherited, others are acquired as a result of the individual's experience and environment. The chief morphological peculiarities of the Kittiwake, such as the black neck-band in the young, are certainly due to inherited differences. The adult's claws are sharper than those of other gulls, which helps the Kittiwake to hold on to small ledges and this may be partly because the tips are not abraded by walking as in the other gulls who spend the whole year close to land. On the other hand this difference is also apparent in the newly born young, so that here too it must have a genetic basis.

As I have shown, some of the differences in behaviour also are innate (the presence or absence of head-turning in the young, their readiness to run away when pecked). Others, for instance the use of mud as nest-material, are presumably also innate. Yet others, such as the social collecting of material, may be due to a complicated interaction of acquired and inherited factors. The existence of inherited factors leading to differences in behaviour should not blind one to the possibility that experience may be able to modify a pattern. This was shown in one of my experiments in which a newly-hatched Black-headed Gull learned after a day to take the food from the throat of its Kittiwake foster-parent in the way a young Kittiwake does from the first. Such modifiability is no objection to the existence of innate differences; it only shows that these differences may not be as rigid as has sometimes been supposed.

With the adaptations described in this paper, and which are summarized in the list below, I hope to have shown how this one change to nesting on tiny ledges on steep cliffs has had repercussions in many aspects of the life of the species and has led to morphological changes as well as a great many alterations in behaviour. In many animals adaptive differences between species have been described but I know of no other case where one relatively simple change can be shown to have been responsible for so many alterations.

GROUND-NESTING GULLS	KITTIWAKE
High predation-rate in nesting colonies.	*Predation pressure relaxed on cliffs.*
Alarm-call frequent.	Alarm-call rarer.
Adults leave nest when predator some way distant.	Remain on nest until predator very close.
Vigorously attacks predator intruding in colony.	Very weak attacks at most at intruding predator.
Brooding birds disperse droppings and carry egg-shells away from nest.	Neither droppings nor egg-shells dispersed.
Young cryptic in appearance and behaviour.	Young not cryptic either in appearance or behaviour.
Clutch-size normally three eggs.	Clutch-size normally two eggs.
Suited to life in colony on ground.	*Adapted to life on cliffs.*
A. Several fighting methods.	More specialized to fighting in one way (grabbing beak and twisting).

Upright threat posture occurs, derived from preparation to peck down at opponent.

No upright threat.

Beak does not specially direct attacks. Not known if it is such a strong releasing stimulus as in the Kittiwake.

Beak releases and directs attacks.

Beak turned away in appeasement but not elaborately hidden.

Beak turned away in appeasement and elaborately hidden.

B. Young run away when attacked.
No head-flagging in young.

Young do not run when attacked.
Head-turning and hiding of beak in young when pecked and appropriate behaviour in attacker.

No neck-band.

Possess black neck-band.

C. Number of nest-sites probably less restricted and therefore probably less competition for nest-sites.

Number of nest-sites restricted, probably more competition.

Often first occupy pairing territories before nesting territories and pairs form away from nest.

Occupy nesting ledges at arrival in breeding area and pairs form on the nest.

Choking not normally used by unmated males as advertisement display.

Choking normal advertisement display of unmated males.

D. Copulation on the ground, female stands.

Copulation on the tiny ledge or nest, female sits on tarsi.

E. Nest-material collected near nest, building not synchronised, individual collecting.

Nest-material collected in unfamiliar places, synchronization of building and social collecting.

Little stealing of nest-material.

Birds very ready to steal nest-material.

Nests often unguarded before laying of first egg.

Nests guarded.

Nest-building technique relatively simple.

Nest-building technique more elaborate.

Mud not used.

Mud as nest-material.

Only one or, at most, very short series of depositing jerks.

Prolonged jerking of head when depositing nest-material.

Only traces of trampling on nest-material.

Prolonged trampling on nest-material.

Nest has relatively shallow cup.

Nest has deeper cup.

F. Young leave nest a few days after hatching.

Young have to stay on nest for long period.

Young fed by regurgitation on the ground.

Young fed from throat.

Nest-cleaning absent or less conspicuous.

Young and adults pick up and throw away strange objects falling into nest.

Parents have feeding call, probably to attract young.

Parents have no feeding call.

Hungry young make themselves conspicuous to parents by head-pumping.

Head-pumping absent in young.

Parents learn to recognise own young in a few days.

Parents do not recognize own chicks at least up to the age of four weeks.

G. Young face any direction. Vigorous wing-flapping in young.

Young face wall much of the time.
Flight movements much weaker.

H. Weaker claws, cannot hold on so well.

Strongly developed claws and toe-musculature.

ACKNOWLEDGEMENTS.

I am grateful to the Janggen-Pöhn-Stiftung, St. Gallen, Switzerland, for a research grant which permitted this work and to the Nuffield Foundation for providing part of the equipment. Acknowledgements are also due to the Natural History Society of Northumberland, Durham and Newcastle-on-Tyne and to the National Trust for amenities for working on the islands. I would like to thank Dr. N. Tinbergen for his encouragement and most helpful advice and also Prof. A. C. Hardy for his hospitality in the Zoology Department in Oxford. Among others to whom I am grateful for helpful criticism are R. E. Moreau and my husband, who also helped in the field and revised the English text.

REFERENCES.

ARMSTRONG, E. A. 1950. The nature and function of displacement activities. Sympos. Soc. Exp. Biol. 4 : 361–384.

BAKER, E. C. S. 1935. The nidification of birds of the Indian Empire. 4. London.

BENT, A. C. 1921. Life histories of North American gulls and terns. U.S. Nat. Mus. Bull. 113.

BUTURLIN, S. A. 1906. The breeding-grounds of the Rosy Gull. Ibis (8) 6 : 131–139.

CULEMANN, H. W. 1928. Ornithologische Beobachtungen um und auf Mellum vom 13. Mai bis 5. September 1926. J. Orn. 76 : 609–653.

DARLING, F. F. 1938. Bird Flocks and the breeding Cycle. Cambridge.

DIRCKSEN, R. 1932. Die Biologie des Austernfischers, der Brandseeschwalbe und der Küstenseeschwalbe nach Beobachtungen und Untersuchungen auf Norderoog. J. Orn. 80 : 427–521.

DROST, R. 1952. Das Verhalten der männlichen und weiblichen Silbermöwen (Larus a. argentatus Pont.) ausserhalb der Brutzeit. Vogelwarte 16 : 108–116.

DWIGHT, J. 1925. The gulls (Laridae) of the world; their plumages, moults, variations, relationships and distribution. Bull. Amer. Mus. Nat. Hist. 52 : 63–408.

FISHER, J. & LOCKLEY, R. M. 1954. Sea-birds. London.

GOETHE, F. 1937. Beobachtungen und Untersuchungen zur Biologie der Silbermöwe (Larus a. argentatus Pontopp.) auf der Vogelinsel Memmertsand. J. Orn. 85 : 1–119.

GOETHE, F. 1955. Beobachtungen bei der Aufzucht junger Silbermöwen. Z. Tierpsych. 12 : 402–433.

HUXLEY, J. S. & FISHER, J. 1940. Hostility reactions in Black-headed Gulls. Proc. Zool. Soc. London (A) 110 : 1–10.

KIRKMAN, F. B. 1937. Bird Behaviour. London and Edinburgh.

LACK, D. 1954. The Natural Regulation of Animal Numbers. Oxford.

LOCKIE, J. D. 1952. The food of Great Skuas on Hermaness, Unst, Shetland. Scot. Nat. 64 : 158–162.

LOCKLEY, R. M. 1947. Letters from Skokholm. London.

LORENZ, K. 1953. Die Entwicklung der vergleichenden Verhaltensforschung in den letzten 12 Jahren. Zool. Anz. Suppl. 17 : 36–58.

MOYNIHAN, M. 1953. Some displacement activities of the Black-headed Gull. Behaviour 5 : 58–80.

MOYNIHAN, M. 1955. Some aspects of reproductive behaviour in the Black-headed Gull (Larus ridibundus ridibundus L.) and related species. Behaviour Suppl. 4 : 1–201.

NAUMANN, J. A. 1820. Naturgeschichte der Vögel Deutschlands. Leipzig.

NICHOLSON, E. M. 1930. Field-notes on Greenland birds. Ibis (12) 6 : 280–313, 395–428.

NOBLE, G. K. & WURM, M. 1943. The social behavior of the Laughing Gull. Ann. N.Y. Acad. Sci. 45 : 179–220.

PALMER, R. S. 1941. A behavior study of the Common Tern (Sterna hirundo hirundo L.). Proc. Bost. Soc. Nat. Hist. 42 : 1–119.

PALUDAN, K. 1951. Contributions to the breeding biology of Larus argentatus and Larus fuscus. Vidensk. Medd. Dansk. Naturh. Foren. 114 : 1–128.

PALUDAN, K. 1955. Some behaviour patterns of Rissa tridactyla. Vidensk. Medd. Dansk. Naturh. Foren. 117 : 1–21.

PAYNTER, R. A. 1949. Clutch-size and the egg and chick mortality of Kent Island Herring Gulls. Ecol. 30 : 146–166.

PERRY, R. 1940. Lundy, Isle of Puffins. London.

PERRY, R. 1948.   Shetland Sanctuary.   London.
PFEFFER-HÜLSEMANN, K. VON. 1955.   Die angeborenen Verhaltensweisen der Sturm-
    möwe (*Larus c. canus* L.).   Z. Tierpsych. 12 : 443–451.
SALOMONSEN, F. 1941.   Tretaaet Maage (*Rissa tridactyla* (L.)) som Ynglefugl i
    Danmark.   Dansk. Orn. Foren. Tidsskr. 35 : 159–179.
SELOUS, E. 1905.   The bird watcher in the Shetlands.   London.
STARK, A. & SCLATER, W. L. 1906.   The birds of South Africa. 4.   London.
TINBERGEN, N. 1952.   " Derived " activities; their causation, biological significance,
    origin and emancipation during evolution.   Q. Rev. Biol. 27 : 1–32.
TINBERGEN, N. 1953.   The Herring Gull's World.   London.
TINBERGEN, N. 1956.   On the functions of territory in gulls.   Ibi' 98 : 401–411.
TINBERGEN, N. & MOYNIHAN, M. 1952.   Head flagging in the Black-headed Gull; its
    function and origin.   Brit. Birds 45 : 19–22.
VENABLES, L. S. V. & VENABLES, U. M. 1955.   Birds and Mammals of Shetland.
    Edinburgh and London.
WATSON, J. B. & LASHLEY, K. S. 1915.   Homing and related activities of birds.
    Pap. Dept. Mar. Biol. 7.   Publ. Carnegie Inst. Washington 21 : 1–104.
WEIDMANN, U. 1955.   Some reproductive activities of the Common Gull *Larus
    canus* L.   Ardea 43 : 85–132.
WITHERBY, H. F. et al. 1944.   The Handbook of British birds. 5.   London.

8

Reprinted from *15th Intern. Ornithol. Congr. Proc.*, K. H. Voous, ed., E. J. Brill, 1972, pp. 424–433

The significance of the pair-bond in the Kittiwake

J. C. Coulson

University of Durham, England

INTRODUCTION

The duration of the pair-bond in birds is extremely variable lasting only a few minutes or hours in some species but persisting for several years in others. In species which tend to retain the mate from one breeding season to another, there are some where the pair remains together throughout the whole year whilst others separate some time after the completion of breeding and re-form the pair again at the beginning of the next breeding season. The Kittiwake *Rissa tridactyla* belongs to the last category and although the temporary break in the pair-bond may last less than three months, in other pairs in the same colony the break may be as long as eight months.

The chance of both members of a pair surviving from one breeding season to the next is directly related to the survival rate of the two sexes. Accordingly, it is to be expected that the retention of the mate from the previous breeding season is more likely to occur in those species with a low annual mortality rate and sea-birds, amongst others, are known to have a low annual death risk.

It is not difficult to appreciate the advantages of retaining the same mate from one breeding season to the next. The successful rearing of young involves a great deal of co-operation between the members of a pair in which both sexes assist in the incubation of the eggs and the care of the young. Experience of the same mate in a previous year probably ensures better co-ordination of parental responsibilities and also it may reduce the time spent in establishing a firm attachment or pair-bond between male and female and allow the next stages in breeding to commence earlier.

However, not all Kittiwakes retain their mate even when their mate is alive and present in the same colony. This type of change of mate, which for convenience will be referred to as 'divorce', would appear to be a disadvantage if one postulates an advantage in mate retention. This paper reports the effects on the breeding biology of Kittiwakes in relation to their age and mate tenacity. In addition, an examination has been made of the effects of changing mate caused by divorce and the death of the previous partner.

Recently, COULSON (1968) reported that Kittiwakes nesting in the central half of a colony showed many differences from those nesting near the edge. Because this effect could lead to erroneous conclusions if not taken into consideration, this factor has been included in the analysis.

METHODS

The Kittiwake colony used in this study is one which became established on a warehouse at North Shields, Northumberland in 1949. Since 1954, almost all of the breeding birds have been ringed with unique combinations of coloured rings using a non-fading, brightly coloured P.V.C. which is sufficiently strong to withstand abrasion (COULSON 1963). As the colony was in active growth until about 1960, only information between 1960 and 1970 inclusive has been used. Observations have been made on the breeding birds in each breeding season and, in particular, accurate records have been obtained of

104

the date of laying of the first egg (error ± 1 day), clutch size and hatching and breeding success. The composition of each pair has been determined in each year.

Previous studies on the Kittiwake (COULSON & WHITE 1958) have shown that there is almost complete colony tenacity in the breeding birds. Although 24% of non-breeding birds ringed at North Shields were seen in other colonies within 120 mile radius, no breeding birds were found breeding away from the colony. Since that time only two adults have been found breeding in another colony and that was within 10 miles of North Shields, well within the area which is repeatedly searched for colour-ringed birds. Accordingly, the movement of two birds in 16 years from a colony of 100 pairs represents a very small loss and it is reasonable to assume that any bird which fails to return to the colony has died and the return rate can be used to estimate the adult mortality rate.

To simplify presentation of results, the age of the birds have been presented in three groups; young birds breeding for the first time, birds breeding for 2nd–4th time and those breeding for more than the fourth year (the oldest breeding for the 16th time).

The colony was divided almost equally into centre and edge. This centre division represents the distribution of nesting birds when the colony was half its present size and therefore divides the colony according to the nest site preference of the colonising birds.

FACTORS AFFECTING PAIR-BOND

DIVORCE

The divorce rate varies with both the age of the bird and also with its position in the colony. The results presented in Table 1. show that there is a significant decrease in the proportion divorced as age increases and also the birds nesting in the centre of the colony are, age for age, less likely to divorce their mates than those birds nesting at the edge. There is also a suggestion that the difference in the divorce rate between birds nesting at the centre and edge increases with the bird's age but this is not significant.

TABLE I

THE PROPORTION OF FEMALES DIVORCED IN THE CENTRE AND EDGE OF THE COLONY

| | female breeding experience (years) | | | |
| | 2–4 | | 5–16 | |
	N	%	N	%
Centre	127	35	194	23
Edge	112	43	120	38

ADULT MORTALITY

There is no indication that the adult annual mortality rate is age specific and all of the age classes can be combined. Table 2 shows that the mortality rate of

TABLE 2

THE MORTALITY RATES OF MALE AND FEMALE KITTIWAKES NESTING IN THE CENTRE AND AT THE EDGE OF THE COLONY

sex	colony position	bird-years at risk	number dying	percentage annual mortality
Males	Centre	555	75	13.6
	Edge	432	87	20.1
Females	Centre	570	66	11.6
	Edge	454	59	13.1

NOTE: The difference between the mortality rates of males in the centre and at the edge of the colony is significant ($P < .01$). The difference between the two groups of females is not significant.

males in the centre of the colony is lower than those at the edge ($P < .01$). No such difference occurs in the female.

The appreciable difference in the mortality rates of males nesting centrally and on the periphery of the colony does not result from predation at the colony, which is negligible. This differential mortality takes place away from the colony and often during the non-breeding season when the birds are fully oceanic.

The greater survival of males in the centre of the colony results in fewer pairs changing mate because their partner has died than at the edge of the colony. Thus both mortality and divorce contribute to an appreciably higher proportion of the birds retaining mates in the centre of the colony. The reason for the difference in the divorce rate is discussed later.

TABLE 3

THE MEAN DATE OF LAYING OF THE FIRST EGG ACCORDING TO BREEDING EXPERIENCE OF THE FEMALE, PAIR STATUS AND POSITION IN THE COLONY

All dates in May. *Sample size in parentheses.*

	breeding experience (years)					
	1		2–4		5–16	
	edge	centre	edge	centre	edge	centre
Same mate			18 (64)	17 (83)	19 (74)	15 (150)
Divorced	21 (93)	22 (99)	20 (48)	16 (44)	18 (46)	16 (44)
Mate died			20 (26)	17 (16)	26 (19)	19 (24)

NOTE: The sample sizes also apply to Tables 5, 7 and 9.

TIME OF LAYING

The time of laying of the first egg of the clutch, that is the time that the pair first indicate that they are in a state which allows reproduction, has been analysed according to the age of the female, the position in the colony and the status of the pair in relation to the previous breeding season (Table 3).

Birds breeding for the first time obviously form only one category, although some may have been in the colony in a previous year and this influences the breeding date (COULSON 1966). On average, these young birds breed latest in the colony and there is no difference in their mean dates at the centre and the edge of the colony.

The remaining results can be analysed with respect to one variable at a time. For example there are six pairs of dates which can be strictly compared to examine effects of edge and centre position and where all other factors are comparable. These results are evaluated in Table 4. There is no significance in the differences

TABLE 4

DIFFERENCE IN DATE OF LAYING ACCORDING TO CONDITIONS

Age:		
Old-young	=	−1.4 days
Position:		
Centre-edge	=	+3.5 days
Mate:		
Same-divorce	=	+0.7 days
Mate:		
Same-dead	=	+3.6 days

in the time of breeding, after the first time. However, comparable birds at the centre and edge of the colony breed an average of 3.5 days later on the edge. Birds which change their mate also breed later but there is an indication that the delay is shorter in birds which are divorced than in those whose mate has died. This difference is not significant.

TABLE 5

THE MEAN CLUTCH SIZE ACCORDING TO BREEDING EXPERIENCE OF THE FEMALE, PAIR STATUS AND POSITION IN THE COLONY

Sample sizes as in Table 3

	breeding experience (years)					
	I		2–4		6–16	
	edge	centre	edge	centre	edge	centre
Same mate			2.06	2.20	2.23	2.31
	1.88	1.86				
Divorce			1.98	2.09	2.13	2.09
Mate died			2.00	2.00	1.95	2.00

TABLE 6

DIFFERENCES IN CLUTCH SIZE ACCORDING TO CONDITIONS

Age: Old-young	=	+0.063 egg	or 3.0% increase
Position: Centre-edge	=	+0.057 egg	or 2.7% increase
Mate: Same-divorce	=	+0.128 egg	or 6.1% increase
Mate: Same-died	=	+0.213 egg	or 10.1% increase

CLUTCH SIZE

Birds breeding for the first time lay significantly smaller clutches than older birds (Tables 5 and 6), but there is again no difference between those nesting centrally and those on the periphery. Thereafter, only those birds which retain their mate show both a progressive increase in clutch size with age and also a difference between the numbers of eggs laid on the edge and in the centre of the colony. Birds which change their mates show a markedly lower clutch size than those birds which retain their mate, and again there is a suggestion that divorced birds lay a slightly larger clutch than those which have changed partners because of the death of their former mate. Thus the death of the mate results in a 10 per cent reduction in clutch size or, perhaps more correctly, that mate retention results in a 10 per cent increase in clutch size.

BREEDING SUCCESS

Just as in the time of laying and in clutch size, there is no significance in the difference in breeding success between birds breeding at the edge and in the centre of the colony (Tables 7 and 8). However, there is a significant increase in breeding success with age in the success of birds which have retained their mates.

TABLE 7

THE BREEDING SUCCESS ACCORDING TO THE BREEDING EXPERIENCE OF THE FEMALE, PAIR STATUS AND THE POSITION IN THE COLONY

Sample sizes as in Table 3

	breeding experience (years)					
	1		2–4		5–16	
	edge	centre	edge	centre	edge	centre
Same mate			66%	65%	68%	70%
	62%	57%				
Divorce			61%	63%	63%	69%
Mate died			58%	58%	63%	67%

TABLE 8

DIFFERENCES IN BREEDING SUCCESS ACCORDING TO CONDITIONS

Age:			
Old-young	=	+4.93%	or increase of 7.7%
Position:			
Centre-edge	=	+2.12%	or increase of 3.3%
Mate:			
Same-divorce	=	+3.45%	or increase of 5.4%
Mate:			
Same-dead	=	+6.08%	or increase of 9.5%

Further, there is also a significant difference between the success of birds which have changed their mates because of death of their partner as compared with those which are divorced (P<.05). Although the birds at the centre have a slightly better success than those at the edge, the difference is not significant.

Birds whose previous partner had died since the last breeding season have a breeding success of the same order as birds breeding for the first time. Breeding success is enhanced by an increase in age and the retention of the previous year's mate.

ANNUAL PRODUCTION OF YOUNG

The interaction between clutch size and breeding success can be expressed in terms of the number of young fledged per pair. The data presented in Tables 9 and 10 show that there is no difference between the number of young fledged

TABLE 9

THE MEAN NUMBER OF YOUNG FLEDGED PER PAIR ACCORDING TO BREEDING EXPERIENCE OF THE FEMALE, PAIR STATUS AND POSITION IN THE COLONY

Sample sizes as in Table 3

	breeding experience (years)					
	1		2–4		5–16	
	edge	centre	edge	centre	edge	centre
Same mate			1.37	1.43	1.51	1.62
	1.16	1.06				
Divorce			1.23	1.31	1.34	1.47
Mate died			1.15	1.29	1.22	1.27

by birds breeding for the first time at the edge and in the centre of the colony. Further, they fledge fewer young than any other age group or any combination of age and mate status. The production of older birds is significantly increased by age, position in the colony and the pair status and there is also a significantly (P<.01) lower production by birds whose previous mate has died as compared with those who have been divorced.

TABLE 10

DIFFERENCES IN YOUNG FLEDGED PER PAIR ACCORDING TO CONDITIONS

Age:			
Old-young	=	+0.11	or increase of 8.3%
Position:			
Centre-edge	=	+0.10	or increase of 7.3%
Mate:			
Same-divorce	=	+0.15	or increase of 11.2%
Mate:			
Same-dead	=	+0.25	or increase of 19.2%

Thus within a single colony the number of young fledged per pair varies, *on average*, between 1.07 in young birds breeding for the first time to 1.62 in old birds which nest in the centre of the colony and retain their previous mate. A young bird which has already bred but whose partner has died fledges only 1.15 young per year.

INTERACTIONS OF AGE, PAIR STATUS AND NEST POSITION

The males nesting centrally in the colony have an appreciably higher survival rate to those on the edge. On average a central male will breed for 10 years whilst one on the edge has only 6 breeding years. Further, the central male changes mate less frequently so that not only does he produce more offspring each year but he lives longer and produces appreciably more young in his lifetime. There is a clear advantage in being a central male but consideration of how this is achieved is beyond the scope of the present paper. However it should be emphasised that one factor in achieving this greater success is his ability to retain the same mate from year to year.

EFFECT OF BREEDING SUCCESS ON THE PAIR-BOND

The advantage in retaining the mate from the previous year has become apparent in the previous sections. It would seem to be a marked disadvantage for a pair to divorce since the marked advantage in pair retention is lost. However, some pairs seemed doomed to failure because, having received enough stimulation to initiate egg laying, the drive to incubate or to care for the chick is not sufficient to cause successful breeding. It would appear that the failure to breed successfully is often the result of inefficiency on the part of one of the partners but this is sufficient to result in a complete failure in breeding. Under such conditions it seems possible that a recombination of partners may give some success, even if below that of established pairs, whereas the continuation of the previous unsuccessful pair-bond is more likely to result in repeated failure.

In Table 11 the effect of the breeding success in the previous breeding season is examined in relation to whether the pair-bond is maintained or not. In the

TABLE 11

EFFECT OF PREVIOUS BREEDING SUCCESS ON THE PAIR-BOND

pair in current breeding season	breeding in previous year		
	successful	failed	total
Same as last	434	60	494
"Divorced"	175	100	275
Total	609	160	769
Percentage "divorced"	29%	63%	

cases where the breeding was successful, 29 per cent of the pairs changed mates whereas more than twice this proportion (63%) changed mate after an unsuccessful breeding season.

It should be pointed out that this effect, although contributing to the success of breeding, accounts only for a proportion of the divorces. This probably arises from the change of mate not being directly linked to failure of breeding but to the early break up of the pair in that breeding season which, in turn, appears to cause considerable variation in the time of return to the colony in the next breeding season. Change of mate is brought about by the other partner not being in the colony at the time the other returns for the next breeding season.

INCUBATION PATTERN IN THE KITTIWAKE

Detailed study of the Kittiwake shows that individuals are considerably more variable in their behaviour than was first believed. This is probably true generally amongst animals with a complex pattern of breeding. In order to evaluate this variation, continuous records of the presence of the male and female at the nest site were obtained by monitoring pairs of Kittiwakes with radio-active leg-bands. The situation was arranged so that the males carried twice the radio-active load of the female, and both sexes could be identified by the amount of radiation received by a Geiger counter placed over the nest. The level of radiation received by the counter was recorded on a chart moving at a constant speed. Figure 1 shows the incubation pattern of two pairs, one successful and the other which failed to hatch their eggs. The successful pair soon established a clear rhythm with both birds taking an equal part. A similar rhythm was established by the other pair but then, half-way through incubation, the male failed to return to relieve his partner on several occasions and finally did not return for 5 days. His partner successfully attempted to fill the gap on the single days but was unable to do so on the long absences of 5 days and it is reasonable to assume that this was when the eggs were chilled and the embryos died. This male showed a poor drive to incubate but the female had a very strong drive and would have suc-

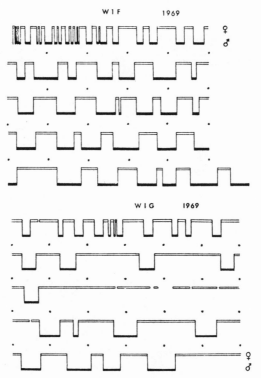

FIGURE 1. The role of male and female Kittiwakes in the incubation pattern of a successful and unsuccessful pair. The dots indicate midnight and the distance between adjacent dots denotes 24 hours. Note the absence of the male on several days and for a continuous period of six days on site WIG.

ceeded in hatching the eggs if paired to a more capable male. Breeding is an integration of a series of very complex interactions between the members of a pair. It is not, perhaps, unreasonable to consider that this female has a better chance of breeding successfully with a different mate. In fact in 1970 she paired with the same mate as in 1969 and once again breeding failed because of failure of the male to take a full share in the incubation, indicating that the cause of failure in one year can persist and result in failure in a future year.

DISCUSSION

The pair-bond in the Kittiwake results in more successful breeding when this relationship has been established for more than one year. This undoubtedly results in a complex pair relationship which becomes more efficient through individual experience of a mate. The long term pair-bond appears to induce a greater reproductive drive resulting in a better all-round breeding performance.

In the Kittiwake, the ability for a single pair to breed in isolation has been lost and the species is an obligatorily colonial. The birds obtain stimulation from their mates and their neighbours, the latter probably coming via the social greeting ceremony (COULSON & WHITE 1958). The more complex the breeding behaviour, the greater the chance that some malfunction can occur. I believe that many Kittiwakes find it difficult to develop into full breeding condition. Few breed as well as the most successful pairs and some birds miss a breeding season while many, usually young birds, pair but fail at some stage which may be at nest building, egg laying, incubation, brooding or feeding of the young.

If this difficulty does occur, then it is clearly an advantage for the potentially good breeding birds to breed together. In the Kittiwake, this appears to be achieved by the best birds returning to the colony first and hence pairing up with potentially good mates; further, the segregation between the centre and the edge birds will enhance this. In addition, the retention of a good mate is clearly an advantage while pairs which have failed to breed successfully are more likely to change mates, and the result of this divorce is likely to increase the overall breeding success of such birds.

The pair-bond in the Kittiwake has been evolved into a highly complex relationship between two individuals which can be more productive than the simpler position seen in many other bird species. Clearly the effect must involve individual recognition but the small but significant differences in the performance of birds which change their mates by divorce rather than due to the death of their previous partner needs cautious interpretation. If divorce was more likely amongst potentially poor breeding birds, then the performance of these birds might, on average, be expected to be worse, not better, than those whose mates have died. It might be tempting to make a comparison between our own society and that of the Kittiwake and suggest that the effects of divorce are, on average, less traumatic than the death of the partner. This type of conclusion would be unwise at the present state of information, but further work on the behaviour of individuals is in progress.

With the evidence so far available, there are strong indications that, accepting the social organisation in the Kittiwake, the retention or the divorce of a mate follows that system which produces the maximum number of offspring for each breeding pair.

REFERENCES

COULSON J. C. & WHITE E. (1958) The effect of age on the breeding biology of the Kittiwake *Rissa tridactyla*. Ibis 100, 40–51.

COULSON J. C. & WHITE E. (1960) The effect of age and density of the breeding birds on the time of breeding of the Kittiwake *Rissa tridactyla*. Ibis 102, 71–86.

COULSON J. C. (1963) Improved coloured-rings. Bird Study 10, 109–111.

COULSON J. C. (1966) The influence of the pair-bond on the breeding biology of the Kittiwake Gull, *Rissa tridactyla*. J. anim. Ecol. 35, 269–279.

COULSON J. C. (1968) Differences in the quality of birds nesting in the centre and on the edges of a colony. Nature 217, 478–479.

Part III

EGG LAYING

Editor's Comments
on Papers 9 Through 13

In general egg laying occurs after the nest is built and at a time when one or both parents are ready to incubate. However, some birds build no nest at all. In other species the male, the female, or both together build the nest. Some birds do not incubate, while in other species the male alone, the female alone, or both participate in incubation. The diversity of behavior patterns, representing adaptations to a wide range of environmental conditions, reflect a diversity of mechanisms which ensure the coordination of behavioral, physiological, and ecological factors.

What stimuli influence the occurrence of egg laying? It has long fascinated scientists that stimuli provided by the courting male (or a suitable substitute), initiate the sequence of events leading to egg laying. This fascination is shown in the paper by Wallace Craig (Paper 9), one of the forerunners of ethology, who preened and tickled the head and neck of a tame, "lonely" pigeon and found that she laid an egg. The notion that stimulation from the male pigeon, could influence egg laying was again taken up by L. H. Matthews (Paper 10), who tried to separate the role of visual, auditory, olfactory, and tactile stimuli.

The third paper in this sequence, by Lehrman et al. (Paper 11), was written fifty years after Craig's original observation. The general question addressed by Craig, Matthews, and Lehrman, is the same: How does the courting male influence ovarian development? By the time Lehrman's work was done in 1961, the relationship between the hypothalamus and the pituitary was well known, and the identity of the hormones which are important in ovulation had been established. Scientific writing had changed dramatically, as had the standards for acceptable experimental procedures. A reading of these three papers provides a sense of the continuity and of the increasing sophistication in behavioral science over the last fifty years.

The previous papers show that stimuli from the male influence ovulation and egg laying. Stimuli from the eggs themselves can also affect egg laying. Thus, birds can be classified as indeterminate or determinate layers. In indeterminate layers the number of eggs laid can be changed by adding or removing eggs just before the time of laying. The female stops laying only when the usual clutch size is reached. Phillips (Paper 12) was able to collect seventy-two eggs from a single Yellow-shafted Flicker (*Collaptes auratus*) by removing the newly laid egg each day. The domestic hen is derived from an indeterminate egg layer. Determinate layers have a fixed clutch size, and the ovary produces only as many yolks as eggs will be laid. Manipulation of eggs in the nest has no influence on the number of eggs laid. Pigeons and doves (Columbidae) are determinate layers with a clutch of two eggs.

The above considerations focus on induction of egg laying. The paper by Lack (Paper 13) is addressed to the question of which factors influence clutch size. First, it is curious that there is great variation in the number of eggs laid by birds. Some birds lay only one egg, others lay more than eight. While a final solution is not available, several theories have been proposed. Some suggest that important limiting factors include the size of the nest, the size of the eggs, the intensity of predation, and demands made on the parents which might influence their later reproductive potential. Lack suggests that the average clutch size of all birds is selected to yield the largest number of young that the parents can successfully feed.

9

Reprinted from *J. Morph.* **22**(2):299–305 (1911)

OVIPOSITION INDUCED BY THE MALE IN PIGEONS

WALLACE CRAIG

Department of Philosophy, University of Maine

The influence of the male upon the time of oviposition is a matter in regard to which pigeons differ from some other birds, notably the domestic fowl. With regard to the fowl I have consulted a number of poultry keepers and experts, chiefly Dr. Raymond Pearl and Dr. Frank M. Surface, of the Maine Agricultural Experiment Station, where the most extensive studies of the egg-laying of fowls have been, and are being carried on. Dr. Pearl and Dr. Surface tell me that the domestic hen, and also the hen of the wild Gallus bankiva so far as can be ascertained, commence their spring laying at an approximately fixed date which can neither be deferred by withholding the cock nor advanced by giving the cock before the usual time.

Pigeons differ widely from poultry in this respect. If, from the winter season onward, an old female pigeon be kept unmated and isolated, she refrains from egg-laying, in evident distress for want of a mate, until the breeding season is far advanced; at length she does begin to lay, but her laying without a mate manifestly partakes of the abnormal. And a virgin pigeon, if kept isolated from other pigeons, may postpone her laying for a still longer period. On the other hand, a female pigeon, young or old, will lay very early in the season if she be early mated. Moreover, there is a pretty definite interval between the first copulation and the laying of the first egg, namely six or seven days; if the egg be delayed much beyond this time, the fact indicates some indisposition on the part of the female. And as the pair rear brood after brood throughout the season, this time-relation between copulation and egg-laying is regularly repeated.

The utility of this time adjustment in pigeons seems obvious. The male pigeon takes his turn daily in the duty of incubation: hence the female must not lay the eggs before he is ready to sit. This aspect of the matter, which has to do with pigeon sociology, has already been treated elsewhere (Craig '08) and will be discussed more fully in a book dealing with pigeon behavior. The present paper is to show, not why the male should determine the time of oviposition, but how he does determine it.

The thesis of the present paper is, that the influence of the male in inducing oviposition is a psychological influence; that the stimulus to oviposition is not the introduction of sperm, for the male can cause the female to lay even though he does not copulate with her. This is easily proven by an experiment, which requires only pigeons, patience, and time, and I shall now recount seven repetitions of such experiment, the first two being accidental cases, the other five being trials designed and carried out on purpose to test the thesis.

Case 1 (1903). In the spring of 1903 I brought together a virgin female dove (individual female no. 7, the species in all these trials being the blonde ring-dove, Turtur risorius) and a young inexperienced male, intending simply that they should mate in the normal manner. The young male played up to the female, but due to his inexperience and to other causes which need not be discussed here, his mating behavior was imperfect and he did not copulate with her. Nevertheless, in due time (six days) she laid an egg, and a second egg, as usual, forty hours later. This was the first intimation to me that a male bird can stimulate the female to lay, without copulating with her. Such an explanation seemed so absurd at that time that I dismissed it with the assumption that the birds must have copulated unobserved, and I did not even test the eggs to see if they were fertile. Looking back on that case now, however, and considering the observed behavior of that male, I feel reasonably certain that he did not fertilize the eggs but simply stimulated oviposition through the psychic (neural) channels.

Case 2 (1904). A female dove (no. 5) had been kept alone ever since her mate had died in November, 1903, and as time wore on she showed intense anxiety to mate. She being a very tame bird, I had often caught and held her gently, but she did not like to be held, so one day in early March I tried tickling her head and pulling the feathers about her neck somewhat as a courting male would do it, and, finding that the poor lonely bird received these attentions with intense pleasure and became still more tame, I continued to preen her neck daily. She now acted toward the hand as if it were a mate, went through a nesting performance in her seed dish, there being no nest in her cage, and to my astonishment laid her eggs in due season. The first egg was laid March 11 and the second March 13. There is no doubt in my mind that the caressing of this bird's head and neck brought on oviposition. I once tried to repeat the experiment with another female dove, but she would not accept the touch of the hand as the former dove had done. Yet there is other evidence indicating that, with a specially tamed bird, this experiment, inducing oviposition by the hand, could be successfully repeated.

This case called to mind that of 1903, and suggested an experiment to determine definitely whether the male dove can stimulate the female to lay, without actual copulation. Opportunity to try this experiment was not found till 1907 and following years, when it was planned as follows.

Method of the regular trials

The experiment requires an unmated female dove that is not laying eggs, preferably a young dove that has never laid. It is best tried early in the season (e.g., in February), especially if an old dove be used, for, as said above, if the female is kept too long without a mate she may lay without one. Side by side with this female, in a separate cage, is placed an unmated male, and the two are given several days to become acquainted. When they act toward one another like mate and mate, the doors separating them are opened and they are allowed to come together for a time, under constant supervision. When they attempt to copulate, a slender rod which can be thrust between the bars of the cage is used to keep them apart. Such attempts are made many times in a day, mostly in the afternoon, and are continued for several days in succession; hence it is best that the experimenter should be able to devote some hours a day for several days in succession to a single pair or at most two pairs of birds. Whenever the birds are not under surveillance they are shut apart, each in his or her own cage. But they should be allowed to come together daily until the egg is laid.

A factor which caused difficulty in one of my trials was the nest. In cases 1, 2, 3 and 6, the bird laid without any nest at all (except that in case 6 a nest was given just a few hours before the egg was deposited). But in case 4 (*q.v.*) the female refused to lay without a nest: it was then necessary to remove the male and make the trial again, first giving the female a nest, and waiting long enough to prove that the nest alone would not cause her to lay.

Results of the regular trials

Case 3 (1907). Female dove, no. 20. This bird had been bought recently from a dealer, and it was not known whether she had laid earlier in the season. But she was kept isolated for some time, during which she showed no inclination to lay. She was then given a male in the manner indicated. No nest given.

June 9. Male allowed in cage of female, and plays up to her.

June 15. First egg.

June 17. Second egg. (The second egg was of no special interest. After the first egg was laid, I generally left the doors open, allowing the pair to come together without surveillance.)

Case 4 (1908). Female, the same. She had not laid since the close of last season. No nest given.

February 4. Male allowed to enter.

The female was unresponsive and showed by her behavior that this time she was holding back for want of a nest. This deficiency was supplied in the following manner (vide ut supra.)

February 8. Male taken away to another building.

March 10. Nest put in cage. Female paid practically no attention to it. Many days were allowed to pass, in order to make sure that the nest alone would not stimulate the female to lay.

121

March 21. Male (after short period in sight of female, that they might become re-acquainted) allowed to enter.

March 27. Egg laid.

Case 5 (1910). Female, the same as in cases 3 and 4. She has laid no eggs since last season (1909.)

January 20. I begin to allow male in cage, at same time putting nest in.

January 29. Egg laid.

Case 6 (1908). Female, no. 19. Virgin, has never laid. No nest given. In this case, the date on which the female was first given the requisite stimulus cannot be stated so definitely as in the other cases.

July 12. Male, in his cage, placed close to cage of female. Cooing commences. Female so excited that she several times assumes, and maintains in extreme degree, the copulation posture.

July 14. Male allowed into cage of female, but he fights her, so that it is necessary to remove him (otherwise the female might be painfully injured), and to allow the pair a few days more of preliminary acquaintanceship.

July 18. Male allowed to begin his series of daily visits.

July 22. Egg laid.

Case 7 (1910). Female no. 19, the same as in case 6. She has laid no eggs since last summer (1909.)

For several days before contact with the male, a nest was kept in her cage; but she paid no attention to it, showing that the nest alone would not stimulate her to lay.

January 20. Male allowed to enter.

January 26. Egg laid.

SUMMARY

1. In six cases, stimulation of a female dove by a male, without copulation, was followed by oviposition; and in one other instance (case 2), stimulation by the hand of man in imitation of a male dove was followed by oviposition

2. In six of the seven cases (being all except case 3, in which the previous history was unknown), it was known that the female

had laid no eggs previously during the current year. In two of these six cases the dove was a virgin and had never laid.

3. It is true that the female may, if left without a mate, begin to lay late in the season. Hence it might be suspected that the sequence of stimulation and egg-laying in the seven cases was mere coincidence. But this is precluded, first of course by the fact that coincidences are not known to happen seven times in succession, and further by the following considerations.

4. In some of the trials it was proven that the female when stimulated by the male laid much earlier in the season than she did when not so stimulated. This is shown in the following table.

Female, no. 20.

1908. (Case 4), stimulated by male, laid March 27.

1909. (Control), without male, began to lay May 13.

1910. (Case 5), stimulated by male, laid January 29.

Female, no. 19.

1909. (Control), without male, began to lay April 26.

1910. (Case 7), stimulated by male, laid January 26.

5. The interval between the first stimulation by the male, and the laying of the first egg, was as follows:

Case 1. 6 days.

Case 2. (Male not used.)

Case 3. 7 days.

Case 4. 6 days.

Case 5. 9 days.

Case 6. 4 to 10 days, depending on what is regarded as the first stimulation in this case.

Case 7. 6 days.

The average and the variation of these intervals tally closely with the average and the variation of the interval in normal breeding, between the first copulation and the laying of the first egg.

6. There were no exceptions. Ovoposition never failed to follow within nine days after the first contact with the male. (The only partial failure was that of the first trial in case 4, which was due to faulty experimental conditions.)

CONCLUSION

These facts make it certain that the male dove can stimulate the female to lay, without copulating with her.

Harper ('04) mentioned the fact that ovulation in the pigeon does not take place until after the bird is mated, but he was in doubt as to how far the influence of mating was a 'mental' one and how far it was a matter of the introduction of sperm. The present paper goes to show that the stimulus to the whole process of egg development and laying is a psychic (neural) stimulus, not dependent upon the introduction of sperm.

BIBLIOGRAPHY

CRAIG, WALLACE 1908 The voices of pigeons regarded as a means of social control. Am. Jour. Sociol., vol. 14, pp. 86–100.

HARPER, EUGENE HOWARD 1904 The fertilization and early development of the pigeon's egg. Am. Jour. Anat., vol. 3, pp. 349–386.

Reprinted from *Proc. Royal Soc. Lond.*, Ser. B, **126**:557–560 (1939)

Visual stimulation and ovulation in pigeons

By L. Harrison Matthews

*(Communicated by F. H. A. Marshall, F.R.S.—
Received 27 September 1938)*

It has long been known that the female pigeon, unlike the carefully selected domestic hen, does not ovulate spontaneously. Harper (1904) showed that in pigeons ovulation occurs not less than 8 days after the introduction of the male to a female when both are ready for mating. He further recorded that two female pigeons when confined together may both take to laying eggs. The act of mating therefore is not necessary to induce ovulation, the mere presence of a companion sufficing. This exteroceptive stimulus to ovulation is usually regarded as exerting its influence through the pituitary.

The experiment described in this paper was designed to show whether the stimulus is visual, auditory, olfactory, tactile or a combination of two or more of these alternatives. The essential part of the experiment carried the confining of two females together a stage further. Two birds were confined in adjacent cages separated by a sheet of glass, and single birds were confined by themselves but provided with a mirror. The birds were thus supplied with companions to which they had no tactile access.

Wooden cages with fronts of half-inch mesh wire netting were used. They measured $18 \times 14 \times 10$ in. Some of them were joined together in pairs, the partition between them being movable, so that it could be removed or a sheet of glass substituted for it. Others were provided with slides at the back so that a sheet of mirror could be slipped in to cover the back wall.

The birds used were dealers' ordinary stock, mongrels showing a predominance of the common blue rock. They were placed in solitary confinement for a month, as a preliminary to the experiment, so that the effect of any previous stimuli might pass. They were then confined as follows:

Double cages.
    (1) Male and female with no partition (control).
    (2) Male and female with glass partition.
    (3) Two females with glass partition.
    (4) Two females with no partition.

Single cages.
    (5) Female alone, with mirror.
    (6) Male alone, with mirror.
    (7) Female alone without mirror (control).

Owing to the difficulty of ascertaining the sex of mongrel pigeons a mistake was made with some of the birds. One of those in cage no. 3 turned out to be a male, so that this cage merely duplicated cage no. 2. No. 6 was confined as a female, but was found to be a male after the experiment had started. Cages nos. 1 and 2 were kept in a different building from the rest. All the cages were arranged so that none of the birds could see any of those in other cages.

The results obtained are shown in Table I, where the dates when eggs were laid are shown.

It will be seen that eggs were obtained from the control pair in 9 days' time, and from other birds after longer intervals. The female of the control pair was isolated on 12 June. She laid again on 14 and 15 June, but thereafter produced no more eggs up to the end of July when the experiment ended. The control female, shut up by herself with no mirror, did not lay at all up to 12 June. On that day she was placed in a double cage with a male, and laid on 23 and 25 June. The male (6), which was mistaken for a female and confined with a mirror, soon started calling and displaying in the usual manner of pigeons (cf. Whitman 1919) to its mirror image.

The results obtained show that the stimulus which causes ovulation in the pigeon is a visual one. The ovulation of birds separated from the male by a sheet of glass, or confined alone with a mirror, shows that the stimulus is not tactile. The absence of ovulation in the control, which was denied

126

sight of another bird but was in the same room with the others, shows that the stimulus is not olfactory or auditory. Were it either, or both, she would have laid. Her subsequent ovulation when confined with a male showed her to have been a normal bird which lacked only the visual stimulus for ovulation.

TABLE I. DATES OF EGG-LAYING

Birds placed in solitary confinement 2 February. Experiment started 3 March

1 control ♂	2 ♂ glass ♀	3 ♂ glass ♀	4 ♀	♀	5 ♀ with mirror	7 ♀, no mirror
12 Mar.	15 Mar.	29 Apr.	23 Apr.		14 May	
13 Mar.	17 Mar.	1 May	25 Apr.		16 May	
30 Mar.	28 Mar.	19 May		27 Apr.	27 May	
1 Apr.	30 Mar.	23 May		29 Apr.	28 May	
16 May	1 June	1 June	8 May		9 June	
18 May		4 June	10 May		11 June	
3 June		15 June		19 May	24 June	
		18 June		20 May		
♀ isolated		4 July	22 May			Placed with ♂
12 June			23 May			12 June
14 June				2 June		23 June
15 June				4 June		25 June
No further			18 June			9 July
eggs to end						10 July
of July						

Although visual stimulation from a companion of the same sex, or even from a mirror image, is sufficient to produce ovulation, it would appear that the response of the female to the behaviour of the male is also self stimulatory. When the birds are denied access to each other, or are of the same sex, the pattern of the mating behaviour cannot be completed, and the lower level of stimulus produced is reflected in the longer interval before the appearance of the first eggs.

It is claimed that this experiment lends further support to the explanation of sexual display in birds and other animals brought forward by Marshall (1936), who concludes that display does not produce a conscious sexual selection as proposed by Darwin, but is essentially a mutual pituitary stimulation. The visual stimulation here shown to exist in pigeons when confined with a single companion, or merely a mirror image, is carried to a higher level in those birds which nest in colonies. Fraser Darling (1938) showed that a definite minimum number of pairs is a requisite in any breeding colony of certain species of birds that nest gregariously, in order to produce the threshold stimulus for ovulation. Visual stimulation has also been demonstrated in pigeons by Patel (1936), who showed that unless the male can see the female sitting on the eggs during incubation, he secretes no milk from the crop gland after the eggs are hatched.

The writer wishes to express his thanks to Dr Richard Clarke, who kindly found accommodation for the experimental birds at the Clifton Zoological Gardens, and to Keeper Victor Jones who looked after them.

REFERENCES

Darling, F. Fraser 1938 "Bird Flocks and the Breeding Cycle." Cambridge.
Harper, E. H. 1904 *Amer. J. Anat.* **3**, 349.
Marshall, F. H. A. 1936 Croonian Lecture. *Philos. Trans.* B, **226**, 423.
Patel, M. D. 1936 *Physiol. Zoöl.* **9**, 129.
Whitman, C. O. 1919 "The Behaviour of Pigeons." Washington.

11

Reprinted from *Endocrinol.* 68(3):507–516 (1961)

THE PRESENCE OF THE MATE AND OF NESTING MATERIAL AS STIMULI FOR THE DEVELOPMENT OF INCUBATION BEHAVIOR AND FOR GONADOTROPIN SECRETION IN THE RING DOVE (*STREPTOPELIA RISORIA*)[1]

DANIEL S. LEHRMAN, PHILIP N. BRODY
AND ROCHELLE P. WORTIS[2]

Institute of Animal Behavior, Rutgers University, Newark, New Jersey

ABSTRACT

We have studied the effects of stimuli provided by the presence of a bird of opposite sex, and by the presence of a nest-bowl and nesting material, on the development of readiness to incubate eggs in male and in female ring doves, and on gonadotropin secretion in females. All subjects had previous breeding experience. The results show that the presence of a male in the cage causes the gradual development, over an 8-day period, of readiness to sit on eggs, and that the additional presence of a nest-bowl and nesting material augments this effect, starting on about the 6th day. The presence of the female, and of nesting material, have similar effects on the incubation behavior of the male, except that the effect of nesting material may be seen earlier (about the 4th day) in the male than in the female. Gonadotropin activity in the female, as indicated by oviduct growth and by incidence of ovulation, is stimulated by the presence of a male, and augmented by the presence of nesting material. The occurrence of incubation behavior in females is closely associated with that of ovulation, and a number of lines of evidence suggest that a progestin is implicated in the instigation of this behavior.

WHEN a pair of ring doves which have had previous breeding experience is placed in a breeding cage containing a nest and eggs, they do not sit on the eggs until after several days, during which they exhibit the progression of courtship and nest-building behavior which normally precedes the laying and incubation of their own eggs (10). Both sexes participate in incubation. It has been shown that association with the mate, and the presence of a nest-bowl and nesting material, facilitate the successive development of the pair's readiness to build a nest and to sit on eggs, although our previous work has not distinguished between the role of the male and that of the female. Nest-building can be induced by exogenous estrogens, and incubation behavior by exogenous progesterone (11, 23).

These findings suggest that the onset of incubation behavior may be

Received September 1, 1960.

[1] Supported by a research grant (M-2271) from the National Institute of Mental Health, United States Public Health Service.

[2] National Science Foundation Cooperative Graduate Fellow.

associated with the presence of gonadal hormones, and that the secretion of these hormones may be elicited or facilitated by stimuli coming from the mate and from the nesting situation. The purpose of the present experiment is to examine: a) the effect of the presence of the mate and of a nesting situation (nest-bowl and nesting material) on the development of readiness to incubate in male and in female ring doves; b) the effect of these stimuli on gonadotropin secretion in the female ring dove, as indicated by ovarian activity; and c) the relationship between ovarian activity and readiness to incubate.

MATERIAL AND METHODS

Animals. The subjects were ring doves (*Streptopelia risoria*) procured from Mr. J. W. Steinbeck of Concord, California, or the first-generation offspring of animals from this source.

Cages and Maintenance. Experimental cages were of wood, each 32″ wide, 18″ deep, and 14″ high, with wire-mesh front doors. They were provided with dispensers for water, food and a grit-and-mineral mixture, and (when required) a glass bowl 4¼″ in diameter, in which the birds build their nests. These cages were illuminated by strips of fluorescent lights mounted on a wall 5′ 6″ in front of the doors.

Stock cages were 35″ cubical cages built with wooden frames and wire-mesh sides.

Isolation cages were metal rat cages 16½″ wide, 9½″ deep, and 7″ high, mounted in racks.

Water, food and grit were continuously available in all cages. Nesting material consisted of pine needles or hay placed on the floor in a corner of the breeding cage.

Breeding cages, stock cages and isolation cages were kept in separate rooms. Lights in all rooms were clock-controlled, being turned on at 6:00 A.M. and off at 8:00 P.M., EST. Temperature in all rooms was 72°–74° F, except for brief and irregularly-distributed periods of malfunctioning of the temperature-control apparatus.

Experience prior to testing. The laboratory-reared birds were hatched in experimental cages. All birds were kept in stock cages after being separated from their parents at the age of 21 days, or after being received from the supplier. The sex and degree of maturity were determined by an exploratory laparotomy, and the birds placed in pairs in experimental cages. They were allowed to breed once, their young being removed when 21 days of age. At the same time, the subjects were removed to individual isolation cages. They were kept in isolation for 3–5 weeks, and then placed in the experimental cages for testing. During the tests, subjects were never paired with birds with which they had previously been mated.

In summary, all subjects were birds with previous breeding experience consisting of one successful breeding cycle, and had been in isolation for from 3 to 5 weeks prior to testing.

Experimental procedure. The general procedure was to place birds, singly or in pairs, in experimental cages for a variable number of days, after which they were tested to determine whether they would sit on eggs, and the females were autopsied to determine oviduct weights, and whether they had ovulated.

Three groups were arranged. *Control:* females were placed alone in breeding cages without nest bowl or nesting material; *Mate:* subjects were placed in the breeding cages in male-female pairs, without nest bowl or nesting material; *Mate+Nesting material:* birds were placed in pairs in the cages, with a nest bowl and a supply of nesting material in each cage.

The birds were taken from the isolation cages and placed in an experimental cage between 9:00 A.M. and 10:00 A.M. (on day 0). Between 0 and 8 days later, at the same hour, a nest bowl containing a nest and two eggs, taken from a breeding pair in the colony, was placed in the cage, the original nest bowl, if any, being removed. The cage was then visited every 30 min. to determine whether either of the birds was sitting on the eggs. If neither bird had sat by the end of 3 hours, both birds were removed. If either bird sat during the 3-hour period, it was removed, and the other bird allowed to remain in the cage until it sat, or until 3 additional hours had passed. All females were killed for autopsy as soon as they were observed sitting, or at the end of the observation period.

In summary, birds kept in test cages for varying periods of time (0 to 8 days), either alone, with a mate, or with a mate and nesting material, were tested for incubation response to eggs, and then killed for autopsy.

Twenty pairs of birds were used for each treatment and each time interval in the experimental groups, and 20 females for each time interval in the control group, a total of 340 males and 400 females.

RESULTS

Incubation behavior

Females left alone in test cages for varying periods of time, up to 7 days, and then tested by being presented with nests and eggs, do not sit on the eggs (Fig. 1, "control"). Previous work has shown that birds of either sex, kept singly in cages with nests and eggs, and with nesting material, do not sit on them during test periods of up to 6 weeks (10).

By contrast, when birds are placed *in pairs* in the cages, and then presented with nests and eggs after periods of up to 8 days, incubation behavior is induced, the number of birds sitting on the eggs gradually increasing from day to day, both in females (Fig. 1, "mate") and in males (Fig. 2, "mate").

The presence of a nest bowl and nesting material in the cage during the period before the introduction of the nest and eggs further stimulates the development of incubation behavior, as shown by the greater numbers of birds of both sexes (Figs. 1, 2, "mate+nesting material") which sit on the eggs under these conditions, compared with birds stimulated only by the presence of the mate.

It may be noted that the two sexes do not respond in quite the same way to the presence of nesting material. P-values for the differences between the incidence of incubation behavior in the two stimulus situations are shown for the various days, separately for males and for females, in Table 1, where it will be seen that the presence of nesting material induces a higher incidence of such behavior (compared with that occurring in birds kept with a mate alone) starting on the 4th day in the males, but not until the 6th day in the females, a difference which is also apparent on visual comparison of Figures 1 and 2. Possible reasons for this difference between the sexes will be discussed below.

131

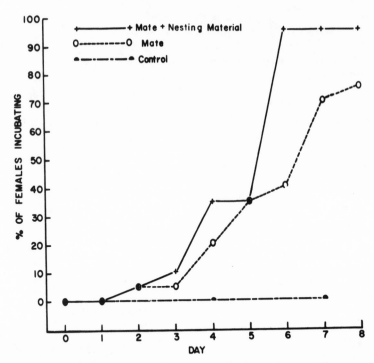

FIG. 1. Effect of association with a mate or with a mate+nesting material on the development of incubation behavior in female ring doves.

Each point is derived from tests of 20 birds. No individual bird is represented in more than one point.

The abscissa represents the duration of the subject's association with a mate or with a mate+nesting material, or (for the control group) the time spent alone in the test cage, just prior to being tested for response to eggs. "Day 0" means that the bird was tested immediately upon being placed in the cage.

Ovarian activity and the relation between ovarian activity and incubation behavior

Figure 3 shows the incidence of ovulation in females autopsied on the same day as that on which they were tested for incubation behavior. The

TABLE 1. P-VALUES FOR DIFFERENCES IN INCIDENCE OF INCUBATION BEHAVIOR BE-
TWEEN "MATE" AND "MATE+NESTING MATERIAL" GROUPS[1]

Sex	Number of days association with mate (or with mate+nesting material)								
	0	1	2	3	4	5	6	7	8
Female	n.d.[2]	n.d.	n.d.	.5[3]	>.3[4]	n.d.	<.001	.05	.08
Male	n.d.	n.d.	n.d.	.12	<.02	<.01	<.001	.02	.01

[1] Incidences taken from Fig. 1 (females) and Fig. 2 (males).
[2] No difference between frequencies of occurrence of incubation behavior with and without nesting material.
[3] P-values in ordinary type are calculated by Fisher's exact method. Conditions for the use of χ^2 not met (25).
[4] P-values in italics are based on χ^2.

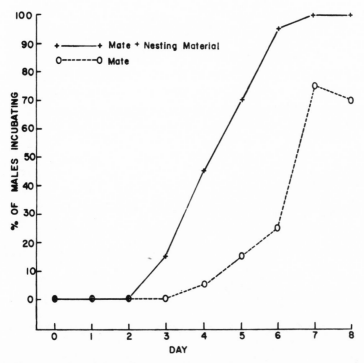

FIG. 2. Effect of association with a mate or with a mate+nesting material on the development of incubation behavior in male ring doves. See caption, Fig. 1.

absence of ovulations in the control group (kept alone in cages), and the sharply rising frequencies of ovulation in the two groups kept with mates and nesting material, indicate that ovulation is induced by stimuli associated with the presence of the mate, augmented by those associated with the presence of the nest bowl and nesting material.

The close similarity between the curves describing the incidence of incubation behavior (Fig. 1) and of ovulation (Fig. 3) in the females of the various groups suggests that the physiological conditions associated with ovulation are the same as those associated with readiness to incubate eggs. Table 2 shows the relationship between the readiness to incubate on the test day and the occurrence of ovulation prior to the test (i.e. up to the time of autopsy), for all females in the groups in which incubation behavior

TABLE 2. COINCIDENCE OF INCUBATION BEHAVIOR AND OF PRECURRENT OR CONCURRENT OVULATION IN FEMALE RING DOVES

Ovarian activity	Behavior	
	Incubated	Did not incubate
Ovulated	111	18
Did not ovulate	14	177

$$\chi^2 = 200.4.$$

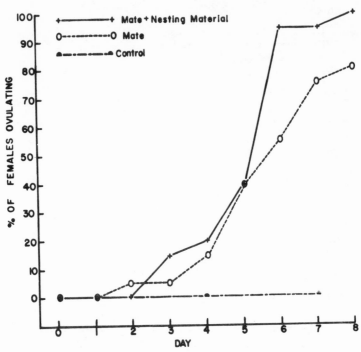

FIG. 3. Effect of association with a mate or with a mate+nesting material on the occurrence of ovulation in female ring doves. See caption, Fig. 1.

and ovulations occurred. The chi-square value of 200.4 indicates a very high degree of association between the occurrence of ovulation and of incubation behavior. The same strong association is shown when the "mate+nesting material" group ($x^2 = 111.95$) and the "mate" group ($x^2 = 86.51$) are examined separately.

Oviduct weights, summarized in Figure 4, show a similar pattern: no increase in oviduct weight occurs while the female dove is alone in the cage; duct weight increases as a result of the presence of a male; and this increase is augmented by the presence of nesting material. The inversion of the oviduct weight curves for the "mate+nesting material" and the "mate" groups between the 6th and 7th day is due to the fact that most birds of the "mate+nesting material" group ovulated earlier than those of the "mate" group, giving up oviduct weight in the form of albumen secretion (21). In each of the two groups, the oviduct weights on day 2 are significantly higher than on day 0 (Mann-Whitney U Test, $P < .05$) (25), indicating that increases in oviduct weights can be detected within 48 hours after the beginning of association with a male.

DISCUSSION

Most species of birds breed seasonally, the timing of the breeding season being influenced by a variety of environmental variables, such as daily light period, weather, availability of nesting sites, presence of other birds, etc. (12, 13). It has long been apparent from naturalists' observations that one of the functions of the courtship singing and posturing of male birds is to stimulate the development of sexual receptivity in the female (8), and F. H. A. Marshall, in his classic papers on sexual cycles (15, 16), suggested that this stimulation occurs *via* the elicitation of the secretion of gonad-stimulating hormones by the female's hypophysis.

Harper (6), Craig (2), and Matthews (17) have shown that in female

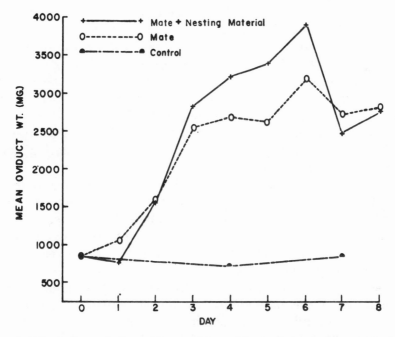

FIG. 4. Effect of association with a mate or with a mate+nesting material on oviduct growth in female ring doves. See caption, Fig. 1.

pigeons and doves, egg-laying may be stimulated by the presence of a male. It has further been demonstrated in English sparrows (19) and in starlings (1) that the ovary-stimulating effects of additional illumination in the non-breeding season may be augmented by the presence of males. In our laboratory, adult females kept in isolation from other birds do not lay eggs, and the present study shows that the presence of the male is necessary to stimulate the secretion of gonadotropins sufficient to elicit ovulation.

This experiment does not demonstrate that *no* gonadotropin secretion occurs in the absence of the male. The oviducts of our control animals are considerably heavier (700–800 mg.) than those of doves taken just before the onset of gonadotropin secretion at sexual maturity (400–500 mg.) (12a). Further, all these experiments were carried out at a constant day length of 14 hours light, 10 hours dark. Unpublished experiments from our laboratory indicate that, if the birds are placed without mates in a shorter day-length (9 hours) the ovaries and oviducts regress considerably below the level found in our control birds. The effect of stimulation from the male on gonadotropin secretion in our birds is, therefore, co-active with stimulation from other sources, and this interaction is now under study here.

Naturalists' observations also suggest that the presence of an appropriate nesting site and of nesting material is a factor in the conditions stimulating normal gonadotropin secretion during the breeding season in birds. A number of field observers have noted that breeding may be delayed by the absence of nesting material during the normal season (9), and Hinde and Warren (7) found that depriving female domestic canaries of nest material and of a nest site delayed the laying of eggs in the spring. Some dry-country tropical birds breed irregularly, always following rainfall, and Marshall and Disney (14) demonstrated experimentally in the case of the African red-billed weaver, *Quelea quelea*, that the presence of green grass, used for nesting material, is the principal stimulus for the beginning of breeding. In the present study it is clearly indicated that the presence of nesting material augments the effect of the male in stimulating gonadotropin secretion in the female.

It is apparent that the onset of the readiness of female doves to incubate is related to ovarian activity. This is indicated in the present study by the close similarity between the incidences of incubation behavior and of ovulation. Further, incubation behavior can reliably be induced in doves by the administration of progesterone (11), which probably appears endogenously at about the time of ovulation (4).

In female doves, it thus appears that the secretion of gonadotropins is in part stimulated by the mate and by the nesting situation (nest bowl and nesting material), and that the ovarian secretory activity thus induced contributes to the occurrence of incubation behavior at about the time of ovulation. The situation in the male is more obscure. We have not yet been able to demonstrate any histological differences between those males which do, and those which do not, sit on eggs; nor between males which have, and males which have not, been placed with females and with nesting material. Nevertheless males, like females, develop incubation behavior as a result of being associated with a mate, and this development is augmented by the presence of nesting material. We are continuing this study, and it is of course possible that we will detect changes in the testes or hypophysis (or elsewhere) associated with the onset of incubation behavior. Another possibility, also now under study, is that the endocrine condition of the male does not change as a result of association with a female, but that his behavior changes in response to the changes in *her* behavior, consequent on the endocrine changes induced in her by association with the male. We cannot at present decide the point. It may be noted, however, that the consensus of data in the literature of sexual photoperiodicity is that males of most wild bird species can be brought to full spermatogenesis by increased light-period alone, while ovum growth brought about solely by this stimulus never achieves ovulation (3).

The fact that the presence of nesting material appears to influence the incidence of incubation behavior in males by the 4th day, while it does not do so in females until the 6th day (Table 1) requires some explanation. In ring doves, as in other species of pigeons and doves (5) the male does most of the gathering of nesting material, which he carries to the nest, where the female does most of the building. It may be that the presence of nesting material alters the reproductive condition of the female partly by altering the behavior of the male toward her.

What of the role of prolactin in initiating incubation behavior in doves? This hormone is traditionally stated to form the principal physiological background of such behavior in chickens (22) and is sometimes assumed to be involved in all types of parental behavior in birds and other animals as well (20). The results reported here suggest that, in doves, ovarian hormones are the principal instigators of incubation behavior. Further, injected progesterone induces incubation behavior in doves, without any evidence of crop growth, and crop growth does not normally begin until after the beginning of incubation (11). There is evidence that participation in incubation induces prolactin secretion (18), and that prolactin so induced maintains incubation behavior (24), but it may be questioned whether the *initiation* of this behavior depends upon this hormone. Unpublished experiments in our laboratory indicate that prolactin does not induce incubation behavior in doves as reliably as does progesterone, even in doses large enough to cause full crop development, and that it does not do so at all when the dosage is small enough so that no appreciable crop growth occurs.

Acknowledgment

The authors are indebted to Mr. Willis Butler for assistance with the autopsies.

REFERENCES

1. BURGER, J. W.: *Anat. Rec.* **84**: 518. 1942.
2. CRAIG, W.: *J. Morph.* **22**: 299. 1911.
3. FARNER, D.: *Recent Studies in Avian Biology,* A. Wolfson, Ed.: 198. Univ. of Ill. Press, Urbana, 1955.
4. FRAPS, R. M.: *Mem. Soc. Endocrin.* **4**: 205. 1955.
5. GOODWIN, D.: *Avicult. Mag.* **1955**: 54. 1955.
6. HARPER, E. H.: *Amer. J. Anat.* **3**: 349. 1904.
7. HINDE, R. A. AND R. P. WARREN: *Anim. Behav.* **7**: 35. 1959.
8. HOWARD, H. E.: *Territory in Bird Life.* Chatto and Windus, London, 1920.
9. LACK, D.: *Proc. zool. Soc. Lond.* **1933**: 231. 1933.
10. LEHRMAN, D. S.: *J. comp. physiol. Psychol.* **51**: 32. 1958.
11. LEHRMAN, D. S.: *J. comp. physiol. Psychol.* **51**: 142. 1958.
12. LEHRMAN, D. S.: *Ibis* **101**: 478. 1959.

12a. LEHRMAN, D. S. AND P. BRODY: *Proc. Soc. exp. Biol.*, N. Y., **95**: 373. 1957.
13. MARSHALL, A. J.: *Ibis* **101**: 456. 1959.
14. MARSHALL, A. J. AND H. J. DE S. DISNEY: *Nature* **180**: 647. 1957.
15. MARSHALL, F. H. A.: *Phil. Trans.* **226B**: 423. 1936.
16. MARSHALL, F. H. A.: *Biol. Rev.* **17**: 68. 1942.
17. MATTHEWS, L. H.: *Proc. roy. Soc.* **126B**: 557. 1939.
18. PATEL, M. D.: *Physiol. Zoöl.* **9**: 129. 1936.
19. POLIKARPOVA, E.: *C. R. Acad. Sci. U.R.S.S.* **26**: 91. 1940.
20. RIDDLE, O.: *Proc. Amer. phil. Soc.* **75**: 521. 1935.
21. RIDDLE, O.: *Endocrinology* **31**: 498. 1942.
22. RIDDLE, O., R. W. BATES AND E. L. LAHR: *Amer. J. Physiol.* **111**: 352. 1935.
23. RIDDLE, O. AND E. L. LAHR: *Endocrinology* **35**: 255. 1944.
24. SAEKI, Y. AND Y. TANABE: *Poult. Sci.* **34**: 909. 1955.
25. SIEGEL, S.: *Nonparametric Statistics*. McGraw-Hill, New York, 1956.

12

Reprinted from *Auk* 4:346 (1887)

EGG-LAYING EXTRAORDINARY
IN COLAPTES AURATUS

Charles L. Phillips

Egg-laying extraordinary in Colaptes auratus.—On May 6th, 1883, I found in a large willow tree, a hole containing two eggs of this bird; I took one, leaving the other as a nest-egg, and continued to do this day after day until she had laid *seventy-one eggs*.

The bird rested two days, taking *seventy-three days to lay seventy-one eggs*. I think this is something very unusual; I have quite frequently heard of from fifteen to twenty-eight being taken from one bird, but this is a large number comparatively. I have the set complete, in my cabinet, and prize it very highly.

This was published in a small journal called the 'Young Oölogist', Vol. I, No. 2, 1884; but it being a rather obscure paper, and not reaching the general public, I concluded to send it to 'The Auk' for publication.— CHARLES L. PHILLIPS, *Taunton, Mass.*

13

Reprinted from *Evolution as a Process*, J. Huxley, A. C. Hardy, and E. B. Ford, eds., George Allen & Unwin, 1954, pp. 143–156

The Evolution of Reproductive Rates

DAVID LACK

THE reproductive rate differs considerably in different species of animals. Some, such as man or a tsetse fly, produce only one young at a birth, while at the other extreme various fish and marine invertebrates lay well over a million eggs at a spawning. The variations are quite large even within one class of animals. In birds, for instance, the average clutch-size of different species varies between 1 and 15. Even within the same species the variations may be far from negligible. Thus the average clutch-size of different populations of the Great Tit *Parus major* varies from 3 to 12 (Lack, 1950), and of the copepod *Arctodiaptomus bacillifer* from 6 to 60 (Hutchinson, 1951).

In the past, variations in the reproductive rate have usually been explained as due to differences in the mortality rate. Migratory birds have larger clutches than resident species because they have to undergo the additional danger of a long journey. Elephants have only one young at a birth because they are long-lived and, if they had more, they would quickly overpopulate Africa. Flatfish lay vastly more eggs than skates to compensate for the vastly heavier mortality among their fry. So runs the argument, which appears in a variety of explicit and implicit forms through the zoological literature. The fact is undoubted: those species with a lower reproductive rate have a lower mortality rate; but the explanation must be wrong. The reproductive rate is a product of evolution, and if the theory of natural selection be accepted, then the number of eggs laid by each species should be that which results in the maximum number of surviving offspring; and those genotypes with a higher egg-number than the normal should be eliminated by selection. Is this statement true? If it is true, what is the selective disadvantage of genotypes with a higher egg-number than the normal? And why should some species lay many more eggs than others?

This question can be answered with some confidence for nidicolous birds. First, it should be made clear that clutch-size is far below the normal limit of egg production; if the eggs are taken when laid, the bird readily lays others in their place, but these other eggs are not laid unless the first are taken. Secondly, the limit is not set by the number of eggs that the sitting bird can incubate, for hatching success is similar for clutches of all sizes found in nature, including those well above the normal (Lack, 1947, 1948a, 1950). The true answer, at least for nidicolous species, is that clutch-size has been adapted through natural selection to correspond with the maximum number of offspring for which the parents can, on the average, find enough food without seriously harming themselves. The evidence for this conclusion will be summarized in succeeding paragraphs. Two types of variation are involved, hereditary variations acted upon by selection, and phenotypic variations adapted to the conditions at the time. It is simplest to consider first some cases in which phenotypic variations can be neglected or excluded.

In England, the Common Swift *Apus apus* lays either 2 or 3 eggs, 2 more often than 3. The parents feed their young on air-borne insects, which almost disappear in continued bad weather, when young Swifts often starve to death. An analysis of the mortality among nestling Swifts in four colonies in southern England in 1946–50 showed that in broods starting with 2 young 82 per cent safely left the nest, whereas in broods starting with 3 young only 45 per cent safely left the nest. Nearly all the losses were due to starvation, broods of 3 suffering more than broods of 2 because the food was at first shared among three mouths instead of two. As a result, broods of 2 gave rise on the average to 1·6, and broods of 3 to 1·4, survivors per brood (Lack and Lack, 1951). Hence 2 was a slightly more efficient clutch-size than 3; and had any Swift laid 4 eggs, one may suppose that a yet higher proportion of the young would have starved, so that fewer, not more, young would have been raised per brood. Broods of 2 were more successful than broods of 3 particularly in wet summers like that of 1948, while in fine summers such as 1949, the average number of young raised per brood was higher from broods of 3 than 2. This makes it understandable how both individuals laying clutches of 2, and others laying clutches of 3, persist in the population. A closely similar result was obtained in a study of the Alpine Swift *Apus melba*, in which the average number of young raised per brood was highest for broods of 3, and the commonest clutch-size found in nature was likewise 3 (Lack and Arn, 1947). In these analyses late broods, which are of smaller size, were omitted.

In passerine birds, unlike Swifts, the nestling mortality proved to be similar in broods of all sizes. But considerable variations were found in the weight of the young when they left the nest, presumably due to differences in the amount of food received, and this might influence their subsequent chances of survival. It was then found that, in the Starling *Sturnus vulgaris*, the mortality-rate after the young left the nest was above average in young from broods of above average size. In first broods in Switzerland, the post-fledging mortality rose with increasing brood-size in such a way that the average number of survivors per brood was similar from broods starting with 5, 6, 7 or 8 young (Lack, 1948a). Hence 5 was the most efficient clutch-size in the Swiss Starling, and it was also the commonest size found in nature, thus supporting the view that clutch-size has been evolved through natural selection.

The above argument presumes the existence of hereditary variations in clutch-size. There is as yet no proof of the latter, though it seems likely on general grounds that specific differences in clutch-size have a hereditary basis. Further, in a breeding colony of the Starling in Holland, the clutches laid in successive years by particular ringed females were less variable in size than those of the colony as a whole, and the same was found in colonies of Common and Alpine Swifts, and in the Great Tits breeding in a Dutch wood (Lack *et al.*, *loc. cit.*, Kluijver, 1951). This suggests that different individuals differ somewhat in the hereditary factors influencing clutch-size. Each individual was not, however, completely consistent; its clutch-size varied within small limits. Hence within the hereditarily determined limits, some of the variations in clutch-size are evidently phenotypic.

In the Song-Thrush *Turdus ericetorum* and Great Tit *Parus major*, unlike the Starling, the mortality-rate of the young after they left the nest did not vary appreciably with brood-size. Hence the average number of survivors per brood was highest from the largest brood-sizes found in nature (Lack, 1949, 1950). Were then the largest brood-sizes the most efficient? In that case, why should not every individual of the species lay a clutch of the largest size? The answer, I suggest, is that, within the specific limits, larger clutches tend to be laid when prospects are more favourable, and smaller clutches when prospects are less favourable, for raising young. If the clutch-size could be perfectly adjusted to the food situation, then the survival-rate would be similar for the young from broods of all sizes. This raises two problems. The first concerns survival value: are the phenotypic modifications in clutch-size adapted to the food prospects for the young? The second concerns the physiological mechanism: How are such modifications brought about? The first question will be considered in some detail, the second only briefly.

There is suggestive evidence in various types of birds that clutch-size can be modified adaptively (within limits) to suit the feeding conditions under which the young will be raised. For instance, in those hawks and owls which prey on voles of the genus *Microtus*, the average clutch-size is half as large again, and sometimes twice as large, during a vole plague as in a year when voles are scarce (Lack, 1947–48). Again, in the arid parts of Africa and Australia, various passerine and other species have smaller clutches in years of low rainfall than in years of high rainfall when food of all kinds is more abundant (Moreau, 1944).

In many European species, clutch-size varies regularly with the time of year. For instance in single-brooded species, any late clutches are usually smaller than the rest. The breeding season of single-brooded species is presumably adapted to coincide with the most favourable time of year for raising young; hence late broods will tend to be raised in poorer conditions, and the smaller size of late clutches may well be adaptive. This probably holds in the Alpine Swift, in which clutches laid at the normal season average 2·7 and late clutches 2·2 eggs, while the mortality from starvation is higher in late than normal broods of the same size (Lack and Arn, 1947).

In some double-brooded species, the average clutch-size declines steadily as the season progresses, and this also may be adaptive. Thus in the Great Tit at Oxford, where the average clutch-size declines from the start to the end of the breeding season, the caterpillars on which the birds feed their young become scarcer from about the time that the first broods leave the nests, while the mortality of the young from starvation is much higher in second than first broods (Gibb, *in progress*). Likewise in the Starling in Switzerland, the commonest size of first broods is 5, but of late broods 4. In second broods, the mortality of the young after they left the nest was higher in broods of 5 and 6 than in broods of 4, and 4 proved to be the most efficient brood-size; whereas in first broods, as already mentioned, 5 was the most efficient size. Similar results were obtained in a comparison of early and late broods in England, where the most efficient brood-size was again the commonest found in nature (Lack, 1948a).

In various other double or treble-brooded species, the average clutch-size rises from April to early June and thereafter declines, hence the second broods are larger than either the first or third (Lack, 1947–48, 1949). The food supply of these species has not been investigated, but another factor, day-length, perhaps has some influence, for with a longer working day the parents can collect more food, and hence, other things being equal, they can raise a rather larger brood. The increase in day-length may also be one of the factors responsible for the marked increase in average clutch-size shown by many species of birds between the tropics and the North Temperate regions (Moreau, 1944; Lack, 1947–48).

Extensive studies of the Great Tit have revealed not only seasonal and regional variations in clutch-size, but also differences between woods in the same region, and differences in the same wood in successive years (Kluijver, 1951; Lack, 1950, 1952). In a wood near Oxford, Gibb (*in progress*) found that the clutch-size of first broods was lower each successive year from 1948 (when the average was 12·3) to 1951 (when it was 7·8 eggs), and that the abundance of the caterpillars on which the Great Tit fed its young also fell off each year. In Holland, Kluijver (1951) found that both the local and the annual variations in clutch-size were partly explicable through an inverse correlation with population density; but the difference was not large, since in any one wood the average was only about ½ egg smaller in years when the population was well above normal than in years when it was well below normal (Lack, 1952). Some other factor must have contributed to the observed differences in clutch-size, and this may have been the food situation, which Kluijver did not study. (Possibly, therefore, the influence of population density was indirect, due to its influence on food.)

To sum up, the available evidence suggests that the variations in clutch-size found in birds are closely adapted to the number of young that the parents can feed. The limits of clutch-size in each species are probably determined by the action of selection on hereditary differences, and some of the regional differences between populations of the same species may also have a hereditary basis. In addition, clutch-size can often be modified phenotypically to suit the immediate conditions. The physiological mechanisms involved have not been studied. In some cases, notably the vole-predators, clutch-size is probably modified by the food situation at the time of laying. In other cases this is not so. Thus the small second broods of the Great Tit are laid when their caterpillar food is most abundant. It should be stressed that, in birds, the clutch is laid some 2–4 weeks before the young require food. To be effective, therefore, any modification in clutch-size must be "anticipatory." The proximate factors involved are not as yet known.

To what extent do the conclusions reached for birds hold in other classes of animals? The closest parallels may be expected in mammals, which resemble birds in having a small family fed by one or both parents. One point is clearer in mammals than birds, since hereditary variations in litter-size have been established for the Rabbit *Oryctolagus cuniculus* and several other species. Since the litters of domestic breeds are larger than those of the wild type, mutations for larger litter-size must be eliminated in the wild by selection. This implies that the number of young surviving from litters above normal size is lower than from litters of normal size, a point which has not been studied under natural conditions. There are, however, suggestive data from the laboratory for the Guinea-pig *Cavia porcellus*. In this species the mortality among the young was found to rise steeply with increasing litter-size, in such a way that the average number of young weaned per litter was highest with a litter of 5; with 6 young in the litter fewer, not more, young were raised (Wright and Eaton, 1929; Lack, 1948b).

The most efficient litter for the laboratory Guinea-pig being 5, it might be thought that this size would have been favoured by natural selection, but in fact the commonest observed size was 3. The discrepancy is probably due to three causes. First, mortality was studied in the laboratory; it would doubtless be higher in the wild. Secondly, mortality was recorded only up to weaning. At weaning, the young from larger litters had a lower average weight than those from smaller litters, hence the subsequent mortality may well have been higher among the young from large than small litters. Thirdly, some of the variations in litter-size were adaptive modifications, smaller litters being produced in poorer, and larger litters in better, feeding conditions. Under these circumstances, the survival-rate from larger litters will be better, and that from smaller litters poorer, than would otherwise be the case. Altogether, the results for the Guinea-pig provide close parallels with those discussed earlier for passerine birds.

Various mammals show seasonal variations in litter-size comparable with those found in birds, though they have not been studied in relation to feeding conditions (Lack, 1948b). An adaptive modification related to food supply is found in the Arctic Fox *Alopex lagopus* in Greenland. The inland populations of this species prey primarily on the Lemming *Dicrostonyx groenlandicus*, and have much larger litters than usual in years when Lemmings are particularly abundant. The coastal populations prey primarily on marine organisms and do not show this variation (Braestrup, 1941).

Outside birds and mammals, the only animals which regularly feed their young are the social insects, in which the number of eggs laid is closely related to the food situation. In both termites and social hymenoptera, the queen lays comparatively few eggs in the early stages of the colony, but the number is vastly increased when there are many workers to help in feeding the larvae (Bodenheimer, 1937).

At first sight it would seem that the factors limiting the reproductive rate must be completely different in those animals which do not feed their young. In fish and invertebrates, it is often claimed that the number of eggs laid is correlated inversely with the degree of safety of the larvae. Thus marine invertebrates with a high degree of brood-protection lay 10–100 eggs, those with a crude type of brood-protection lay 100–

1,000 eggs, and those with no brood-protection lay 1,000–500,000,000 eggs (Thorson, 1950). Similar correlations hold in fish and in insects, but as mentioned earlier, the conventional deduction from these facts is wrong. Natural selection cannot favour the evolution of a smaller egg-number as such. There must be some compensating advantage, and this, I suggest, is a larger size of egg. As between different species, a given food intake by the adult can be utilized to produce many small eggs or a few large ones, and in each species the compromise actually evolved will be that which leads to the maximum number of surviving offspring.

In fact, a smaller egg-number is usually associated with eggs of larger size. Thus in marine invertebrates, the species with brood-protection produce a few large eggs rich in yolk, while those with pelagic larvae produce many small eggs poor in yolk (Thorson, 1950). The advantage of pelagic plankton-feeding larvae is that, being in a medium rich in food, they do not need a large internal food-store, hence the parents can produce a large number of small eggs; the disadvantage is the heavy predation. The advantage of brood-protection is the comparative safety from predators, but the embryos or larvae subsist on internal food-stores, hence the eggs must be large, and this means that the parents can produce fewer. Parallel considerations apply in fish and in insects, in which large eggs are found in viviparous species and others in which a larva of large size at hatching is advantageous. In phanerogamic plants, likewise, some species have many seeds with small food-stores, while others have few seeds with large food-stores.

In birds, clutch-size is well below the physiological limit of egg-production, and the limit is set by the amount of food which the parents can supply to their young. In fish and invertebrates (also in higher plants), on the other hand, the number of eggs laid is probably the physiological maximum, but this maximum is greatly influenced by the size of the eggs, and in particular by the size of the food-store, hence the amount of food that the adults can supply for their offspring again plays a vital role, the difference being that in these other groups, unlike birds, the food is provided wholly beforehand. When comparing different types of invertebrates or fish, an exact correlation between egg-number and egg-size cannot be expected, as the number of eggs laid is probably influenced by several other important factors. These include the ability of the adult organism to obtain food before laying, and also its subsequent survival; proportionately more eggs may be expected from species which die soon after laying than from those which survive to breed again.

In birds it was found that, because of the many variables involved, a study of interordinal or interspecific differences in clutch-size was not nearly so illuminating as a study of intraspecific differences (Lack, 1947–48). In other groups of animals, unfortunately, intraspecific variations in egg-number have been little studied. An exception is the recent review by Hutchinson (1951) of researches on freshwater copepods, from which the following facts are taken. In the Scheinsee, Germany, *Eudiaptomus gracilis* has an average "clutch" of 11 in April, which falls gradually during the summer to 3 in early August, then rises again to 8·8 in early November, and falls again to 5–6 during the winter. In Denmark, the same species usually carries 25–30 eggs in spring but only 6–8 in summer and autumn, both these values being smaller in one particular lake. Hence this species shows both local and seasonal variations in egg-number, and the same holds for other species in northern Europe, U.S.A. and Ceylon.

The reasons for these variations have not been established, but, at least in some cases, the food supply apparently varies in a similar way to the clutch-size, in northern Europe, for instance, reaching a peak in spring and declining in summer. The position is not, however, as simple as this might suggest. Thus in three species of copepods studied by Wesenberg-Lund in Denmark, not only does the egg-number decrease, but the size of the eggs increases, between spring and summer, and Hutchinson calculated that in *Eudiaptomus graciloides*, the summer clutch of about 4 eggs represents a similar egg-volume to the spring clutch of 9–18 eggs. The larvae hatch from these large summer eggs at a later stage of their development. It may therefore be suggested that the large egg-size is adaptive and due to the fact that, when food is scarce, the larvae have a greater chance of survival if they have a larger internal food-store and hatch out at a later stage; but the production of eggs of larger size inevitably means a smaller number of eggs. Here, then, within one species, is an inverse correlation between egg-number and egg-size similar to that found earlier when different families or orders of marine animals were compared, and with the same suggested significance. A similar case occurs in *Daphnia*, in which the larger number of small "summer" eggs poor in yolk contrasts with the smaller number of large "winter" eggs rich in yolk. It may be noted that in fresh-water crustacea, as in birds, the clutch is formed in anticipation of the food situation for the young. Such "anticipatory" adaptation is a characteristic result of natural selection.

This relationship between egg-number and egg-size helps to explain an otherwise puzzling situation found in a very different group of animals, the lizards. In the Wall Lizard *Lacerta sicula*, Kramer (1946) found that the number of eggs laid per clutch on islands off the Italian coast was 2–4, whereas on the adjacent mainland, subspecies of the same species usually laid 4–7 eggs. Genetic crosses between the insular and mainland subspecies showed that this difference was hereditary, the hybrids having clutches of intermediate size. Now the insular lizards not only laid smaller clutches, but had eggs of larger size than those of the mainland lizards. On the islands, the lizards have no predators so far as known, but food and water are sparse, while on the mainland both enemies and food are abundant. A larger egg is probably advantageous for the insular forms, as the young are larger at hatching, and so are better able to withstand food and water shortage. The smaller clutch of the insular lizards is, in my view, an incidental consequence of natural selection favouring eggs of larger size.

A similar difference has been found in California for lizards of the genus *Sceloporus*. High in the mountains, where enemies are absent but food is short, *S. graciosus* has an average of 3·3 eggs per clutch, but in the plains, where enemies abound, the related *S. occidentalis* has an average of 8·5 eggs (Stebbins and Roberts, 1946; Stebbins, 1948). The reason is not that *S. occidentalis* lays more eggs because its young have more enemies, but that, where enemies are absent, the lizard populations are limited by food shortage, and so eggs of larger size are at a selective advantage; and larger eggs mean fewer. (This interpretation is mine, not that of Kramer or Stebbins.) Once again, therefore, there is an inverse correlation between egg-number and egg-size, and the size of the eggs has been evolved in relation to the food situation for the young.

One further, and quite different, factor is claimed to influence the reproductive rate of lower animals. Fecundity, it is said, varies inversely with population density. This claim is based on population studies made in the laboratory with cultures of *Lucilia* (Nicholson, 1950) and *Drosophila* (Pearl, 1932; Sang, 1950) among diptera, of *Tribolium* and other flour-beetles (Park, 1941; Boyce, 1946; Crombie, 1947) and of *Hyalella* (Wilder, 1940) and *Daphnia* (Pratt, 1943) among crustacea.

In *Lucilia*, Nicholson clearly showed that the females laid fewer eggs when the cultures were crowded. He also showed that with a limited food supply the number of larvae which survived to become adults was larger when the culture was started at a low than a high density. Indeed, if the initial density of larvae was too high, none survived, as the food supply became exhausted before they could complete their development. Now *Lucilia*, the Sheep Blowfly, is a mobile animal which, in nature, lays its eggs in a circumscribed habitat such as the flesh of a living or dead sheep. Since, as shown by Nicholson, the larvae die when over-crowded, there must be strong survival value in any behaviour which prevents the adult flies from laying their eggs where too many eggs or larvae are already present, and under natural conditions *Lucilia* presumably reacts to overcrowding by suspending laying and flying off to seek some other place in which to lay.

I would suggest, therefore, that the crowding in Nicholson's experiments was unnatural, and that their true significance is rather different from that claimed. Similarly in *Drosophila*, I would interpret the decline in fecundity with rising population density in the culture bottles as part of the response which, in nature, leads the females to cease laying where the food supply for the larvae is dwindling, and to go elsewhere. In support of this view, the natural breeding-places of *Drosophila*, sap-smears and the like, are evidently much more circumscribed than a laboratory culture bottle, or they would have been found more often; yet as Sang (1950) points out, stunted adults (indicating undernourishment as larvae) are rarely if ever found in nature, though they are common in the laboratory. Further, Sang has shown that the decline in fecundity with rising population density is due, not to population density as such, but partly to a change in the nature of the culture medium through the action of the larvae already present, and partly to a decline in the quantity of food present.

There is a parallel to these findings in the experiments of Salt (1936) on the chalcid *Trichogramma evanescens*, which parasitizes the eggs of the flour-moth *Sitotroga cerealella*. Normally *Trichogramma* lays one egg in each host. When it was supplied with many hosts, egg-laying proceeded apace. If, however, only a few hosts were provided, these were parasitized, but egg-laying was then restrained for about eight hours. After this, the parasite began to lay more than one egg in each host, but had it been unconfined in nature, it would presumably have moved elsewhere in search of unparasitized hosts. The survival value of the behaviour is clear, since Salt found that if 2 or 3 eggs were laid in one host, only one adult usually emerged, while if more than 4 eggs were laid in one host, only one stunted adult, or none at all, emerged.

In flour-beetles, part of the apparent decline in fecundity with rising population density is due to a rise in egg-mortality through accidental cannibalism, but there is also a genuine decrease in the number of eggs laid. This, also, is due not to a high population density as such, but to a resulting shortage sometimes of food, and sometimes of places in which to place the eggs (Crombie, 1947).*

Similarly in crustacea, the data reviewed by Hutchinson (*loc. cit.*) indicate that clutch-size is influenced by food supply, but not by popu-

---

* In a paper seen only after this MS. was completed, Voute (1937–38) showed that the fecundity of the rice-eating weevil *Calandra oryzae* decreased above a certain population density, and that at this same density the beetles started to emigrate. This supports my interpretation of the situation in *Lucilia* and *Drosophila*.

lation density as such. In the copepod *Phyllodiaptomus annae* in Ceylon, for instance, clutch-size is highest at the start of the reproductive season in June and thereafter falls steadily, but population density first rises and then falls during this time. Likewise in European freshwater copepods, clutch-size falls from April onwards, but population density first rises and then falls. Hence clutch-size bears no constant relationship to population density. Hutchinson also cites Slobodkin (*unpublished*) that food supply controls fecundity in *Daphnia obtusa*.

Summarizing, the decline in fecundity as the population increases in a laboratory culture is really a response which, in nature, prevents or reduces egg-laying where or when there will be insufficient food for the larvae to develop. It is not primarily a response to a high population density as such, though the latter is often associated with a dwindling food supply. It seems very possible that the inverse correlation mentioned earlier between clutch-size and population density in the Great Tit *Parus major* has a similar explanation. Provided that the population is not restricted by shortage of nest-holes, the Great Tit breeds at a much higher density in broadleaved than coniferous woods (Kluijver, 1951). This indicates that broadleaved woods provide it with much more favourable conditions, so the question arises as to why any Great Tits breed in conifers. The answer is, I suggest, that broadleaved woods become progressively less favourable for raising young as the tit population increases, until a point is reached at which an empty coniferous wood provides better conditions. This view is supported by Kluijver's finding that the average clutch-size is higher in coniferous woods (the less favoured habitat) than in *crowded* broadleaved woods.

The situation in the Great Tit is much complicated by local and annual fluctuations in the food supply (Lack, 1952); the available evidence suggests that clutch-size is influenced by the food situation, rather than by population density as such. This also applies to those hawks and owls which prey on *Microtus*. Where there is a vole plague, these species not only have larger clutches than usual, but they congregate to breed in much larger numbers than usual. Hence their clutch-size varies directly, not inversely, with population density. The relevant point, of course, is that both their population density and their clutch-size are greatly influenced by the food supply, and any direct or indirect influence of density on clutch-size is negligible.

The vole-predators, also cone-crop feeders like the Crossbill *Loxia curvirostra* and locust-predators like the Rose-coloured Starling *Pastor roseus*, are unusual among birds in that, in any given locality, the food for their young varies greatly in abundance from year to year. Correlated with this, the species concerned are nomadic, moving out of areas where food is sparse, and congregating to breed where food is abundant. In most other birds, the food for the young varies much less, and such species tend to be much more evenly dispersed when breeding. This dispersion was formerly attributed to territorial behaviour (Howard, 1920), but there are grave objections to this view. In particular it is necessary to postulate, as Howard saw, that territory-size is specific. Indeed Howard claimed that it was specific, but actually it varies with the conditions, as shown above for the Great Tit.

Further, a rather even dispersion for breeding is characteristic not only of territorial but also of colonial species, such as the Heron *Ardea cinerea*. In the Upper Thames valley, for instance, there is a small heronry about every five miles. Despite the attraction for this species of colonial breeding, the birds do not collect into one large colony, and there is evidently a severe restriction on the size of each colony. Further, the upper limit of size varies markedly in different parts of England and the Continent, presumably in accordance with food supply.

There is no reason to think that, in the Heron, some individuals expel others from the colonies. The simplest explanation of the dispersion of the colonies is that those individuals seeking breeding sites avoid settling where there are already many pairs present in relation to the food supply. The same explanation may well hold in territorial species, with the hostility of settled pairs perhaps playing a subsidiary role. Bird-ringing shows that, except in the few nomadic species, adult birds tend to breed where they bred in the previous year, probably because the advantage of returning to a known place outweighs the possible chance of finding a new place that is richer in food. In species of this type, the only individuals seeking breeding sites are those breeding for the first time, together with a few older individuals which have found their site of the previous year untenable. Hence dispersion is brought about mainly by individuals breeding for the first time. Kluijver's findings for the Great Tit show that the attractiveness of any particular place depends not only on the nature of the habitat, but on the number of pairs already present, and the observed decline in the reproductive rate with increasing population density shows that there will be survival value in avoiding crowded places. To conclude, therefore, the significance of breeding dispersion in birds is probably similar to the tendency among invertebrates to avoid laying eggs where the chances of larval survival are reduced.

SUMMARY

(1) The thesis advanced here is that the reproductive rate of animals, like other characters, is a product of natural selection, hence that each

species lays that number of eggs which results in the maximum number of surviving offspring.

(2) In nidicolous birds, clutch-size is limited by the number of young that the parents can feed. With broods above normal size, fewer, not more, young are raised per brood. In many species, clutch-size can be modified within limits to suit the particular feeding conditions. Similar considerations hold for mammals; and in social insects the number of eggs laid is adapted to the number of workers, i.e. to the food supply.

(3) In animals which do not feed their young, the number of eggs laid is probably the maximum that the parent can produce, but this is greatly influenced by the size of the eggs, in particular by the size of the food-store. An inverse correlation between egg-number and egg-size is found not only when comparisons are made between different types of animals, but also in the seasonal variations in egg-number within one species of freshwater crustacean, and in the local hereditary differences between lizard populations.

(4) The decline in fecundity with rising population density, found in both birds and invertebrates, is primarily a response to a dwindling food supply for the young, and not to population density as such. As part of this response, birds, diptera and other mobile animals move away from places where conditions are less favourable, and so tend to be dispersed rather evenly for breeding.

ACKNOWLEDGMENTS

I am very grateful to Mr. R. E. Moreau and Professor G. C. Varley for their valuable criticisms of the manuscript.

REFERENCES

BODENHEIMER, F. S. (1937): "Population problems of social insects," *Biol. Rev.*, 12, pp. 393–430.

BOYCE, J. M. (1946): "The influence of fecundity and egg mortality on the population growth of *Tribolium confusum* Duval," *Ecology*, 27, pp. 290–302.

BRAESTRUP, F. W. (1941): "A study on the arctic fox in Greenland," *Medd. Grønland*, 131, No. 4.

CROMBIE, A. C. (1947): "Interspecific competition," *J. Anim. Ecol.*, 16, pp. 44–73.

HOWARD, H. E. (1920): *Territory in Bird Life.*

HUTCHINSON, G. E. (1951): "Copepodology for the ornithologist," *Ecology*, 32, pp. 571–77.

KLUIJVER, H. N. (1951): "The population ecology of the Great Tit *Parus major L*," *Ardea*, 39, pp. 1–135.

KRAMER, G. (1946): "Veränderungen von Nachkommenziffer und Nachkommengrösse sowie der Altersverteilung von Inseleidechsen," *Zeits. Naturforsch.*, 1, pp. 700–10.

LACK, D. (1947): "The significance of clutch-size in the Partridge," *J. Anim. Ecol.*, 16, pp. 19–25.

Lack, D. (1947–8): "The significance of clutch-size," *Ibis*, 89, pp. 302–52; 90, pp. 25–45.

(1948a): "Natural selection and family size in the Starling," *Evolution*, 2, pp. 95–110.

(1948b): "The significance of litter-size," *J. Anim. Ecol.*, 17, pp. 45–50.

(1949): "Family size in certain thrushes (*Turdidae*)," *Evolution*, 3, pp. 57–66.

(1950): "Family-size in titmice of the genus *Parus*," *Evolution*, 4, pp. 279–90.

(1952): "Reproductive rate and population density in the Great Tit: Kluijver's study," *Ibis*, 94, pp. 167–73.

Lack, D., and Arn, H. (1947): "Die Bedeutung der Gelegegrösse beim Alpensegler," *Ornith. Beob.*, 44, pp. 188–210.

Lack, D. and E. (1951): "The breeding biology of the Swift *Apus apus*," *Ibis*, 93, pp. 501–46.

Moreau, R. E. (1944): "Clutch-size: a comparative study, with special reference to African birds," *Ibis*, 86, pp. 286–347.

Nicholson, A. J. (1950): "Population oscillations caused by competition for food," *Nature*, 165, pp. 476–7.

Park, T. (1941): "The laboratory population as a test of a comprehensive ecological system," *Quart. Rev. Biol.*, 16, pp. 274–93, 440–61.

Pearl, R. (1932): "The influence of density of population upon egg production in *Drosophila melanogaster*," *J. Exper. Zool.*, 63, pp. 57–84.

Pratt, D. M. (1943): "Analysis of population development in *Daphnia* at different temperatures," *Biol. Bull.*, 85, pp. 116–40.

Salt, G. (1936): "Experimental studies in insect parasitism. 4. The effect of super-parasitism on populations of *Trichogramma evanescens*," *J. Exper. Biol.*, 13, pp. 363–75.

Stebbins, R. C. (1948): "Additional observations on home ranges and longevity in the lizard *Sceloporus graciosus*," *Copeia* (1948): pp. 20–2.

Stebbins, R. C., and Robinson, H. B. (1946): "Further analysis of a population of the lizard *Sceloporus graciosus gracilis*," *Univ. Calif. Publ. Zool.*, 48, pp. 149–68.

Thorson, G. (1950): "Reproductive and larval ecology of marine bottom inverte-brates," *Biol. Rev.*, 25, pp. 1–45.

Wilder, J. (1940): "The effects of population density upon growth, reproduction and survival of *Hyalella azteca*," *Physiol. Zool.*, 13, pp. 439–61.

Wright, S., and Eaton, O. N. (1929): "The persistence of differentiation among inbred families of guinea-pigs," *U.S. Dept. Agric. Tech. Bull.*, 103, pp. 1–45.

ADDITIONAL REFERENCE

Voute, A. D. (1937–8): "Bevolkingsproblemen II, III," *Natuurk. Tijdschr. Ned.-Ind.*, 97, pp. 210–13; 98, pp. 97–102.

Part IV

POST-LAYING BEHAVIOR

Editor's Comments
on Papers 14 Through 24

24   LORENZ and TINBERGEN

Taxis and Instinctive Behaviour Pattern in Egg-rolling by the Greylag Goose

The incubation phase of the breeding cycle is marked by the appearance of the egg(s) in the nest. There has been confusion in the literature on whether incubation time should be considered from the point of view of the egg or of the parent. Study of the egg requires analysis of when optimal incubation temperature is maintained. Study of the parent requires analysis of the special behaviors associated with warming the eggs. The problem was evaluated by Swanberg (Paper 14). As Swanberg points out, the incubation period can be defined as the time between laying and hatching of an egg, or as the length of time the parents sit on the nest. In order to compare the records of different observers, starting and end points used in estimating incubation time must be known.

When one surveys the division of labor of the parents during incubation, and the patterns of attentiveness (intervals on the nest) and inattentiveness (intervals off the nest) almost all possible arrangements are seen. Skutch (Paper 15) has provided a valuable synopsis of a large body of data and suggests that because incubation by both sexes is the most widespread method seen among birds, this is likely the most primitive mode. Variations on this basic arrangement, with one or the other parent incubating alone, presumably evolve from this more primitive arrangement.

In some birds incubation behavior has been completely suspended. The mound-building birds (Megapodiidae) of the Malayan-Australian region keep their eggs warm by using the heat of decaying vegetation. Frith (Paper 16) has observed that the birds are capable of regulating mound temperature by altering the depth.at which the eggs are buried and the rate of fermentation of decaying organic material. Brood parasites lay their eggs in others' nests and in this way they avoid nest building, incubation, and rearing young. The occurrence of brood parasitism raises interesting questions about the evolution of behavioral and physiological control mechanisms. Brood parasitism has evolved independently in five families: Anatidae, Cuculidae, Indicatoridae, Icteridae, and Ploceidae. The possible evolutionary origin of this behavior is suggested by comparing parasitic species with living closely related nonparasitic species. Thus, Bay-winged Cowbirds (*Molothrus badius*) use the nest of other birds, but incubate their own eggs and raise the young.

The Shiny Cowbirds (*M. bonariensis*) occasionally build their own nests and usually lay in nests of other species. Friedmann was among the first to document brood parasitism extensively. He suggests (Paper 17) that the loss of territorial defense may have led to the loss of nest building and subsequently to brood parasitism.

The evolution from parasitism to symbiosis, seen in the Giant Cowbird (*Scaphidara oryzivora*) of South and Central America has been worked out in intricate detail by Smith (Paper 18). The cowbirds fall into two groups, "skulkers," with mimetic eggs (eggs similar to those of their hosts), or "dumpers," with nonmimetic eggs (eggs not similar to those of their hosts). The hosts fall into two groups: "discriminators" which reject nonmimetic eggs and "nondiscriminators" which accept eggs that differ from their own. Smith shows that where wasps and bees are present, the cowbirds are parasites while where these insects are absent, the cowbirds live symbiotically with their hosts. This poly- are absent, the cowbirds live symbiotically with their hosts. This polymorphism (coexistence of two or more forms) in behavior of parasites and the hosts is adaptive because of the uncertainty of whether or not cowbirds will be necessary to protect against a major common enemy, the botfly. Either social insects or cowbirds can protect against botflies. The presence or absence of host specificity of the parasites and the development of discrimination of their eggs by the host can be understood only in the context of the relationship of the birds to botfiles, bees, and wasps.

During incubation, not only is the egg temperature regulated, but also the eggs are turned periodically. The importance of egg turning was noted in the poultry industry with the introduction of artificial incubators. It was noted that the mortality rate was far higher in incubator-reared eggs than in hen-reared eggs. The paper by New (Paper 19) shows that there are critical periods in embryonic development during which the eggs must be turned.

Several investigators have been interested in the role of hormones during incubation. That these were involved was suggested in 1927 by Lienhart (Paper 20), who showed that plasma from an incubating chicken could induce incubation in a recipient of the plasma. The study of hormone effects has clarified the controversy on whether behavior is the result of hormone effects on the brain alone or whether it is the result of hormone-produced changes in peripheral structures and their sensitivity. Evidence indicates quite clearly that both are important. Thus, Fisher (1956) first showed that implants directly into the brain of minute amounts of hormones are effective in eliciting hormone dependent sexual behavior in the rat. As far as avian parental behavior is concerned, only one experiment has been done using brain implants of hormone. Komisaruk (Paper 21) showed that progesterone implants in the hypothalamus of doves was effective in eliciting incubation behavior. Hormones also act by changing sensitivity and responsiveness to peripheral external stimuli. Changes in brood patch sensitivity occur during breeding season (Hinde et al., 1963) and these can be mimicked by exogenously administered steroids (Hinde and Steel, Paper 22). Note that it remains to be demonstrated that these changes in sensitivity influence the quantitative or qualitative aspects of the response to the eggs.

Another way to approach the question of hormonal influences on behavior is to correlate behavioral and endocrine events. In parasitic birds which show little or no incubation and in birds where male and female roles are apparently reversed, a question of some interest is whether the peculiar breeding behavior is the result of, or the cause of a hormone aberration. To this end Höhn (Paper 23) analyzed the prolactin content of the pituitary of a brood parasite, the cowbird (Höhn and Cheng, 1965). The research suggests a hormonal change accompanies the behavioral changes. Far more research would be required to establish the causal and evolutionary mechanisms involved.

It is fitting to end this section in incubation behavior with the classic paper by Lorenz and Tinbergen (Paper 24) on the egg-rolling behavior of the Graylag Goose. Many ground-nesting birds return eggs to their nest by rolling them back towards the nest with the beak. Starting with a beautiful description of this simple response, Lorenz and Tinbergen laid out a number of fundamental ideas of ethological theory. The *fixed action pattern* is illustrated by the stereotyped egg-rolling movements. They noted that the egg-rolling movement occurs even when the egg slips away, suggesting that once the behavior is initiated, the fixed action pattern occurs without feedback from the environment. Lorenz and Tinbergen also show that very simple cues which they called *sign stimuli* or *releasers* and which can differ from the real egg cause the goose to perform the retrieval response. The *innate releasing mechanism* is the hypothesized physiological mechanism which receives neural information from the releasers and issues motor impulses to produce the fixed action pattern.

REFERENCES

Fisher, A. E. (1956). Maternal and sexual behavior induced by intracranial chemical stimulation. In *Hormones and Sexual Behavior*, C. S. Carter, ed., Vol. 1 (1973), Benchmark Papers in Animal Behavior, Dowden, Hutchinson & Ross, Inc.: Stroudsburg, Pa.

Hinde, R. A., Bell, R. Q., and Steel, E. A. (1963). Changes in the sensitivity of the canary brood patch during the natural breeding season. *Anim. Behav.*, *11*, 553–560.

Höhn, E. O., and Cheng, S. C. (1965). Prolactin and the incidence of brood patch formation and incubation behaviors of the two sexes in certain birds with special reference to Phalaropes. *Nature, 203,* 197–198.

14

Reprinted from Vår Fågelvärld 9:63-77, 80 (1950)

On the concept of »incubation period»

(Med svenskt sammandrag: Om begreppet ruvningstid)

By

P. O. SWANBERG

To many the concept of incubation period seems both simple and obvious. In recent years a considerable diversity of opinion as to the definition of the incubation period has, however, become evident, and a good deal of confusion unintentionally has arisen in the literature. No matter which definition we prefer, we must agree on the desirability, not to say the necessity, of basing the notion of »incubation period», »Brutdauer», »ruvningstid», »rugetid» etc. in the different handbooks on the same concise definition. Since the handbooks are based on special investigations of different species it is desirable that even the same uniform definition should principally be used in bird monographs, although the author may occasionally hold special views of the problem. It should not be necessary to complain of the fact, as does VAGN HOLSTEIN (1942: 71), when defining the incubation period, that »such definition is still missing in the authors who have hitherto studied the subject» (i. e., the incubation of the goshawk). It goes without saying that the different authors should not have to give such a more or less detailed definition in every single instance. In that case so many different definitions would probably be given that the views of one author sometimes could not be compared to those of others.

Before proceeding with the discussion some principal questions should be answered:

What is understood by incubation (»ruvning» in Swedish)?

What is the theoretical and practical meaning of giving the incubation period of the respective species, or, in other words, what do we want to know when we take a handbook or a special paper to find out the incubation period of a certain species of bird?

As regards the Swedish term »ruvning» (incubation) Y. LÖWEGREN (1935: 477) has given the scientific definition in the following words: »Incubation is the process during which the bodies of the parents supply the heat necessary to the development of the embryo in the egg after it has been laid» (the original Swedish). B. H. RYVES (1943: 10) has given an adequate definition of the English parallel.

As regards the second question O. HEINROTH (1922: 173) has given a clear-cut answer in the words: »Ich möchte ... unter Brut-dauer die Zeit verstehen, die bei regelrechter, d. h. ungestörter Bebrütung eines frischen Eies bis zum Auskriechen des jungen Vogels verstreicht, d. h. also die Entwicklungsdauer des Keimlings bis zum Ausschlüpfen». HEINROTH thus says that the incubation period is the period of development of the egg. For a scientific verdict on this matter within poultry farming I approached Prof. NILS OLSSON, chief of Husdjursförsöksanstalten vid K. Lantbrukshögskolan (Ex-perimental Station of Domestic Animals of the Royal Agricultural

High School), who replied: »Most correctly the incubation period is held to be the time necessary for the egg to develop into a full-hatched chick». If we consult a handbook to find out the incubation period of a certain species of bird we probably, in most cases, wish to compare it with that of other birds which we are studying, or to be able to undertake certain practical calculations on the basis of the time of development of the egg. In both cases a uniform definition of the term incubation period is highly desirable. As far as I can see the definition formulated by Heinroth is the only one hitherto given which is in reality u n i v e r s a l l y v a l i d.

HEINROTH (1922: 174) continues by giving a rule for a practical determination of the incubation period, which is said to be the time elapsed from the laying of the last egg until the last young has left its egg shell. This practical definition seems to have been generally used and has on the whole been considered obvious. It has been quoted many times. B. H. RYVES (1934: 13) gives a corresponding definition, and also M. M. NICE (1937: 122) who in her own investigation on the song sparrow, *Melospiza melodia*, further elaborates it by calculating the length of incubation »from the laying of the last egg to its hatching», and R. E. & W. M. MOREAU (1940: 313) follow HEINROTH; R. E. MOREAU repeats this in 1946·(p. 67), likewise E. SUTTER (1946: 56).

A number of other definitions which have been used, appear to be, plainly spoken, meaningless, e. g., the so-called »Kurzbrutdauer». It has been used for »incubation period» in many cases and represents the time from the laying of the last egg to the hatching of the first one. A closer motivation of this special basis of calculation is generally not given.

An explanation of the use of this »Brutperiode im engeren Sinne», as it is called by other authors, is, however, given by KUERZI (1941: 28). He maintains that he has to use it to make his observations on the incubation period of the tree swallow, *Iridoprocne bicolor*, directly comparable to those of S. H. Low, who previously applied the method! But KUERZI does not give any information of the length of development of the egg in his otherwise valuable paper.

When we come to the deviations from HEINROTH's definition caused by various observations of birds which are sitting on their eggs throughout the egg-laying period or during part of it, the problem, however, becomes much more interesting. It has been generally supposed that in such cases incubation begins when the bird begins to stay on the nest longer than is necessitated by the egg-laying proper.

The authors of the great faunistic handbooks are generally not led astray but have applied HEINROTH's definition of the incubation period, although it had been stated just before that incubation begins with the first egg or the like. Regarding the incubation period HEINROTH (1922: 174) himself pointed out: »This period of time need not correspond exactly to the length of time during which the parent incubates, at least not in the forms whose clutches consist of several or even many eggs. It is known that the first eggs are often incubated before the last ones are laid.» G. H. RYVES (1934) ascertained that the incubation of the corn bunting (*Emberiza calandra*) begins with the last egg but one, but it is just in this connection that he declares that he reckons the incubation period in accordance with HEINROTH.

Unfortunately, in the last few decades it is not unusual to find records in the literature as. e. g., those of P. RUTHKE (1941: 90) who in slightly more than one line both states that he recorded an incubation period of 38 days in a nest of the mute swan, *Cygnus olor*, and adds: »HEINROTH states $35\frac{1}{2}$ days». The unsuspicious reader must get the impression that the incubation period of the mute swan is highly variable. If, however, one takes the time to study RUTHKE's statement more closely one will find that if he had used the same calculation basis as HEINROTH he would have come to the result, $36\frac{1}{4}$—$36\frac{1}{2}$ days. (That his statement, nevertheless, should be worthless because four eggs, obviously of unknown identity, were rotten, is another story!)

VAGN HOLSTEIN is one of the authors who has given the problem most attention. In his work on the goshawk, *Accipiter g. gentilis*, he says (1942: 71, 134): »incubation time means the period reckoned from the first time one of the hawks goes on the eggs to incubate — which takes place as soon as the first egg has been laid — until the last young leaves the egg. Thus the incubation time is not identical with the incubation time of the individual egg except in cases when the batch contains one egg only.» A corresponding definition is repeated in his work on the honey buzzard, *Pernis apivorus*, (1944: 96, 169) and on the great spotted woodpecker, *Dendrocopos major*, (1945: 76). HOLSTEIN calls this »the total brooding time». STEINFATT was more cautious when he wrote on the great spotted woodpecker, since he stated (1937: 51) i n f a t t y p e s that the actual incubation period was 12 days, but in order to pay regard to all views even stated »Kurzbrutdauer» and »Gesamtbrutdauer».

The method of stating the incubation period as »the total brooding time», i. e., the time during which a bird may sit on its nest before the last egg has been laid plus the time which elapses until the last egg is hatched, certainly appears attractively correct to many. But can it stand a closer examination?

Two questions now present themselves:

1. Can it be decided whether there actually is incubation or not when we see a bird sitting on its nest, before there is a completed clutch?

2. Even if this be the case, is it then actually the »total brooding time» of the species, »Gesamtbrutdauer» or what it may be called, which primarily is of general interest when calculating the incubation period?

Question No. 1 can at once be answered in the negative. Some examples may be given. The hooded crow, *Corvus corone cornix*, according to The Handbook of British Birds as well as to NIET-HAMMER (1937), begins incubation with the first egg. Yet the incubation period, wisely enough, has been stated by applying HEINROTH's definition. But if there is a clutch of five eggs of hooded crow, will the first egg be hatched four days before the last one? This does not seem to be the general rule (OLSTAD 1935: 10).

As an explanation of the fact that some birds which incubate with the first egg, nevertheless, hatch a larger or smaller clutch within a period which is much shorter than the egg laying period, it has sometimes been assumed that the first eggs laid require a longer time for development than the last ones. HEINROTH (1922: 174) says regarding such cases, when a bird incubates with the last egg but one and the young nevertheless hatch simultaneously, that this may be due to the fact that in the first place the last egg is the freshest and in his opinion thus requires a shorter time of development; secondly, that it seems as if it is more advanced in development when it is laid than the first ones. GROEBBELS (1937: 397) even gives the same explanation of the fact that *Rhea americana* hatches its eggs simultaneously, in spite of the fact that brooding begins six days before the last egg has been

164

laid! But a real support of theories of a s i g n i f i c a n t, regular
difference in the time required for development, which for the
different eggs should counterbalance the possible differences in
»incubation time», could hardly be given. On the contrary, there are
numerous examples showing that the opposite is the case. and
experiments can give further evidence of this. The thick-billed
nutcracker, *Nucifraga c. caryocatactes*, is a fine example of this.
This bird sits hard on the nest from the first egg onwards but all
the eggs generally hatch in the course of 24 hours, in spite of the
fact that the time of development of egg No. 1 is approximately
the same as for egg No. 4. This will, however, be discussed in a later
paper.

In the spring of 1942 during my military service I had to stay
in a place where I could not study the thick-billed nutcracker, which
I had planned to do. By way of a substitute I made up my mind
to examine more closely the hooded crow during the egg-laying
period. Towards such a bird one has no scruples and the nests were
visited several times per day. In one nest the first egg was laid
on April 20. Next day at 3.55 a. m., before it was light, I climbed
the tree. The crow fluttered away. The egg was, however, completely
cold. Unfortunately. forest work was in progress around the nest tree
all the day, but next morning on April 22, at 3.55 a. m. the crow was
sitting on the nest and fluttered away as when incubating. The
egg was, however, still cold. At 6.00 I could feel by touching the
egg that the bird had been covering it. And at 8.05 a. m. the crow
was sitting on the nest, egg No. 2 had just been laid, and No. 1
was still cold.

On April 23 the nest was watched from 3.35 a. m. The male
was roosting right below the nest and flew away at 3.55 a. m. I
then climbed the tree, flushed the female and found that one egg
was cold, but not »ice-cold», the other was cool to the touch.

This indicates that the crow had been sitting on the nest for three
nights without giving the egg, or eggs, any effective incubation heat.
She had not been incubating, only protecting the eggs.

Such procedure may be more usual than has hitherto been believed. In Lapland where the climate is inclement one may find,
especially by night, but also in the daytime, birds which sit on
their eggs although the clutch is not completed. In many cases this
can be ascertained by immediately taking an egg and checking the

temperature by placing it against a sensitive portion of the skin, e. g., the upper lip. If actual incubation has commenced, they are very warm to the touch, almost hot, all over; if incubation has not commenced, a perceptibly lower temperature generally will be felt, and often the eggs are cool on the underside.

Some recent examples can be given: At the time in the summer of 1949 when in Northern Lapland numerous red-necked Phalaropes, *Phalaropus lobatus*, still could be seen in pairs, indicating that to a great extent incubation had not yet commenced, we found a number of completed clutches of red-necked Phalaropes, as well as one with three, and one with two eggs on the night between June 18 and 19. The three eggs were what in my notes I usually call temperate, i. e., the temperature could be estimated at 22°—25° C.,[1]) while all three were cool on the underside (about 20° C.). In the nest with two eggs one was »temperate», the other »faintly tepid» (somewhat more than 25° C.), and both were cool on the underside. For the sake of clarity it should be mentioned that in both cases it was the males that stayed on the nest and came back there, which was stated by KAI CURRY-LINDAHL concerning the nest with three eggs.

The thermal threshold for the beginning of the embryonic development of red-necked Phalaropes is not known, but it seems obvious that the eggs before the clutches were completed were only kept near this temperature. It can be mentioned, by way of an example, that in the domestic hen this lies around 24° C., but still at 28° C. the embryonic development remains unprogressive (STRESEMANN 1934: 292).

At that time of the Lapland summer when only a small part of the small passerine birds on the high plateau near Sautso NE Torne Träsk had commenced incubation, a red-spotted bluethroat (*Luscinia s. svecica*) left its nest just in front of me on June 16, 1949 at 5 p. m., and I could put down in my notes »five eggs, the clutch c e r t a i n l y n o t completed, since the female was sitting on the nest, but the eggs were cool to the touch of the lip». The temperature of the eggs presumably was somewhat above 20° C. The next day at 8 p. m. a Lapland bunting, *Calcarius lapponicus*, left its nest right in front of my feet. There were only three eggs in the nest, and they were cool! The normal clutch of a Lapland bunting is five eggs. The temperature of the eggs confirmed that the clutch was not completed and

---

[1]) According to experiments with hen's eggs.

showed that the bird was sitting on them without supplying the heat necessary for the embryonic development.

The observation that birds can sit on their eggs without actually incubating them is very old. Already 200 years ago MOORE (1735: 11, according to COLE & KIRKPATRICK) said about his pigeons: »They generally stand over the first Egg, which if you please, you may call an improper Incubation, till the next is laid, and then sit close, that both young ones may be hatch'd at once, or pretty nearly; tho some will sit close on the first, and by that means hatch one young one two Days before the other». WHITMAN (1919: 53) during his pioneer behaviour studies made the same observation more than 40 years ago, and COLE & KIRKPATRICK on the basis of notes on several thousands of pigeon eggs have given a diagram and a table showing clearly how the first egg of the pigeon hatches on the average only 11 hours before the second one, in spite of the fact that the bird sits on it about 44 hours, before No. 2 is laid (1915: 495, 497).

Regarding birds which do not incubate until the last egg is laid GROEBBELS (1937: 397) says that it must be assumed that the female when she lays the later eggs incubates the first ones beforehand (vorbebrütet), and that this preliminary incubation increases with the number of eggs and with the time demanded for the egg-laying, but that the eggs nevertheless hatch almost simultaneously. He explains this by the fact that the last egg laid is supposed to have a shorter time of development. This is, however, obviously a treacherous assumption. It is by no means certain that any incubation of the eggs previously laid takes place, when the bird stays on the nest to lay another egg. In the above mentioned crow's nest on April 22 at 8.05 o'clock, the first of the numbered eggs was cold, while the other egg was still quite warm. Other examples could be given for the thick-billed nutcracker.

Regarding the white-rumped swift, *Apus caffer streubelii*, the MOREAUS (1940: 313) report that both parents stay at least two, generally three, twelve-hours nights in the nest hole before the second egg follows the first. Nevertheless, it happens sometimes that the second egg hatches before the first one. The explanation is, no doubt, that they do not incubate during the 24—36 hours they s t a y in the nest hole before the clutch is completed.

As early as 1913 L. J. BONHOTE set forth the view that the procedure of the pigeons, known for centuries, to sit at the beginning only covering the eggs could be used by other species of birds too.

In 1929 B. H. Ryves followed the same line of thought: ». . . when does incubation actually begin? . . . Does this change necessarily take place, with all birds, soon after brooding is in process? Is it not possible that, in some cases, the eggs may be covered for a time without producing any definite change?»

And in 1938 John R. Arnold presented his thesis for the degree of doctor of philosophy, »The systematic position and natural history of the northern blue jay, *Cyanocitta cristata bromia* Oberholser», with many interesting observations of which the following are abstracted with kind permission of the author: In one nest the six young all hatched within 24½ hours in spite of the fact that »the bird sat on the nest from the time of the laying of the first egg», in another nest the five young all hatched in one day. The explanation is obviously this: »The incubating bird was noted to hold the body feathers much closer to her body in the early days of sitting on the nest. In three cases the bird was urged to rise from the nest slowly (by approaching the nest slowly) and the feathers were seen to be between the body and the eggs. The eggs were cold in each of these cases.»

Holstein (1942) who studied the goshawk very thoroughly, among other things the temperature conditions in the first part of the incubation period, says that all his goshawks commenced incubation with the first egg, but the table on page 72 shows that the eggs of pair No. 1 hatched in the course of about 48 hours, although 9 days passed between laying of the first one and the last. In pair No. II 6 days passed between the first and the last eggs, but they hatched in only 24 hours. For pair No. IV the figures are 6 days and about 48 hours respectively. So much can be seen from the tables. Even this tells us quite a lot.

If we now turn to Wingstrand's paper (1943: 3) we see that he found that the time of development of the f i r s t e g g of a goshawk in an incubator proved to be 37.5 days. If we compare this result with Holstein's above mentioned table we find a striking similarity to the time of development of t h e l a s t e g g in the respective clutches which were $35 \pm \frac{1}{2}$, $37 \pm \frac{1}{2}$, $37 \pm \frac{1}{2}$ in I, II, and IV respectively, whereas the table shows that 41 days passed in all these three nests between the laying of t h e f i r s t e g g and its hatching, if the goshawks were left undisturbed. This strongly bears out the supposition that the first and the last eggs in a goshawk clutch have a fairly equal time of development and that, therefore, actual incubation does not take place until several days after the first egg was laid. And HOLSTEIN (1942: 72, 135) himself states: »Only on the supposition that the incubation warmth of the hawks in the egg-laying time is so low that it is unable to start the embryonic development can the phenomenon be explained . . .»

HOLSTEIN's temperature registrations also support this view. It would be very interesting if they had been made continuously during the entire egg-laying period. It is this period which in this connection offers the greatest interest, but it is to be regretted that no temperature curve with corrected values has been given for this period, in which as a matter of fact it is most difficult to get an opportunity of making such registrations. The curve given (1944: 85) has its first really registered value, when three days had passed from the laying of the last egg. It is true that the curve shows a fine rectilinear course from the laying of the first egg, marking a gradual rise of temperature which is believed to be approximately correct, but there are no data supporting the course of the curve between the first and the fourth egg. If a purely physical process were involved it might be justifiable under certain suppositions to start the curve with this straight line, but here a biological process is involved which is difficult to judge, and which may even be influenced by psychical factors.

I, for my part, believe that the curve covering the period from the first egg till what in the diagram is called »incubation day No. 12» actually ought to have had a far more interesting course. This view is supported even by Holstein's tables for the goshawk. It is true that in these tables we do not get the corrected values, nor the air temperature, but in incubation table I for pair I 1938 there is a sudden rise of temperature of no less than 10° C. from the day when the second egg was laid to the next day, and in table II for pair K. I 1939 a rise in the mean temperature of 7° C. took place on the day when the last egg but one was laid. This gives a different picture from the curve just mentioned, which in the corresponding period shows a rise of temperature of only 1° C. per day.

The importance of the development of the brood patches in relation to the incubation temperature need not be discussed here. But in order to avoid any misunderstanding the following should be mentioned. HOLSTEIN (1942: 87, 137) holds that in the goshawk the hens' brood patches do not develop properly in »moulting» until the clutch is complete, hence the incomplete clutch is not raised to a temperature sufficiently high for the development to take place. That the brood patches gradually develop during the egg-laying period is beyond doubt. But it is uncertain whether the theory to the effect that »the stage of moulting» is the actual cause why effective incubation does not begin until the clutch is complete will hold. It is, however, obvious that the more or less developed stage of the brood patches does not in any considerable degree influence the above mentioned circumstance that the thick-billed nutcracker hatches all its eggs within a very short period of time, sometimes only a few hours. If she is compelled to lay a second clutch, the incubation and hatching conditions will mainly be the same as for the ordinary clutch, although the brood patches are now fully developed already before the eggs are laid.

We are fully aware that in certain species embryonic development begins before the last egg has been laid. But w h e n a n d h o w t h i s d e v e l o p m e n t c o m m e n c e s i n t h e o n e o r t h e o t h e r s p e c i e s w e d o n o t k n o w a s y e t with sufficient exactitude.

One practical example showing the risk of reckoning the incubation period from the first egg is found in HOLSTEIN, who made conclusions from the weighings of the embryos from different eggs from two goshawk clutches. A f t e r t h e l a p s e o f 5 d a y s the last egg laid was found to contain an embryo weighing 0.002 g. As this egg was laid six days after the first one, HOLSTEIN reckons that the weight of an embryo with an »incubation period» of eleven days is 0.002 g. And so on. And from this is concluded (1944: 87, 167): »In all probability a quite minimal embryonic development takes place in the eggs of the Goshawk during the first nine days of incubation». And another inference is made from this conclusion: »the lowest temperature limit for the possibility of embryonic development in this bird should be considerably higher than in the domestic hen, viz. at about 30° C.» From this hypothesis still another conclusion is made: ». . . that the embryonic development of all the eggs of a batch does not begin for good until the 10th day . . .».

What happened is consequently this: On the basis of the hypothesis that the embryos in all three eggs have an equal development from the day which is called »the 7th incubation day», i. e., that egg No. 1 practically did not commence its development until egg No. 3 had been laid, and that the subsequent development is largely parallel in all three eggs, the author makes certain calculations and corrections. The values thus obtained, which are based on the view that no considerable embryonic development takes place before at the earliest six days after the first egg was laid, are meant to show that embryonic development does not progress perceptibly until nine days after the first egg has been laid. Or in other words: A theory is used for proving its own correctness and a little more.

Just this procedure of giving the embryonic weight after five days as being valid for eleven days and besides comparing the values thus obtained with figures for other species of birds with a completely different initiation of incubation clearly shows how dangerous it is to apply such a specific and highly varying notion as »the total brooding time». The author even runs the risk of tripping himself up.

Some formalist may hold that if it has been possible to ascertain that a certain bird sits on the nest from the first egg, people say in common usage that it incubates, therefore it is most correct to give as incubation period the »total brooding time». In that case it seems to be the period of time in which the b i r d i s s i t t i n g o n t h e n e s t w i t h e g g s u n d e r i t, which is of interest. But then, what is the interest of drawing the limit just at the hatching of the last egg? The bird — in case a goshawk or other nidicolous birds are involved — will generally remain on the nest several days after! Why then draw the limit just at the hatching of the last egg? If an author is interested in the duration of the impulse to brood in the adult bird he ought to realize like GROEBBELS (1937: 248) that »the brooding bird is not an automaton, in which the impulse to brood suddenly stops at a certain time».

No, what we want to clear up regarding the incubation period is the i n c u b a t i o n p e r i o d o f t h e e g g, not the more or less indistinctly defined duration of the impulse to brood in the birds. And in order to refute the constructed concept of »Brutperiode im engeren Sinne», which several authors have used: we wish to clear up the length of development of the egg, n o t t h e t i m e d u r i n g w h i c h a l l t h e e g g s *together* h a v e b e e n l y i n g u n h a t c h e d i n t h e n e s t.

To this is added the above indicated argument which in my

opinion is of vital importance: the demand of absolute uniformity of the basis of calculation, i. e., a method yielding comparable values in different species, which may have a different biology and also live in different environmental conditions. If »the total brooding time» is used for the »incubation period» it may occur that the brooding conditions of conspecific birds cannot be compared since one of them may have only one egg, while the other may lay, e. g., four eggs in nine days. And in some species it may happen that certain individuals sit on the nest from the first egg even in the case of large clutches, while other individuals only incubate from one of the last eggs laid. Indeed, if the wide definition of incubation period was used one might get, e. g., for the moorhen, *Gallinula chloropus*, a variation of more than 12 days, since in this species there are records both that all eggs hatched on the same day (BONHOTE, 1913. COLLENETTE, 1913), and that in another case the hatching of twelve eggs took at least twelve days (v. ZIEGLER, 1909)!

We need not therefore disregard the problems of the ways in which the brooding impulse manifests itself both before a clutch is completed and after the hatching of the eggs.

Information of the behaviour of the bird on the nest before the last egg has been laid, comparisons between the length of the egg-laying period and the interval during which hatching is commenced and finished are of vital interest. Such information, however, should be discussed separately, and ought to be used in order to complete information on the actual incubation period and should not be confused with it.

For the sake of a clear and unambiguous distinction and, above all, for the sake of uniformity we ought to remember and adopt HEINROTH's rules, which run as follows: *By incubation period is understood the time which with regular, uninterrupted incubation of a newly laid egg elapses until the young has left the egg. And further: In nature it is generally possible to ascertain the incubation period with satisfactory exactitude by checking the time from the laying of the last egg to the hatching of the last young.*

Here a practical rule can be mentioned. In the case of a bird which is suspected of having commenced brooding before the last egg was laid and, in addition, has not succeeded in hatching all the eggs, no reliable registration of the incubation period can be obtained if it has not been possible in some way to ascertain that the last laid egg belongs to those which hatch. People have often sinned against this simple rule.

R. E. MOREAU (1946: 67) says that exceptions can be made from HEINROTH's rule in the cases in which we know that incubation starts earlier. As has been said above, the observer very rarely has so good evidence that he can say definitely whether incubation at the beginning of the egg-laying period actually has been effective or whether it has only been apparent. If I am able to ascertain that a common buzzard, *Buteo buteo*. sits on the nest from the first egg. and if I mark the eggs successively and write down when they hatch, I may maintain that I have ascertained that the bird has been sitting on the first egg for 37 days (since it hatches after 37 days), on the second egg for 36 days. and on the third for 34 days. But I cannot allege that I know that the incubation period o f t h e f i r s t t w o e g g s was 37 and 36 days respectively.

It should be added here that, although it is of rare occurrence, there are several examples where highly varying values of the »incubation period» have been obtained, owing to the fact that certain individuals apparently do not devote themselves completely to brooding until several days after the laying of the last egg. RYVES (1929, 1946) has given a number of such examples. Two categories could be mentioned: in the first place merely accidental deviations, abnormities; secondly, as far as certain species are concerned »normally» fluctuating values in the period from the egg-laying until the hatching is finished.

One might try to neutralize gradually in different ways the values of the first category which are more or less misleading — though in a way very suggestive. In the first place, as exact observations as possible regarding the beginning of incubation should be made. RYVES (1946: 51) showed how in that way he succeeded in ascertaining the length of incubation for a pair of *Emberiza cirlus*. In the first nest it took 16 days from the laying of the last egg until the hatching was finished, in the second nest it took 14 days. But in reality the period of incubation was 14 days in both nests. Another obvious method is that an extensive material is procured, whereupon certain extreme values gradually can be eliminated as mere abnormities. As regards the other category, species with notoriously individually »unreliable» incubation which gives highly varying values, it seems justified for practical reasons to adhere to HEINROTH's rule for determination of the incubation period in the field, since the first purpose of this rule is to make it possible for us to estimate in advance when a certain clutch is going to hatch. It is desirable, however, for these few species to try gradually to add to the information of the ascertained incubation period particulars of the actual time of development of the eggs by way of experiments. One method to effect regular incubation of the eggs by way of experiment is to transfer them to nests of other individuals with continuous incubation going on, or to nearly allied species with a reliable continuous incubation behaviour. If practical arrangements have been made the eggs can subsequently be returned to the nests of the proper parents.

There are two reasons why I do not mention hatching in incubators here. In the first place the field ornithologists, on whom we have to rely to a very great extent here in Sweden, very rarely have the opportunity of using a good incubator; in the second place, it is not only the primary — but difficult — factors, such as temperature and air humidity, which are of importance, but there is probably a number of other factors influencing incubation, of which our knowledge is not yet sufficient. In my opinion we can fully agree with HUGGINS (1941: 156) who says: »These other factors must be known and controlled before we can expect to·artificially incubate wild bird eggs with any degree of success.»

It goes without saying that the high figures for the incubation period discussed above cannot be just neglected. On the contrary, they are indispensable as practical information, and, when compared with the proper time of development of the eggs, they become excellent indicators of the degree of »inattentiveness» which is accidental or distinctive for certain individuals or species and which may be of interest in connection with environmental factors.

[*Editor's Note:* The foreign language summary has been omitted at this point.]

REFERENCES

BONHOTE, I. L. 1913. On incubation. letter to the Editors. Brit. Birds, 6: 29.

COLE, L. J. & KIRKPATRICK, W. F. 1915. Sex Ratios in Pigeons, together with observations on the laying. incubation and hatching of the eggs. Agr. Exp. Station Rhode Isl., Bull. 162.

COLLENETTE, C. L. 1913. On incubation. letter to the Editors. Brit. Birds, 6: 29.

GROEBBELS. F. 1937. Der Vogel. II. Berlin.

HEINROTH, O. 1922. Die Beziehungen zwischen Vogelgewicht, Eigewicht, Gelege-gewicht und Brutdauer. J. f. Orn., 70: 172—285.

HOLSTEIN. V. 1942. Duehøgen. Copenhagen.

— — 1944. Hvepsevaagen. Copenhagen.

— — 1945. Træk af Flagspættens Ynglebiologi. Dansk Orn. For. T., 39: 65—87.

HUGGINS, R. A. 1941. Egg temperatures of wild birds under natural conditions. Ecology. 22: 148—157.

KUERZI, R. G. 1941. Life History Studies of the Tree Swallow. Proc. Lin. Soc. New York, No. 52—53: 1—51.

LOW, S. H. 1933. Further Notes on the Nesting of the Tree Swallows. Bird-Banding. 4: 76—87.

LÖWEGREN, Y. 1935. Ruvning. Sv. Uppslagsbok. bd 23. Malmö.

MOORE, J. 1735. Columbarium: or, the pigeon-house. Being an introduction to a natural history of tame pigeons. London. (Quoted by COLE & KIRKPATRICK, 1915)

MOREAU, R. E., & W. M. 1940. Incubation and Fledging Periods of some African Birds. Auk, 57: 313—324.

MOREAU, R. E. 1946. The recording of Incubation and Fledging Periods. Brit. Birds, 39: 66—70.

NICE, M. M. 1937. Studies in the Life History of the Song Sparrow, I. Trans. Lin. Soc. New York, 4: 1—247.

NIETHAMMER. G. 1937. Handbuch der deutschen Vogelkunde. I. Leipzig.

OLSTAD, O. 1935. Undersøkelse over kråkens forplantningsforhold. Skr. Norske Vid.-Akad. Oslo. I. Matem.-Nat. Kl.. no. 3.

RUTHKE, P. 1941. Beobachtungen am Horst des Höckerschwans (Cygnus olor GMELIN'). Beitr. Fortpfl.-biol. Vögel. 17: 86—90.

RYVES, B. H. 1929. Variability in Incubation- and Fledging-Periods. Brit. Birds, 22: 203—205.

— — & Mrs. 1934. The Breeding-Habits of the Corn-Bunting . . . Ibid., 28: 2—26.

— — 1943. An investigation into the Rôles of Males in relation to Incubation. Ibid., 37: 10—16.

— — 1946. Some criticisms on the recording of Incubation-Periods of Birds. Ibid., 39: 49—51.

STEINFATT, O. 1937. Aus dem Leben des Grossbuntspechtes. Beitr. Fortpfl.-biol. Vögel, 13: 45—54.

STRESEMANN, E. 1934. Aves, in KÜKENTHAL, W.. Handbuch der Zoologie. Bd 7, Teil 2. Berlin.

SUTTER, E. 1946. Zur Bestimmung der Brutdauer. Orn. Beob., 43: 54—57.

WHITMAN, C. O. 1919. The Behaviour of Pigeons. Posthumous work. Washington.

WINGSTRAND, K. G. 1943. Zur Diskussion über das Brüten des Hühnerhabichts Accipiter gentilis (L.). Kungl. Fysiograf. Sällsk. i Lund Förh., bd 13, no. 23.

WITHERBY, H. F., JOURDAIN, F. C. R., TICEHURST, N. F. & TUCKER, B. W. 1945. The Handbook of British Birds, I. London.

VON ZIEGLER, V. 1909. In: Verh. Ornith. Ges. in Bayern 1908, Bd 9: 125.

15

Reprinted from *Ibis* **99**:69-93 (1957)

THE INCUBATION PATTERNS OF BIRDS.

By ALEXANDER F. SKUTCH.

Received on 7 December 1955.

CONTENTS.

Introduction.

Part 1.   A classification of incubation patterns.
  Synopsis of incubation patterns.
  Notes and references.

Part 2.   What determines the participation of the sexes in incubation ?
  The evolutionary sequence.
  Incubation and brood-patches in male passerines.
  Incubation habits in relation to colouration and nidification.
  Influence of the environment.
  Incubation in relation to song and territory.
  Incubation and the sex-ratio.
  Conclusions.

Summary.

References.

INTRODUCTION.

More than a decade ago, I began to collect material for a general survey of the modes of incubation of birds as a whole, giving special attention to the division of labour between the sexes, the rhythm of sitting, and the factors which determine the pattern adopted by each species or group.   But the task proceeded slowly and meanwhile Kendeigh (1952) has performed a great service to all ornithologists by giving us in a single volume a summary of nearly all that is known of the modes of incubation and other aspects of parental care in birds.   There would be no point in attempting to duplicate this vast and laborious undertaking.   Yet I believe that a condensed synopsis of the known patterns of incubation, in the form of a " key ", will be interesting and instructive to naturalists, as revealing at a glance the vast diversity of behaviour which birds have evolved in this single activity, and enabling them to place any observed pattern in the general scheme.   Moreover, an analysis of the facts now available seems to permit the drawing of a few conclusions which appear to be new, or at least not sufficiently emphasized by earlier writers.

PART 1.

A CLASSIFICATION OF INCUBATION PATTERNS.

The following classification is logical rather than phylogenetic.   It does not pretend to trace the course of evolution of incubation patterns, but offers an arrangement based upon differences actually observable.   The primary divisions are determined by the number of adults incubating, whether one, two, or as in a few exceptional groups, more than two at a single nest. When the two parents incubate, we obtain a further division by considering whether they sit alternately or (as rarely occurs) simultaneously. Alternate care of the eggs by male and female is the method most widespread among birds; and in this great division obvious subdivisions are suggested by the occupancy of the nest by night. Whether the male or female is in charge during the hours of darkness seems more significant in strictly diurnal birds than in those which are active both day and night. But in a number of families, especially among seabirds, either sex may cover the nest through the night. Whether this alternation occurs, or whether the same sex is always in charge by night, seems to be determined primarily by the length of the sessions or the frequency of changes in occupancy; for obviously when the sexes replace each other at intervals consistently longer than twelve hours, it will fall now to one, now to the other, to sit at night. Hence the frequency of change-overs forms the basis of further subdivisions in the great group of birds of which the sexes alternate on the eggs.

In the section in which a single parent takes charge of the eggs, the major subdivisions are determined by the sex to which this duty falls. In each of the two resulting classes, further subdivisions are determined by the length of the sessions, or the frequency of coming and going. Since there is a graded series from birds which sit for only a few minutes at a stretch to those which remain continuously on the nest for all or nearly all of the incubation period, further divisions must be somewhat arbitrary. A parent which sits continuously for many days may either fast or be fed by its mate, and this suggests a final dichotomy in the key (IIA3).

When more than two birds take charge of the eggs, we wish to know how many of the coöperating partners are parents and how many are voluntary assistants. Among the anis *Crotophaga* several females lay their eggs in a common heap and the parents of both sexes take turns in covering them. In Bush-tits *Psaltriparus minimus* three birds sometimes incubate in one nest. In Black-eared Bush-tits *P. melanotis*, of which the sexes are more readily distinguishable, supernumerary males help the mated pair to attend their nest and young and many take tufts of down into the pouch while incubation is in progress; but I could not make sure that any of these assistants actually warmed the eggs. Simultaneous incubation by two females of the same or even different species in a nest where both have laid has from time to time been reported as an abnormality, but our classification is concerned only with more constant behaviour.

Finally, to complete our conspectus of the modes of incubation, we have included certain expedients whereby the parents avoid the possibly tedious occupation of applying to their eggs the heat of their own bodies. They may either drop their eggs into the nests of other birds, as with a number of cowbirds, cuckoos, and a few ducks and others; or they may utilize the heat of the sun, decaying vegetation, or volcanoes, as with the megapodes.

In examining the following classification, a certain symmetry is evident. In division I (incubation by both parents) B2a, in which the female covers the eggs by night and the male takes one long session each day, has its counterpart in B3a, in which the male incubates by night and the female takes one long session each day. Similarly B2b, in which the female sits by night and the sexes alternate several times a day, has its counterpart in B3b, where the situation is reversed. In division II (incubation by a single parent), the three subclasses under A (incubation by the female) are almost exactly mirrored by the three subclasses under B (incubation by the male). But there seems to be no known instance of a female bird feeding her continuously incubating partner, to match the nourishing of the immured female hornbill by her mate.

Many birds whose method of incubation is known in a general way, as whether one or both parents sit on the eggs and even which is in charge through the night, have not been studied in sufficient detail to permit their allocation to the finer subdivisions of our table. It is to be hoped that this lack will little by little be remedied. The examples cited under each ultimate division are intended to be illustrative rather than exhaustive; and I have in most instances limited their number to four, even when scores of examples might have been given. Also in order to keep the " key " compact, I have relegated all comments and citations of literature to the notes which follow immediately after it. When no authority is given, either the mode of incubation was learned by my own observations, or it is generally known and recorded in works of reference. An extensive bibliography of pertinent studies, with brief summaries and analyses, is to be found in Kendeigh's monograph (indicated by K in these notes).

SYNOPSIS OF INCUBATION PATTERNS.

PARENTAL INCUBATION.

I. Incubation by two parents.

    A. By both sexes simultaneously.

        Red-legged Partridge *Alectoris rufa.*

    B. By both sexes, alternately.

        1. Either sex may cover eggs by night.

            (a) Change-overs at intervals of about 24 hours, so that the same parent does not often sit for the whole of successive nights.

                Diving Petrel *Pelecanoides urinatrix*, Sooty Tern *Sterna fuscata*, Bridled Tern *S. anaethetus*, Ringed Kingfisher *Ceryle torquata.*

            (b) Change-overs at intervals of much more than 24 hours, so that the same parent may sit for several nights in succession.

Adélie Penguin *Pucheramphus adeliae*, Manx Shearwater *Puffinus puffinus*, Royal Albatross *Diomedea epomophora*, and many other Procellariiformes.

(c) Change-overs at intervals of less than one day (sometimes by both daylight and dark).

Semipalmated Plover *Charadrius hiaticula*, Herring Gull *Larus argentatus*, Cape Wagtail *Motacilla capensis*.

2. Female covers eggs by night.

(a) Male takes one long session each day.

Pigeons and doves, Citreoline Trogon *Trogon citreolus*, Black-throated Trogon *T. rufus*.

(b) Sexes alternate on nest several times a day, the diurnal sessions therefore usually shorter than in 2a.

Amazon Kingfisher *Chloroceryle amazona*, Rufous-tailed Jacamar *Galbula ruficauda*, antbirds (Formicariidae), many warblers (Sylviidae).

3. Male covers eggs by night.

(a) Female takes one long session each day.

Ostrich *Struthio camelus*, Pale-billed Woodpecker *Phloeoceastes guatemalensis*.

(b) Sexes alternate on nest several times a day, the diurnal sessions therefore usually shorter than in 3a.

Many woodpeckers, anis *Crotophaga*, American Coot *Fulica americana*.

II. Incubation by one parent only.

A. Incubation by female.

1. One recess each day, usually long.

Bobwhite Quail *Colinus virginianus*, Marbled Wood Quail *Odontophorus gujanensis*, Guan *Pauxi pauxi*.

2. Several or many recesses each day, sometimes also by night.

Red-head Duck *Nyroca americana*, most hummingbirds, manakins (Pipridae), American flycatchers (Tyrannidae), most song-birds.

3. Female sits continuously for many days.

(a) She fasts while incubating.

Golden pheasant *Chrysolophus pictus*, Great Argus Pheasant *Argusianus argus*, Eider Duck *Somateria mollissima*, Blue Goose *Chen caerulescens*.

(b) She is fed by male.

Hornbills.

B. Incubation by male.

1. One long recess each day.

Little Tinamou *Crypturellus soui*, Bonaparte's Tinamou *Nothocercus bonapartei*.

2. Several or many recesses each day.

Pheasant-tailed Jacana *Hydrophasianus chirurgus*, Ornate Tinamou *Nothoprocta ornata*.

3. Male incubates continuously for a number of days, fasting.

Emperor Penguin *Aptenodytes forsteri*, Kiwi *Apteryx*, Emu *Dromiceius n. hollandiae*.

III. Incubation by more than two birds at a single nest.
     A. Eggs laid by one female.
         1. Several adults assist the female.

Bush-tit *Psaltriparus minimus*.

     B. Eggs laid by two or more females.
         1. Several birds of both sexes participate in incubation.

Anis *Crotophaga*, ?Acorn Woodpecker *Melanerpes formicivorus*.

SUBSTITUTE INCUBATION.

IV. Eggs incubated by birds of other species.

Many cuckoos, cowbirds (Icteridae), honey guides (Indicatoridae), Black-headed Duck *Heteronetta atricapilla*.

V. Eggs incubated without animal heat.

Megapodes, Egyptian Plover *Pluvianus aegyptius*.

NOTES AND REFERENCES.

IA.   I have placed first this extraordinary mode of incubation to emphasize by juxtaposition its contrast with the following heading—incubation by both sexes alternately. The Red-legged Partridge is the only example I know. The female lays in close succession two sets of eggs in separate nests; she incubates one set while her mate takes charge of the other (Goodwin 1953). In the Bobwhite Quail, according to Stoddard (1946), either the cock or the hen may take exclusive charge of the nest, and more rarely the two take turns on the same set of eggs, or in captivity sit side by side in the same nest. Apparently there is no record of the male and female of a pair incubating simultaneously separate sets laid by the latter; but it seems that this custom might easily develop from the situation found in the Bobwhite. In *Lophortyx* the male may take full charge of incubation if the female deserts or is killed (K.).

     In a number of rails, motmots, barbets, woodpeckers, swifts, titmice, etc. both sexes sleep in the nest cavity; but we have no evidence that they divide the eggs between them and incubate simultaneously, by night if not by day.

IB 1a.   Diving Petrel—Richdale (1943); Terns—K.

IB 1b.   Male Adélie Penguins may fast up to 40 days, females 21—Sladen (1953). Albatrosses fast up to 32 days—K. Manx Shearwater—Lockley (1942).

IB 1c.   Semipalmated Plover—Spingarn (1934); Herring Gull—K. The only instance of a passerine in which either sex may incubate by night known to me is the Cape Wagtail, in which Skead (1954) found sometimes the male, sometimes the female sleeping on the nest. This irregularity becomes more understandable in the light of the statement by Moreau (1949) that in the African Mountain Wagtail *Motacilla clara* both parents slept on or alongside the nest. In cases like this, it is usually scarcely possible to tell which actually sits upon the eggs, and whether there is constancy in this detail. A similar uncertainty exists with all those birds, mentioned above, in which both parents sleep with the eggs in well closed nests. When this happens with passerines, such as some titmice and swallows, it is more likely to be the female who maintains close contact with the eggs, since in the Passeres males usually lack brood patches for the efficient transfer of heat.

IB 2b.   In addition to many certain examples of this class, there are many more that are probable; but the similarity of the sexes makes it difficult to determine which is in charge of the nest by night. It is likely that many toucans, motmots, woodhewers (Dendrocolaptidae), and ovenbirds (Furnariidae) belong here; but the sex which incubates through the night is not definitely known.

IB 3a.   Ostrich—K. The Pale-billed Woodpecker is included here provisionally on the strength of a single day's observation at a nest which was prematurely lost.

IB 3b.   Anis often nest communally, but nests of single pairs fall here. American Coot—Gullion (1954).

IIA 1.   Bobwhite—Stoddard (1946); *Pauxi*—Schäfer (1953, 1954a).

IIA 2.   The Red-head Duck takes recesses by night as well as by day (Low 1945).

IIA 3a.   Golden Pheasant and Eider—Goodwin (1948); Argus—Delacour (1951); Blue Goose—Manning (1942).

IIA 3b.   As is well known, the female hornbill encloses herself in the nest cavity by walling up the doorway, leaving only a small aperture through which her mate passes food for her and later for the nestlings as well (Moreau 1936, 1937; Moreau & Moreau 1940, 1941). Although no other instance of uninterrupted incubation by a female sustained by her mate is known to me, more or less close approach to it is made by a number of other species, in which the sitting female is so well nourished by her partner that her absences are infrequent or short, or both. Among these are owls and parrots (K), and such passerines as the Cedar Waxwing *Bombycilla cedrorum* (Putnam 1949) White-throated Magpie-Jay *Calocitta formosa* (Skutch 1953), and numerous cardueline finches, such as the Goldfinch *Carduelis carduelis* (Conder 1948).

IIB 1.   Bonaparte's Tinamou—Schäfer (1954 b).

IIB 2.   Pheasant-tailed Jacana—Hoffmann (1949); Ornate Tinamou—Pearson and Pearson (1955).

IIB 3.   Emperor Penguin—Stonehouse (1953); Emu and Kiwi—K.

IIIA 1.   Bush-tit—Addicott (1938); (compare Skutch 1935).

IIIB 1.   Anis—Bent (1940), Davis (1940). Four or five Acorn Woodpeckers may attend one nest, but the number of layers is uncertain.

IV.   Cowbirds—Friedmann (1929); parasitic ducks—Friedmann (1932).

V.   Megapodes—Wallace (1872), Sibley (1946), K; Egyptian Plover—K.

PART 2.

WHAT DETERMINES THE PARTICIPATION OF THE SEXES IN INCUBATION ?

The type of the incubation pattern is determined largely by innate factors, so that it persists through fairly wide fluctuations in weather, although it may break down in extreme conditions. Hence the position of a species in the foregoing classification is rarely changed by the normal, day to day variations in temperature, rainfall, and the like, which, however, may markedly affect the length of the birds' periods on and off the nest. An exhaustive study of incubation patterns would include these variable features along with the more constant features. At present, however, we shall confine our attention to the latter, especially the role of the sexes, reserving the former for future discussion.

The evolutionary sequence. In a number of families placed lowest in current schemes of classification, as the rheas, emus, and kiwis, the male takes care of the eggs and chicks with little if any help from the female, and in a great many families considered to be more advanced he has a share in incubation equal to, if not greater than, that of his mate, while in the highest group, the Passeriformes, the female is largely and in many species wholly responsible for keeping the eggs warm. When one recalls these facts, it is tempting to postulate that incubation by the male alone is the primitive condition in the avian stock. The fact that in a number of fishes, including Stickle-backs, Fighting Fish, and Pipe Fish, and in such batrachians as the Nurse-frog of Europe, the male alone takes charge of the eggs and young, makes this hypothesis more attractive. One might suppose that, with the psychic development of birds, as the affective bonds between the sexes grew stronger, the female's closer association with her mate led her to share increasingly in his parental duties, until the participation of both parents in all phases of the nesting operations became widespread among birds. Finally, with the evolution of song, territorial defence, or ornamental plumage along with elaborate courtship displays, the male came to neglect domestic pursuits for these more exciting occupations, until he quite ceased to incubate or, as in many pheasants, hummingbirds, cotingas, manakins, and birds of paradise, he lost all interest in the nest and young.

Weighing against this attractive speculation is the fact that in a number of orders of birds, especially those occupying intermediate positions in current taxonomic systems, are families in which the female takes the sole or at least the leading role in incubation, along with families in which the male is largely or wholly responsible for the eggs. Thus in the Gruiformes, in the majority of whose families the sexes share parental duties rather equally, the female alone is said to incubate in the bustards, whereas in the hemipodes it is chiefly if not solely the male who attends the eggs. In the Charadriiformes, where again incubation by both sexes is the prevailing mode, the male takes charge of the eggs in the painted snipes, the jacanas, and some of the phalaropes and sandpipers; while the female does most of the incubation in a number of other species of Scolopacidae, so that in this single family both trends are evident. When we recall how, through much of the avian class, the weight of incubation shifts now to the female, now to the male sex, we must agree with Kendeigh (1952) that incubation by both sexes is most probably the primitive, as it is the prevailing, method among birds, and from this basic arrangement various stocks have diverged in the direction of greater male or greater female participation. For this, too, we can find analogies among the cold-blooded vertebrates from which the warm-blooded birds and mammals have descended. In some of the cichlid fishes the sexes alternate in the care of the eggs and newly hatched young in a manner that reminds one strongly of birds (Lorenz 1952).

The constancy of the incubation pattern in each species and even in whole families of birds, leaves little doubt that it is genetically determined ; and it is highly probable that its physical basis lies in the structure of the nervous system and the functioning of the endocrine glands. It will help us to understand the evolution of incubation patterns if we bear in mind that the habit of incubation is a secondary sexual, if not a sex-linked, character, like song and ornamental plumage in birds, horns and manes or beards in mammals. As a result of genetic " accidents ", sometimes no doubt the crossing-over of genes from one to another of the sex chromosomes, these secondary sexual characters often pass from male to female, or vice versa, in a way for which no good reason in the form of a recognized utility can be assigned. The fact that in some deer and cattle the females as well as the males bear horns, in others the male alone, seems to be largely unrelated to the usefulness of these weapons to the females. In many families of birds, there are species in which both sexes are brightly coloured along with those in which the male alone is brilliant; and in many instances it seems impossible to correlate these differences with the habits and needs of the birds. Similarly, song is on the whole a male character which may or may not pass to the female, and it is often difficult to find a reason for the singing of the female in one species and her failure to do so in a related species. Many of the fluctuations in incubation habits seem to be òf the same irrational or non-adaptive character, the result of genetic accidents uncontrolled by the needs of the birds.

In certain families or wider groups, incubation habits are in a particularly unstable condition, so that we find species in which the male incubates along with those in which he rarely does so. For example, in the vireos the male's part in incubation is most irregular. The most careful studies of the Red-eyed Vireo *Vireo olivacea* (Lawrence 1953) and the closely related or possibly conspecific Yellow-green Vireo *V. flavoviridis* (Bent 1950) revealed incubation by the female alone; but statements that the male of the former sits on the eggs are not lacking (*e.g.* Forbush 1929), and in view of the variability of this feature in the genus as a whole are not to be lightly brushed aside as unreliable. On the other hand, in the Philadelphia Vireo *V. philadelphicus* (Lewis 1921), Warbling Vireo *V. gilvus* (Rust 1920), Hutton's Vireo *V. huttoni* (Van Fleet 1919), and Bell's Vireo *V. bellii* (Nice 1929), the males seem regularly to incubate, often singing freely while they sit in the nest. So, too, in the warblers (Sylviidae) there are a number of species in both the Old and New World in which the male incubates, along with many, at least in the Old World, where the family is best represented, in which he fails to do so (Witherby *et al.*, 1938). The swallows show the same inexplicable mixture of species in which both sexes incubate and those in which only the female does so (Allen & Nice 1952).

The great suborder of the Tyranni is generally held to contain some of the more primitive families of the order Passeriformes, although many of its members exhibit amazing specializations in various directions; and in view of its taxonomic position the study of the wide fluctuations in incubation habits and other aspects of parental care to be found within it should be particularly instructive. Although in the antbirds incubation by both sexes seems general, the ovenbirds and woodhewers exhibit cases of non-participation by the male which are difficult to explain. In the American flycatchers and cotingas the male is not known to incubate and may or may not feed the young; in the manakins he is not known to attend the nest at any stage in any species. The failure to incubate, or even to feed the young, probably preceded rather than followed the amazing developments in plumage and courtship habits in many manakins and cotingas, and in less conspicuous degree in some of the American flycatchers. It is far more likely that a genetically determined release from attendance upon the eggs left the males free to acquire colours, ornamentation, and modes of life which make them unfit for parental duties, than that the acquisition of these peculiarities was followed by a growing aloofness from the nest. So, too, in the suborder Passeres, or song-birds proper, elaborate song and great preoccupation with territory by the males may have followed upon rather than preceded absten-tion from incubation.

Incubation and brood-patches in male passerines. The apparently non-adaptive character of many variations in incubation habits, at least within the passerines, is further attested by the lack of correlation between incubation and the presence of a brood-patch in the males. While it is true that even without an incubation patch a bird can retard the cooling of already warmed eggs by sitting on them as a sort of heated blanket, it is through the area of highly vascularized bare skin which most incubating birds apply directly to the eggs that heat is most effectively transferred to them. Although the males regularly take turns on the nest in such oscinine groups as bush-tits *Psaltriparus*, wren-tits *Chamaea*, and grosbeaks *Pheucticus* or *Hedymeles*, they fail to develop highly vascularized incubation-patches, and it is doubtful if these are present in typical form in any males of the suborder Passeres (Bailey 1952). Yet Smith (1950) saw the male Yellow Wagtail *Motacilla flava* apply bare ventral skin to the eggs as he settled down for his regular turns at incubation, and found the eggs warm after he had been sitting. Mewaldt (1952) found incubation patches on a number of male Clark Nut-crackers *Nucifraga columbiana*, and in some cases they seemed as well deve-loped as in nesting females. Likewise Amadon and Eckelberry (1955) collected in Mexico a male Magpie-Jay *Calocitta formosa* with a well developed brood-patch; although at a Guatemalan nest of this species that I watched carefully I failed to find evidence that the male covered the eggs, and incubation by males is unusual in the Corvidae.

In the Tyranni or non-oscinine passerines the situation is equally confusing; for incubation patches are sometimes present even in families and genera in which males do not incubate, at least in the species which have been watched at the nest (e.g. *Tityra* and *Myiodynastes*—Davis 1945). In American flycatchers of the genus *Empidonax* the males of some species do show an incubation patch while those of other species lack it (Parkes 1953); yet in no member of this genus or even family is the male known to incubate. Except in the tropical antbirds, it is not easy to find a nest of a passerine at which the male incubates and which is at the same time favourably situated for observing whether he applies a patch of bare skin to the eggs, in the manner of Smith's Yellow Wagtail; but anyone who has an opportunity to determine this point may without any sacrifice of life make an important contribution to ornithology.

It appears, then, that the pattern of incubation, and especially its major feature the participation of the sexes, is genetically controlled, and like other heritable characters may be altered by mutations which are not called forth by the needs or mode of life of the race in which they occur. So long as these changes do not seriously detract from reproductive efficiency, they may persist even if they are no improvement over the earlier mode of incubation; but where they result in a marked decrease in hatching success they will be eliminated by selection. Thus there is no reason to suppose that every modification in incubation patterns is adaptive. The only condition for its persistence is that it is not detrimental. Can we detect definite trends in the evolution of incubation patterns in relation to the structure, coloration, or mode of life of birds ?

Incubation habits in relation to colouration and nidification. It was long supposed that there is a close correlation between the colouration of a bird and its part in incubation. Thus Wallace (1871) wrote : " when both sexes are of strikingly gay and conspicuous colours, the nest is . . . such as to conceal the sitting bird; while, wherever there is a striking contrast of colours, the male being gay and conspicuous, the female dull and obscure, the nest is open and the sitting bird exposed to view ". Since colouration is so variable a character in animals, and birds are on the whole conservative in their habits and architecture, Wallace supposed that colouration was determined by the form of the nest, rather than the reverse. If a species nested in a hole or burrow, both sexes could don brilliant plumage without jeopardizing eggs and young by their conspicuousness; if it used an open cup, the sitting female was kept dull by natural selection and the male alone was free to become colourful. Wallace appeared to be ignorant of the prevalence of incubation by males in many avian orders.

In view of our vastly amplified—although still far from adequate— knowledge of the life histories of birds, it may be worth our while to re-examine the facts and try to discover some correlation between the three variables, colouration, nidification, and incubation pattern. In some of the most colourful of the non-passerine families, as kingfishers, motmots, jacamars, bee-eaters, toucans, and woodpeckers, male and female are nearly or quite equal in brilliance and share fairly equally all the labours of the nest, which is nearly always in a cavity of some sort, so that the colour of the incubating bird is not likely to catch the eye of a predator. But in some families in which the females of North American species are far duller than the males the Central American representatives have the sexes alike. In my early years in Central America, I was eager to learn whether, when the sexes were similar in appearance, they would take more equal parts in the nesting duties than in the north where they are strikingly different. But as the years passed I became convinced that, among the passerines at least, colouration has little to do with the participation of the male in incubation, and shows almost as little correlation with the form of the nest. No family displays a greater variety of bright colours than the tanagers, and in some of the most brilliant genera (e.g. *Tangara*) the sexes scarcely differ in appearance, yet only the female sits in the open nest. These are balanced by other genera (e.g. *Ramphocelus*), also with open nests, in which the male is far more brilliant than the female, which alone sits. When there is a closed nest combined with striking sexual differences in colour, as in the euphonias and chlorophonias, the situation is the same, only the female incubates. In the Icteridae, some of the most brilliant of the tropical orioles exhibit slight sexual differences in their yellow and black attire; yet whether the nest is a shallow open cup or a deep closed pouch, the male, so far as known, fails to incubate. So, too, with the wood warblers, in some of the most colourful tropical species the females hardly differ from their mates; but the male is not known to incubate, even when the nest is roofed, as in *Myioborus* and *Ergaticus*.

No family exhibits a more wonderful diversity of architecture than the Tyrannidae, yet whether the nest is open or well closed the male has never been seen to sit in it. On the other hand, the antbirds, without being brilliant, often exhibit pronounced sexual differences in colouration ; yet male and female participate equally in incubation in their usually open nests. Among the passerine birds of tropical America, there is scarcely any correlation between colouration, nidification, and incubation pattern, which on the whole vary quite independently of each other. It is true that in some of the manakins and cotingas the ornate male leaves to the far duller female all the parental offices; but the same is true of some members of these families which exhibit little or no sexual dimorphism; and it is probable that the males relinquished incubation long before they acquired their colourful attire, their bizarre wattles, or their elaborate courtship antics.

In the most lavishly adorned of birds, as the peacocks, Argus pheasants, and the most ornate of the birds of paradise, the abundant, brightly coloured nuptial plumes of the males would not only jeopardize the nests by their conspicuousness, but by their size and unwieldiness might seriously impede the birds in the performance of parental chores. It is quite understandable why in the more ornate species of these families the males remain aloof from the nest, although in the more plainly attired members, as the quail and the manucodes, the males not only take an interest in the young but sometimes incubate, or at least brood the nestlings (Rand 1938). But even in these families it is probable that the males ceased to sit in the nest before rather than after acquiring their splendid adornments. As Huxley (1938) pointed out, the evolution of exaggerated courtship characters is facilitated by the higher reproductive potential of males which do not pair but use all their arts to attract the greatest number of females. A strictly monogamous male can father only as many offspring as he and his single mate can adequately nourish; but a promiscuous male, who proves especially attractive to the females, can fecundate the eggs of many and so each season leave a numerous progeny. Such an outstanding advantage in reproduction accelerates the evolution of the bright colours, elaborate plumes, and striking postures which attract the females, bringing about results which could hardly be attained by monogamous birds. But the reproductive advantage which leads to these arresting products of sexual selection is enjoyed only *after* a male becomes polygamous and ceases to attend the nest; so that the most elaborate nuptial adornments are on the whole consequences rather than antecedents of the males' abstention from incubation. Yet in the Quetzal *Pharomachrus mocinno*, one of the most beautiful birds of the western hemisphere, the male takes his full share in incubating and attending the young, as in the less ornate trogons. The long, graceful plumes of his train are badly frayed or even broken off short by constant flexure and abrasion on his innumerable passages in and out of the doorway of the rough hole in a decaying trunk; but at the following moult he grows them afresh.

While on the subject of the influence of colouration on incubation patterns, we might consider the colours of eggs. The usually white, unmarked eggs of pigeons and doves are of a type common enough among birds which nest in holes or build covered structures but rare in such frail, open nests as most pigeons build. These birds sit on the first egg much of the time before the second is laid, after which male and female together keep the usually two eggs almost constantly covered, thereby reducing the probability that their gleaming whiteness will attract hostile eyes. If an enemy approaches the nest, the parent, as I have observed from concealment, crouches motionless and does not flee so long as there is a chance of escaping detection. Although it is not impossible that the conspicuousness of the eggs brought about the pigeons' habit of keeping them continuously covered, it is more probable that the birds' close sitting is reponsible for their remaining unpigmented; for with them cryptic colouration lacks the selective advantage it enjoys in eggs more frequently exposed.

With the Pauraque *Nyctidromus albicollis* of tropical America, I have repeatedly noticed that the eggs, although buffy and mottled, are more conspicuous against the browns of the ground litter on which they lie than is the plumage of the birds themselves. Since male and female together keep the eggs rather constantly covered, natural selection again has not produced that close resemblance to the substratum which might have resulted if they were exposed for longer periods. On the other hand, the anis and other cuckoos which only gradually work up to full constancy in incubation leave their chalky white eggs exposed for considerable periods. But the green leaves which anis and Squirrel Cuckoos *Piaya cayana* bring to their nests stain the eggs which lie upon them, and the scratches which soon disfigure the soft outer layer of anis' eggs allow the blue of the inner shell to show. Together the dark, irregular stains and scratches gradually reduce the conspicuousness of the originally white eggs.

Influence of the environment. When I began to study birds in the highlands of tropical America, I asked whether the more severe conditions there prevailing would not lead to incubation and brooding by the males, even in families of which the female alone sits on the nest in the warm lowlands. At high altitudes, low temperatures, chilling winds, cold rains, and frosts make it far more important to keep eggs and nestlings well covered; and male participation in these duties would seem to be highly advantageous. I found a number of nestlings which died apparently of cold and exposure. Many hummingbirds, and honeycreepers of the genus *Diglossa*, nested on the high mountains at a season when severe frosts were of almost nightly occurrence and penetrating winds sometimes blew by day. But I could find no instance of male incubation in hummingbirds, honeycreepers, wood warblers, wrens, thrushes, and other families in which the males fail to incubate in milder climates.

Although in spite of the cold the nests of the White-eared Humming birds *Hylocharis leucotis* which I watched in the high mountains of Guatemala were not attended by males, at least one member of the family which breeds at high altitudes has made the adjustment which the climate seems to demand. When Moore (1947) announced that in the highlands of Ecuador he had collected a male Gould's Violet-ear *Colibri coruscans* which had been sitting on eggs, ornithologists familiar with the habits of humming birds hesitated to conclude from this single observation that male incubation is customary even in this species. However, Schäfer (1954 c) found and photographed a male Gould's Violet-ear incubating eggs and feeding young in the mountains of Venezuela. In view of this confirmation of Moore's discovery, his observations on certain other Andean humming birds, whose males showed greater interest in the nest than one expects in this family, should stimulate further investigations. If a harsh environment is capable of radically altering the incubation pattern, one might expect this to occur among the small birds of the high Andean páramos; and it is to be hoped that before long evidence will be accumulated on this point.

In the far north, severe conditions seem as little effective in inducing incubation by the male in passerine birds as on high tropical mountains. The male Snow Bunting *Plectrophenax nivalis* is as careless of incubation as most other male finches (Tinbergen 1939; Sutton & Parmelee 1954). Although Blair (1936) reported incubation by the male Lapland Longspur *Calcarius lapponicus* in arctic Norway, in arctic and subarctic America females alone were found on the eggs by Grinnell (1944), Wynne-Edwards (1952), and Sutton & Parmelee (1955). Similarly with the Greater Redpoll *Acanthis flammea*, Wynne-Edwards found only females incubating on Baffin Island.

In the Shetlands and other northern islands where food is relatively scarce, the Wren *Troglodytes troglodytes* is monogamous and the male takes a good share in feeding the nestlings, although under the milder conditions prevailing farther south polygyny is frequent in this species and the female often attends the young alone (Armstrong 1952, 1955). This climatically induced change in breeding behaviour might, if carried far enough, lead to incubation by the male; but except in the case of Gould's Humming bird, I am aware of no instance where such a departure from the system prevalent in the family has actually occurred, apparently in response to ecological conditions.

Our second example of a major change in an incubation pattern which we may with some confidence attribute to the environment is an instance of the loss rather than the acquisition of the habit of incubation by one of the sexes. In the penguins, rather equal participation by male and female appears to be general; but in the Emperor Penguin *Aptenodytes forsteri* the female goes off to the sea soon after laying her single egg on the Antarctic ice and leaves the male to take full charge of it through the cold and darkness of midwinter. This midwinter incubation brings the young out of the shell at such a time that the more favourable period of the year is available for caring for them through their five months of helpless dependency and they will be ready to go to sea while open water is close by. But during the coldest months, open water, where alone the penguins can find food, is so far distant from the breeding ground that change-overs on the egg, such as are practised by other penguins, could be effected, if at all, only by means of exhausting journeys over the sea ice, and consequently have been abandoned by this species. When the females at length return, it is to take charge of the newly hatched chicks, leaving the emaciated males free to march off to the water and recover from their long fast (Stonehouse 1953). In the Adélie Penguin *Pucheramphus adeliae* the male takes charge of the newly laid eggs while his mate goes off to fish in the sea, returning about two weeks later. Since he had been ashore several weeks before laying began, the male's period of fasting may extend to 40 days (Sladen 1953). From such an arrangement, the Emperor Penguin's peculiar method of incubation might be derived simply by extending the male's initial session to cover the two months required for hatching the eggs.

We are able to cite more numerous examples of less radical modifications of the incubation pattern, not involving a change in the participation of the sexes, which seem to have been made in response to fairly constant features of the environment. The infrequent change-overs and consequent long sessions, often amounting to many days together, of Adélie Penguins and many albatrosses, shearwaters, and petrels, are in at least many instances caused by the wide separation of the breeding ground on dry land from the foraging area on the high sea. When the penguins' nests are miles inland, the walking journey is time-consuming. The peril that the smaller petrels run from the big gulls whenever they approach land by day, or even on moonlit nights, has also been an important factor in reducing the number of changes of duty on the nest and in making the meals of the nestlings infrequent but copious.

In numerous other instances the environment, living and lifeless, has made attendance at the nest more constant, even if it has not lengthened the separate sessions. That male and female may share incubation, yet by no means keep the eggs constantly covered, is clear to anyone who has studied toucans, puffbirds, barbets, ovenbirds, and even some of the antbirds. When the eggs are hidden in a hole or closed nest there is no pressing need for one partner to remain in attendance until its mate comes to relieve it; in mild weather the periodic cooling of the eggs is no more harmful when both parents are responsible for them than when a single parent incubates taking alternating sessions and recesses. But when the nest is in a crowded colony whose members covet the sticks or stones which compose it, it becomes imperative to keep it constantly guarded lest it be carried off by thieving neighbours in the absence of the proprietors. Since these conspicuous colonies are often frequented by predators alert to snatch up unattended eggs or chicks, and Herring Gulls *Larus argentatus* may even eat the eggs and young of neighbours of their own kind (Tinbergen 1953), there is an additional motive for maintaining constant guard. Thus in colonial gulls, terns, herons, and other species, each parent remains on duty until its mate returns to relieve it, with a fidelity conspicuously absent among many birds whose nests are not exposed to these perils.

The surprisingly long sessions of diminutive manakins, antbirds, and other species of the tropical forest may be of value in reducing the number of comings and goings, each of which may reveal to the watchful eyes of snakes and other nest-robbers the position of a nest in itself inconspicuous. Far from the tropics, bitter weather may make continuous incubation impera- tive lest the eggs freeze. Although some northern birds nest in winter or early spring while snow still covers the ground, the male does not on this account take a share in incubation, but rather supplies the female with so much food that she can sit almost uninterruptedly, as happens with the Pine Siskin *Spinus pinus* (Weaver & West 1943) and the Red Crossbill *Loxia curvirostra* (Lawrence 1949). Since other species of the Carduelinae which breed in midsummer or even in the tropics incubate in much the same fashion, with the female sitting for hours together while her mate regurgitates food to her, it may be doubted whether this system is really an adaptation for nesting in cold weather. The generally northern distribution of this division of the Fringillidae makes it probable that its mode of incubation is a response to nesting at low temperatures and has persisted even in species which later came to breed in milder months or lower latitudes. None of the obviously tropical branches of the Fringillidae, nor indeed any other tropical passerine known to me, exhibits comparable behaviour.

Another northern bird which nests amidst snow and ice is the Canada Jay *Perisoreus canadensis*. Lawrence (1947) watched two birds, presumably male and female, sit simultaneously and almost continuously in the nest, now one and then the other on top of its partner, in snowy weather in Canada; but Warren (1899) found only the female incubating while snow lay deep in Michigan. Incubation by the male is exceptional in the Corvidae. The behaviour of Mrs. Lawrence's jays may have been a response to severe weather; but in view of Warren's conflicting report, it looks like an abnormality rather than the custom of the species; and the fact that an egg was broken suggests that such double incubation is not a favourable arrangement. We must await further observations on this bird of the far north before drawing conclusions.

Incubation in relation to song and territory. Of all the great divisions of birds, the Passeres is the group in which song reaches its highest development, and incubation by the male, who chiefly sings, is most conspicuously absent, although it does indeed occur in scattered families and species. We know that song is of importance in relation to the advertisement and defence of territory, and in this suborder we have the most numerous and careful studies of territorial behaviour. It is tempting to look upon the waning of incubation in the male as an adaptation which permits him to devote more attention to song and territorial defence. There is a rather widespread view that singing and non-incubation are causally related, and at one time I inclined to this opinion. But more mature reflection makes it appear untenable. The Tyranni are generally held to be more primitive than the Passeres and few of its members rank as songsters. Nevertheless, many of the American flycatchers and a few of the cotingas sing freely and even sweetly in the morning twilight, but through the rest of the day their songs are rarely heard except at moments of great excitement. A few renew their singing in the evening twilight. In these families, spells of incubation in broad daylight would certainly not interfere with the males' vocal exercises.

Even in the Passeres, many good and persistent songsters incubate, and many which are practically songless fail to do so. Among the latter are crows, jays, and many tanagers, and honeycreepers. Some of the males that sing well and loudly continue to do so while taking their turns on the eggs. This has been frequently recorded for vireos of a number of species; and I have known a Pepper-Shrike *Cyclarhis gujanensis* to pour forth his song at the top of his strong voice while sitting in his open nest, continuing this for two hours. In the Rose-breasted Grosbeak *Pheucticus ludovicianus* (Ivor 1944) and the Black-headed Grosbeak (*P. melanocephalus*) males take a large share in incubating and frequently sing on the nest, in the latter species so loudly that Weston (1947) used the song, audible at several hundred feet, for locating nests. The boldly coloured and tuneful male grosbeaks are about the last birds one would expect to find incubating in a family in which even the most soberly attired male sparrows rarely sit on the eggs. Even female birds of many kinds often sing rather loudly while incubating. Laskey (1944) reported that female Cardinals *Richmondena cardinalis* sing loudly in the nest; and among those which I have heard are the Yellow-tailed Oriole *Icterus mesomelas*, Melodious Blackbird *Dives dives*, Blue-black Grosbeak *Cyanocompsa cyanoides*, Orange-billed Sparrow *Arremon aurantiirostris*, Highland Wood Wren *Henicorhina leucophrys*, and others.

In many habitats silence at the nest does not appear to be essential to safety. Not only do sitting birds sometimes sing loudly, or shout out in answer to their mates, or murmur special low nest-songs; the nestlings themselves are frequently noisy when taking food and even while awaiting their meals. Many Tyranni and non-passerines practise the sweet but simple songs or calls of the adults before leaving the nest. Even when approaching the nest with the utmost hesitancy and circumspection, the parents are often exceedingly noisy, presenting a mixture of excessive caution and apparently foolhardy recklessness, which is understandable only if their behaviour has reference to predators which hunt with their eyes or noses rather than with their ears. The noisiness of these otherwise so discreet birds may not be prejudicial if their chief enemies are snakes, which have undeveloped hearing. Among the birds about me as I write, two members of the Sylviidae, the Long-billed Gnatwren *Ramphocaenus rufiventris* and the White-browed Gnatcatcher *Polioptila plumbea* are fine songsters and the males regularly incubate; while most of the male tanagers and flycatchers neither have a song worthy of the name nor share in incubation.

As to defence of territory, this can be satisfactorily attended to (with or without song) even when both sexes incubate, as in the Wren-tits *Chamaea fasciata* studied in California by Erickson (1938), the Graceful Warbler *Prinia gracilis* watched in Egypt by Simmons (1954), the Black-headed

Grosbeak (Weston 1947), barbets, woodpeckers, kingfishers, and many other birds.

Incubation and the sex-ratio. Monogamy, with the equal participation of the sexes in all reproductive activities after laying, is ideally suited to a species in which the sexes are present in about equal numbers. When there is an excess of males or females, either some must refrain wholly from reproduction, or the sexes must take unequal parts in nesting. In the Black-eared Bush-tit *Psaltriparus melanotis* supernumerary males aided mated pairs in caring for the young and sometimes even in building, but I found no evidence of actual polyandry (Skutch 1935). The attendance of extra males or females at the nests of mated pairs has been recorded as an exceptional occurrence in a few species; but usually a strongly unbalanced sex-ratio is associated with polygamy, which normally gives a larger proportion of the population some share in perpetuating the kind. When males are more numerous than females, as in the tinamou *Crypturus variegatus* (Beebe 1925), the jacana *Hydrophasianus chirurgus* (Hoffmann 1949), and apparently also the painted snipe *Rostratula* (Mayr 1939), each female may lay eggs in a number of nests, leaving to a male the whole responsibility of hatching them and caring for the young. By putting the males to work in this fashion, a female may produce more eggs and young than if she laid only as many eggs as she and a single male could together hatch and attend. But in the phalaropes, where also the male takes over the domestic duties, there appears not always to be a preponderance of this sex (Mayr 1939). Likewise in the tinamou *Nothoprocta ornata*, of which the male alone incubates, Pearson & Pearson (1955) saw no indication of either polygyny or polyandry; while in the tinamou *Nothocercus bonapartei* Schäfer (1954 b) found the two sexes in about equal numbers, although one male incubated the eggs which three females laid in a single nest. Such an arrangement is hardly advantageous for the species and doubtfully so for the male who shoulders all the responsibility of reproduction; and it is difficult to explain except as an inheritance from an ancestral stock in which females predominated. Likewise the male Rhea *Rhea americana* takes charge of the eggs laid in a common heap by several females, which suggests an excess of the latter sex.

Usually, however, the presence of several females to each male leads the latter to desist from all participation in nesting and confine his attention to courtship alone. Unfortunately, we know very little about the sex-ratio in pheasants, birds of paradise, manakins, humming birds, and other groups with well developed arena displays or courtship assemblies along with the male's complete withdrawal from nesting activities. In some of the colonial, polygynous passerines, including species of Icteridae and Ploceidae, the relative paucity of males is well attested (Moreau & Moreau 1938; Skutch 1954). It is probable that the colonial oropéndolas, grackles, caciques, and other Icteridae sprang from monogamous ancestors in which the males had already lost the habit of incubation, and that the incidence of a strongly unbalanced sex-ratio, with accompanying polygyny, led the males to desist also from feeding the young, as they do in monogamous members of this family.

When there is a strongly unbalanced sex-ratio, the failure of the less numerous sex to incubate has an advantage which is not obvious in most species in which a single parent sits, for it permits breeding by many individuals which would otherwise remain childless. The aberrant behaviour of the males of some normally monogamous species suggests how polygyny might arise. A male Starling *Sturnus vulgaris* shared incubation at two nests but later fed the young at only one (Schüz 1943). Male Song Sparrows *Melospiza melodia* and Robins *Erithacus rubecula* do not incubate; but a bigamous male of the former attended the young of both his wives (Nice 1937); and a bigamous male of the latter might have done so if he had not disappeared before the brood of his second mate hatched (Lack 1953). In this connection, Brewer's Blackbird *Euphagus cyanocephalus* is of special interest because of the family in which it belongs. In California, Williams (1952) found that the sex-ratio of this icterid varied from year to year; and the greater the excess of females, the more instances of polygyny he observed. Unlike the typical polygynous or non-pairing Icteridae, the male blackbirds fed the young in the nests of several mates at the same time. In cases like this, a sustained increase in the number of females might lead the male so to divide his time between them that he would in the end quite fail to attend eggs and young, as in those passerines which exhibit the most highly evolved polygyny.

The evolution of polyandry from a stock in which the female alone attends the nest might be difficult; but if we assume equal participation by the sexes or exclusive charge by the male as the primitive condition, it is easy to imagine how a situation such as we find in the tinamous and jacanas arose. In the first case, we might suppose that the female on her absences from incubation is courted by an unoccupied male and, deserting her nest, goes off with him to start another brood, while her first mate continues to care for her first set of eggs and the young he hatches from them. With an increasing number of males to distract her from domestic obligations, this desertion of nests and mates would happen more frequently, until the female laid sets of eggs only to abandon them to the care of her latest associate. If, however, we assume full charge of the nest by the male as the primitive arrangement, we need only to suppose that as males became relatively more numerous the females laid more often in order to provide work for all of them. These probable routes of the evolution of both polygyny and polyandry require some individual adjustments or original behaviour by the birds themselves; or at least the process would be accelerated by such spontaneous responses to the novel situations which a growing disparity in the numbers of the sexes would create. But if we adopt the concept of "organic selection" (Huxley 1942; Skutch 1955), which postulates the gradual replacement of individually acquired behaviour by supporting mutations, we need not have recourse to Lamarckism.

Conclusions. The rarity of incubation by the female alone in the more primitive groups of birds makes it highly improbable that this was the original condition in the avian stock. A survey of incubation habits in the whole avian class leads to the conclusion that incubation by the male, or by both sexes, was the ancestral method. If we assume that male incubation was the original mode, we must further suppose that in the evolution of such families as jacanas and phalaropes, in which the male is largely or wholly responsible for the eggs, the female acquired and then lost the habit of incubation; for these families belong to an order in which incubation by both sexes is the prevailing method. On the whole, the probable view is that in ancestral birds both sexes shared incubation, and that from this original and still predominant condition some stocks diverged in the direction of greater participation by the male, others in the direction of greater participation by the female.

Where there is great disparity in the numbers of the sexes, the advantage of shifting all or at least most of the labours of the nest to the more abundant sex is fairly obvious, because it permits fuller participation of the whole adult population in reproduction. In the Emperor Penguin, abstention from incubation by the female is an adaptation to breeding by an unusually large bird with a very long period of helpless dependency in an extra-ordinarily rigorous environment. But with this outstanding exception, and that of Gould's Hummingbird which nests at high altitudes in the Andes, the incubation pattern in its major feature, the participation of the sexes, is surprisingly unresponsive to ecological conditions, although many less radical changes may be traced to such influences. In many situations where incubation by both sexes would seem to be highly advantageous, as among birds which nest on high tropical mountains or amid snow and ice in the north, a single sex takes charge of the eggs if this is the family tradition.

On the other hand, in most monogamous birds in which only the female incubates, it is hardly possible to point to any advantage which the species has derived from the male's aloofness from the eggs. When I reflect upon a great variety of birds that I have studied and found only the female incubating, I can think of exceedingly few in which the male's sitting would be detrimental. In areas where snakes with poor hearing or nocturnal mammals are the chief predators on eggs and young, he might even sing while incubating without jeopardizing the eggs. Possibly in some of the biggest and most brilliant tanagers, as the Scarlet-rumped Black Tanager *Ramphocelus passerinii*, the Scarlet Tanager *Piranga olivacea*, or the bright red Summer Tanager *P. aestiva*, the male would imperil the nest by his eye-catching attire; yet after hatching, these males often visit it with food, and in movement they are even more conspicuous than when quietly sitting. And in many other tanagers the female, who alone incubates, is hardly less colourful than the brightest males. No male bird whose nest I have studied in the tropics has seemed to be so burdened with the maintenance of territory that he could not well have spared four or five hours a day for incubation; and indeed the incubating females were in some cases as much concerned with territorial defence as their idle mates. In sum, I believe that neither the safety of the nest nor the integrity of the territory would be diminished if the males of the majority of passerine species with which I am familiar took to incubation.

On the other hand, when I recall how many manakins, cotingas, American flycatchers, icterids, and other birds of tropical America raise without the least male assistance broods of the same size as most of their neighbours in which pairs are regularly formed and the males help at the nest, I am led to believe that few passerine species, at least in the lowlands, would be adversely affected if the males, which now feed the nestlings but do not incubate or brood, were to relinquish feeding as well. In the more favourable tropical environments, a single parent seems in many instances quite capable of adequately reproducing the species, and the assistance of the male is not strictly necessary, except for fecundation. Hence incubation or non-incubation by the second parent seems in many instances to be non-adaptive. When as a result of mutations a male bird loses the incubation habit, the mutant persists simply because this new departure is not injurious, not because it adds to the efficiency of reproduction. In migratory species breeding at high latitudes, the urgency of rearing in a short season many offspring to replace high annual losses might place a premium upon the specialization of functions which results when the female alone incubates and the male is free to devote all his attention to the establishment and maintenance of territory; but in these birds with their larger broods the male's aid in feeding the young is imperative. But tropical birds, like tropical people, manage to survive without being so efficient as the inhabitants of more rigorous northern lands. It is probable that, in the Passeriformes at least, non-incubation by the male began in the tropical groups, as a mutation neither useful nor harmful, and later happened to give some slight advantage—if it is an advantage—to species which undertook a long and perilous annual journey to breed at high latitudes.

As we have seen, incubation by a single parent is advantageous to species with a strongly unbalanced sex-ratio if it is combined with habits which permit a fuller participation in reproduction by the more numerous sex, as in polyandrous tinamous and jacanas and polygynous Icteridae and Ploceidae. But an unbalanced sex-ratio is itself a misfortune which can only weaken the monogamous species in which it arises by diminishing its reproductive capacity. Such a disparity in the numbers of the sexes should give an advantage to any subsequent mutation that threw the weight of family cares on the more abundant sex. Thus non-incubation by one of the sexes seems in many instances to be either useless, or if advantageous, it is so merely because it serves to compensate for other genetic accidents which upset the sex-ratio.

It is instructive to compare incubation habits with the use of distraction displays. On the whole, the latter are given only by birds whose nests are so situated that they are likely to be of some use in luring away predators (Armstrong 1954; Skutch 1954). But the male's participation in incubation varies in a largely capricious fashion, so that usually we can point to no good reason for his behaviour in this respect. The reason why distraction displays are usually adaptive and incubation patterns, at least in the matter of the participation of the sexes, appear rarely to be so, seems obvious. It is not beyond the mental powers of birds to learn by experience that they can entice an enemy from the nest by a slow and conspicuous retreat in front of the animal; but we can hardly expect them to understand the relation between temperature and the development of an embryo unseen within the shell. Hence birds' intelligence has had some influence upon the evolution of distraction displays by the method of organic selection; whereas, since in many environments the participation of both sexes in incubation is not essential, incubation patterns have been free to vary more or less at random, without the initiative of mind nor close control by natural selection.

Still, as ornithologists we have no reason to regret the accidental loss of the incubation habit by so many male birds, because it has added greatly to the diversity of bird behaviour and consequently to the fascination of our studies. Whether, as Selous thought, there is greater charm in watching the female build and incubate while her mate sings close by, or in watching the two take equal shares in all the activities of the nest, is a question of taste which admits no argument. We should be grateful that both arrangements are available for our contemplation. Moreover, the release of the male from incubation, and finally from all association with the nest, opened the way for the evolution, largely under the influence of sexual selection, of marvellous courtship habits and wonderful adornments, which in some instances have become so lavish that the bird who wears them is physically incapacitated for attending eggs or young.

SUMMARY.

A classification of the known methods of incubation of the birds of the world, on the basis of the division of labour between the sexes and the apportionment of time on the eggs, shows that practically every feasible pattern has been adopted by one species or another.

A review of the evidence suggests that incubation by both sexes was the primitive method among birds.

Among passerines, there appears to be little correlation between incubation by the male and his possession of a brood-patch. Some males which regularly incubate lack it, whereas others which fail to incubate develop it.

There is also poor correlation between the form of the nest (whether open or covered), the coloration of the two sexes, and their participation in incubation.

Participation of the sexes is not generally correlated with the environment. The non-incubation by the female Emperor Penguin and the sharing of incubation by the male Gould's Hummingbird, seem to be instances of response to rigorous climates.

Except possibly in migratory birds whose breeding season is short, incubation by the male seems compatible with efficient territorial defence. Some males and even females sing loudly from the nest, apparently without betraying it to predators.

When the sex-ratio is strongly unbalanced, the failure of the less numerous sex to incubate makes it possible for a larger proportion of the adults to reproduce. But in a number of species in which polygamy or polyandry occurs, the sex-ratio is not known to be unbalanced.

Many modifications in the incubation pattern, especially those involving the participation of the sexes, appear to be non-adaptive, and to persist because they do not decrease reproductive efficiency.

REFERENCES.

ADDICOTT, A. B. 1938. Behavior of the Bush-tit in the breeding season. Condor 40 : 49–63.

ALLEN, R. W. & NICE, M. M. 1952. A study of the breeding biology of the Purple Martin (*Progne subis*). Am. Midl. Nat. 47 : 606–665.

AMADON, D. & ECKELBERRY, D. R. 1955. Observations on Mexican Birds. Condor 57 : 65–80.

ARMSTRONG, E. A. 1952. The behaviour and breeding biology of the Shetland Wren. Ibis 94 : 220–242.

ARMSTRONG, E. A. 1954. The ecology of distraction display. Brit. J. Anim. Behav. 2 : 121–135.

ARMSTRONG, E. A. 1955. The Wren. London.

BAILEY, R. E. 1952. The incubation patch of passerine birds. Condor 54 : 121–136.

BEEBE, W. 1925. The Variegated Tinamou *Crypturus variegatus*. Zoologica 6 : 195–227.

BENT, A. C. 1940. Life histories of North American cuckoos, goat-suckers, humming-birds, and their allies. U. S. Nat. Mus., Bull. 176.

BENT, A. C. 1950. Life histories of North American wagtails, shrikes, vireos, and their allies. U.S. Nat. Mus., Bull. 197.

BLAIR, H. M. S. 1936. On the birds of east Finmark. Ibis (13) 6 : 280–308, 429–459, 651–674.

CONDER, P. J. 1948. The breeding biology and behaviour of the Continental Gold-finch *Carduelis carduelis carduelis*. Ibis 90 : 493–525.

DAVIS, D. E. 1940. Social nesting habits of the Smooth-billed Ani. Auk 57 : 179–218.

DAVIS, D. E. 1945. The occurrence of the incubation patch in some Brazilian birds. Wilson Bull. 57 : 188–190.

DELACOUR, J. 1951–52. The Pheasants of the World. 2 Vols. London & New York.

ERICKSON, M. M. 1938. Territory, annual cycle and numbers in a population of Wren-tits (*Chamaea fasciata*). Univ. Calif. Publ. Zool. 42 (5) : 247–334.

FORBUSH, E. H. 1929. Birds of Massachusetts and other New England States 3. Boston.

FRIEDMANN, H. 1929. The Cowbirds. Springfield & Baltimore.

FRIEDMANN, H. 1932. The parasitic habit in the ducks, a theoretical consideration. Proc. U.S. Nat. Mus. 80 (18) : 1–7.

GOODWIN, D. 1948. Incubation habits of the Golden Pheasant. Ibis 90 : 280–284.

GOODWIN, D. 1953. Observations on voice and behaviour of the Red-legged Partridge *Alectoris rufa*. Ibis 95 : 581–614.

GRINNELL, L. I. 1944. Notes on breeding Lapland Longspurs at Churchill, Manitoba. Auk 61 : 554–560.

GULLION, G. W. 1954. The reproductive cycle of American Coots in California. Auk 71 : 366–412.

HOFFMANN, A. 1949. Ueber die Brutflege des polyandrischen Wasserfasans *Hydro-phasianus chirurgus* (Scop.). Zool. Jb. 78 : 367–403.

HUXLEY, J. S. 1938. Darwin's theory of sexual selection and the data subsumed by it in the light of recent research. Am. Nat. 72 : 416–433.

HUXLEY, J. S. 1942. Evolution, the modern Synthesis. London.

IVOR, H. R. 1944. Bird study and semi-captive birds : the Rose-breasted Grosbeak. Wilson Bull. 56 : 91–104.

KENDEIGH, S. C. 1952. Parental care and its evolution in birds. Illinois Biol. Mon. 22.

LACK, D. 1953. The Life of the Robin. Rev. ed. London.

LASKEY, A. R. 1944. A study of the Cardinal in Tennessee. Wilson Bull. 56 : 27–44.

LAWRENCE, L. de K. 1947. Five days with a pair of nesting Canada Jays. Canadian Field-Nat. 61 : 1–11.

LAWRENCE, L. de K. 1949. The Red Crossbill at Pimisi Bay, Ontario. Canadian Field-Nat. 63 : 147–160.

LAWRENCE, L. de K. 1953. Nesting life and behaviour of the Red-eyed Vireo. Canadian Field-Nat. 67 : 47–77.

LEWIS, H. F. 1921. A nesting of the Philadelphia Vireo. Auk 38 : 26–44, 185–202.

LOCKLEY, R. M. 1942. Shearwaters. London.

LORENZ, K. Z. 1952. King Solomon's Ring. London.

LOW, J. B. 1945. Ecology and management of the Redhead, *Nyroca americana*, in Iowa. Ecol. Mon. 15 : 35–69.

MANNING, T. H. 1942. Blue and Lesser Snow Geese on Southampton and Baffin Islands. Auk 59 : 158–175.

MAYR, E. 1939. The sex ratio in wild birds. Am. Nat. 73 : 156–179.

MEWALDT, L. R. 1952. The incubation patch of the Clark Nutcracker. Condor 54 : 361.

MOORE, R. T. 1947. Habits of male hummingbirds near their nests. Wilson Bull. 59 : 21–25.

MOREAU, R. E. 1936. The breeding biology of certain East African hornbills. (Bucerotidae). J. E. Africa Uganda Nat. Hist. Soc. 13 : 1–28.

MOREAU, R. E. 1937. The comparative breeding biology of the African hornbills (Bucerotidae). Proc. Zool. Soc. London, 1937 (A) : 331–346.

MOREAU, R. E. 1949. The African Mountain Wagtail *Motacilla clara* at the nest. Ornithologie als biologische Wissenschaft : 183–191.

MOREAU, R. E. & MOREAU, W. M. 1938. The comparative breeding ecology of two species of *Euplectes* (Bishop Birds) in Usambara. J. Animal Ecol. 7 : 314–327.

MOREAU, R. E. & MOREAU, W. M. 1940. Hornbill studies. Ibis (14) 4 : 639–656.

MOREAU, R. E. & MOREAU, W. M. 1941. Breeding biology of the Silvery-cheeked Hornbill. Auk 58 : 13–27.

NICE, M. M. 1929. The fortunes of a pair of Bell Vireos. Condor 31 : 13–18.

NICE, M. M. 1937. Studies in the life history of the Song Sparrow I. Trans. Linn. Soc. N. Y. 4 : i–vi, 1–247.

PARKES, K. C. 1953. The incubation patch of males of the suborder Tyranni. Condor 55 : 218–219.

PEARSON, A. K. & PEARSON, O. P. 1955. Natural history and breeding behaviour of the tinamou, *Nothoprocta ornata*. Auk 72 : 113–127.

PUTNAM, L. S. 1949. The life history of the Cedar Waxwing. Wilson Bull. 61 : 141–182.

RAND, A. L. 1938. Results of the Archbold Expeditions. No. 22. On the breeding habits of some birds of paradise in the wild. Am. Mus. Novit. 993 : 1–8.

RICHDALE, L. E. 1943. . The Kuaka or Diving Petrel, *Pelecanoides urinatrix* (Gmelin). Emu 43 : 24–48, 97–107.

RUST, H. J. 1920. The home life of the Western Warbling Vireo. Condor 22 : 85–94.

SCHÄFER, E. 1953. Estudio bio-ecológico comparativo sobre algunas Cracidae del norte y centro de Venezuela. Bol. Soc. Venezolana Cien. Nat. 15 : 30–63.

SCHÄFER, E. 1954a. Der Vogel mit dem Stein auf dem Kopf. Kosmos 50 : 9–13, 118–124.

SCHÄFER, E. 1954b. Zur Biologie des Steisshuhnes *Nothocercus bonapartei*. J. Orn. 95 : 219–232.

SCHÄFER, E. 1954c. Sobre la biología de *Colibri coruscans*. Bol. Soc. Venezolana Cien. Nat. 15 : 153–162.

SCHÜZ, E. 1943. Brutbiologische Beobachtungen an Staren 1943 in der Vogelwarte Rossitten. J. Orn. 91 : 388–405.

SIBLEY, C. G. 1946. Breeding habits of megapodes on Simbo, Central Solomon Islands. Condor 48 : 92–93.

SIMMONS, K. E. L. 1954.   The behaviour and general biology of the Graceful Warbler *Prinia gracilis*.   Ibis 96 : 262–292.
SKEAD, C. J. 1954.   A study of the Cape Wagtail *Motacilla capensis*.   Ibis 96 : 91–103.
SKUTCH, A. F. 1935.   Helpers at the nest.   Auk 52 : 257–273.
SKUTCH, A. F. 1953.   The White-throated Magpie-Jay.   Wilson Bull. 65 : 68–74.
SKUTCH A. F. 1954.   Life histories of Central American Birds.   Pacific Coast Avifauna 31 : 1–448.
SKUTCH, A. F. 1954–55.   The parental stratagems of birds.   Ibis 96 : 544–564, 97 : 118–142.
SLADEN, W. J. L. 1953.   The Adelie Penguin.   Nature 171 : 952.
SMITH, S. 1950.   The Yellow Wagtail.   London.
SPINGARN, E. D. W.   1934.   Some observations on the Semipalmated Plover (*Charadrius semipalmatus*) at St. Mary's Islands, Province of Quebec, Canada.   Auk 51 : 27–36.
STODDARD, H. L. 1946.   The Bobwhite Quail.   New York.
STONEHOUSE, B. 1953.   The Emperor Penguin *Aptenodytes forsteri* Gray.   I.   Breeding behaviour and development.   Falkland Islands Dep. Surv. Sci. Rep. 6 : 1–33.
SUTTON, G. M. & PARMELEE, D. F. 1954.   Nesting of the Snow Bunting on Baffin Island.   Wilson Bull. 66 : 159–179.
SUTTON, G. M. & PARMELEE, D. F. 1955.   Summer activities of the Lapland Longspur on Baffin Island. Wilson Bull. 67 : 110–127.
TINBERGEN, N. 1939.   The behaviour of the Snow Bunting in spring.   Trans. Linn. Soc. N. Y. 5 : 1–94.
TINBERGEN, N. 1953.   The Herring Gull's World.   London.
VAN FLEET, C. C. 1919.   A short paper on the Hutton Vireo.   Condor 21 : 162–165.
WALLACE, A. R. 1871.   Contributions to the Theory of Natural Selection.   2 ed. London & New York.
WALLACE, A. R. 1872.   The Malay Archipelago.   4 ed.   London.
WARREN, O. B. 1899.   A chapter in the life of the Canada Jay.   Auk 16 : 12–19.
WEAVER, R. L. & WEST, F. H. 1943.   Notes on the breeding of the Pine Siskin. Auk 60 : 492–504.
WESTON, H. G., Jr. 1947.   Breeding behavior of the Black-headed Grosbeak. Condor 49 : 54–73.
WILLIAMS, L. 1952.   Breeding behavior of the Brewer Blackbird.   Condor 54 : 3–47.
WITHERBY, H. F. *et al* 1938.   The Handbook of British Birds.   London.
WYNNE-EDWARDS, V. C. 1952.   Zoology of the Baird Expedition (1950).   I. The birds observed in central and south-east Baffin Island.   Auk 69 : 353–391.

16

Reprinted from *CSIRO Wildl. Res.* 2(2):101–110 (1957)

EXPERIMENTS ON THE CONTROL OF TEMPERATURE IN THE MOUND
OF THE MALLEE-FOWL, *LEIPOA OCELLATA* GOULD (MEGAPODIIDAE)

By H. J. FRITH*

[*Manuscript received July 22, 1957*]

Summary

The mallee-fowl, *Leipoa ocellata* Gould, is shown to be capable of detecting
artificially induced changes in the internal temperature of its incubator mound,
and of altering its behaviour to deal with them. It is suggested that the tongue is
the organ involved in the perception of mound temperature, and that the temperature-regulation behaviour is based on certain responses.

I. INTRODUCTION

The mallee-fowl, *Leipoa ocellata* Gould, tends its nesting mound throughout
the breeding season, and regulates the temperature of the egg chamber. There
are eggs present in this chamber from September to March or April, and during
this period there are marked seasonal variations of air and soil temperatures which
affect the mound from the outside, as well as changes within it due to the first
increasing, then declining, heat output from the fermentation of the buried organic
matter. The mallee-fowl is capable of balancing these two changing sources of
heat so that the eggs are maintained at a near-constant level. The behaviour pattern
necessary for this regulation has been seen to be sufficiently adaptable to deal with
unusual air temperature conditions and unseasonal weather (Frith 1956). In 1955
some experimental work was undertaken in an attempt to determine the factors
which governed the temperature-regulation behaviour of the mallee-fowl.

II. METHODS

Two nearby mounds similar in size were selected. In one a 50-ft coil of resistance
cable was buried 12 in. below the egg chamber whilst the mound was being prepared
for the approaching breeding season. The cable was connected to a mobile 240 V
generating plant placed 300 yd from the mound. This equipment was capable of
causing a measurable increase in temperature in the egg chamber within 10 min
of being switched on, and thereafter of raising its temperature up to 20°F an hour.
This mound will be referred to as the heated mound.

At the other mound, when two eggs had been laid the fermenting organic
matter was removed and replaced by mallee sand. This was covered and disguised
by a small amount of organic matter; the eggs were then returned and the mound
restored to its original contour. As this treatment was carried out in spring it
resulted in the removal of the chief source of heat for that time of the year. This
mound will be referred to as the cooled mound.

One bulb of a Negretti and Zambra recording soil thermograph was buried
in the egg chamber of each mound, and the other bulb of each was buried in the

* Wildlife Survey Section, C.S.I.R.O., Canberra.

soil 6 in. below the surface and close to the mound. Air temperatures were recorded 3 ft from the ground.

As only two soil thermographs were available it was not possible to record continuous temperatures in a control mound, but observations and spot temperature readings at eight other mounds in the same paddock were used. as controls in interpreting the results in the heated mound. The heated mound itself was used as control for experiments with the cooled mound. Hides were erected and the birds watched at the experimental mounds. General observational data reported in this paper were collected at other mounds and in other years.

III. RESULTS

(a) Control of the Mound

Table 1 shows the average time per day spent at a mound by the male and female. The data are confined to those occasions on which the mound was watched from before dawn until after dusk.

TABLE 1

ACTIVITY AT THE MOUND IN DIFFERENT SEASONS (1953–55)

Season	Number of Observations	Time per Day Spent at Mound (hr)		Time per Day Digging at Mound (hr)	
		Male	Female	Male	Female
Spring	7	7·0	4·2	3·0	1·0
Summer	10	13·0	3·1	6·2	1·2
Autumn	12	10·0	6·5	7·3	4·3
Mean	—	10·0	4·6	5·3	2·2

It is clear from Table 1 that the female spends much less time throughout the season working at the mound than does the male. The disparity is greatest in spring and summer when the female is burdened with egg production. The time-table of activity at the mound also varies greatly between the sexes. The male usually arrives at the mound alone at dawn, but he is sometimes accompanied by the female. The female is seldom seen to visit the mound alone. The male, who invariably begins the work, never loses any time, but the female frequently does— she may first move about feeding and working spasmodically for up to one hour before beginning constant and effective digging. Even then her activities are often misdirected, and frequently she has been watched kicking soil back into the hole in the bottom of which the male is steadily digging it out. Frequently during the course of the work she wanders away and sometimes does not return.

The work completed, the female, if still present, departs and is not seen again; but the male stations himself nearby and remains there throughout the day ready to deal with chance changes in weather. His actions on such an occasion were observed at midday on March 10, 1953, when the sky clouded over and a storm threatened after the mound had been opened for sunning purposes; immediately the shadow fell on the nest the male left his station and mounded up the excavation. He is also ready to deal with other eventualities, such as the repairing of any damage to the soil cover of the eggs, even in the presence of man (Frith 1956, p. 89). Plate 1, Figure 1 illustrates this response on the part of a normally shy and elusive bird.

A further example of the male's control of the mound and its temperature was seen on December 4, 1955. On this occasion an egg was due to be laid but the day dawned dull, cloudy, and unsettled. At first light, 4.30 a.m., the male walked onto the mound, scratched into the surface for a few inches, and then retired. The female was present, but did not dig and finally wandered away. At 8 a.m. the clouds began to clear and bright sunshine fell on the mound. At 8.06 a.m. the female returned to the mound, and both birds began to open it rapidly. At 8.30 a.m. the rift in the clouds closed and the sunshine disappeared; at this moment the female was 50 yd from the mound. As the sunshine faded the male, with wings and crest raised, rushed at the female attacking her and driving her off. He then returned and refilled the partly opened mound. The female did not return that day, and presumably the egg was laid elsewhere. This behaviour is similar to that of *Alectura* where the female is not allowed near the mound except when an egg is to be laid (Fleay 1937). In normal circumstances the female mallee-fowl is allowed to participate in the maintenance of the mound; but at critical moments, when its internal temperature may be jeopardized, the male assumes complete control and will even drive her away. On December 4 continued exposure of the interior of the mound would have meant a decrease in its temperature, and presumably this consideration led to the attack on the female.

(b) Temperature Regulation

The responses stimulated by the artificial manipulation of mound temperature can only be interpreted by reference to the normal behaviour of the birds in the different seasons of the year. In the spring, fermentation of the organic matter produces more than enough heat to maintain the egg chamber at the required temperature. The mound is opened quickly, in the early morning. On refilling, the cool sand from the surface is mixed with the warm sand from the deeper layers. When fermentation is at its height, the mound is opened daily; when it begins to slow down, the attendance of the birds becomes less regular.

In the autumn, when the fermentation heat has become inadequate, the mound is opened for an extended period in the middle of the day. A wide, saucer-shaped depression is dug out, leaving the eggs covered by only 2–3 in. of sand. The frequency of the opening depends on conditions, and varies from mound to mound.

In midsummer the mound is increased in height, to prevent too much solar heat reaching the egg chamber. Attendance by the birds is relatively infrequent, and the opening of the mound is probably done mainly to test the internal temperature. though it will have the effect of reducing the temperature gradient that has become established. "Mounding up" is also the normal response to unseasonable cold temperatures.

(i) *Cooled Mound*

The organic matter was removed from the cooled mound on September 13, 1955. At this time openings typical of spring were being performed by the birds. The removal of the organic matter led to an immediate reduction in mound temperature from 92°F to 60°F.

On the following day the springtime cooling activities were abandoned. The birds altered their behaviour completely and began the typical autumn-type openings designed to increase the amount of solar heat reaching the eggs. They did not visit the mound until 10 a.m. and then, having opened it, allowed the interior to remain exposed to the sun until 4 p.m. When the mound was restored the height, when fully heaped, was reduced from 20 in. to 6 in. All other mounds at this time were maintained 18–24 in. above ground level. On September 15 an egg was laid; but whereas the previous eggs had been laid 12 in. from the mound surface, on this occasion the birds dug to 30 in. before the female laid—presumably they were searching for warmer soil.

The autumn-type openings were continued throughout the breeding season, but without the assistance of organic heat they were not capable of raising the temperature above 85°F except for one period of very hot weather in January, when the mound temperature reached 92°F. When this happened the mound was built up to 18 in., but when the internal temperature was not maintained it was reduced to 6 in. and the autumn-type openings were resumed.

Figure 1 shows the temperature of the egg chamber of the cooled mound during the period September 6–17, 1955, and also the effect on it of the sudden change in the birds' activities. Clearly the birds could detect the unsatisfactory temperature conditions within the mound and were able to alter their behaviour in an attempt to deal with it. The failure to secure the desired temperature emphasizes the importance of the organic matter as a source of heat for much of the breeding season.

(ii) *Heated Mound*

Spring.—Figure 2 shows the temperature conditions within the egg chamber of the heated mound during a typical experiment in October 1955, when the birds were behaving as is normal for spring. The male visited the mound at dawn on October 21, but did no digging; the internal temperature was satisfactory. At 10 a.m. the artificial heat was switched on and the egg chamber temperature was raised to 116°F. The following day the male came to the mound at 7 a.m. and immediately opened the egg chamber and allowed it to cool to 88°F. During the night October 22–23 the egg chamber temperature increased sharply owing, no doubt, to conduction from the lower layers of the mound that had been artificially heated. At 5 a.m. on October 23 the male again opened the mound and reduced the temperature to 86°F. Although complete digging-out on consecutive days is common in late August and early September, when the fermentation of the organic matter is at its maximum, it is most unusual in October, when the fermentation heat output has declined.

On October 24 the egg chamber temperature was 92°F. The bird visited the mound at 6 a.m. and dug about 1 ft into it, then refilled it, and remained in the vicinity until midday; he did no sustained digging. At 11 a.m. the mound was

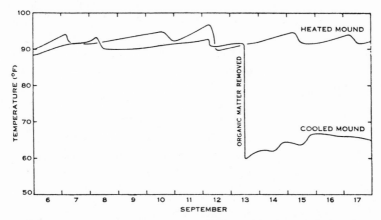

Fig. 1.—Temperature in the egg chamber of the cooled mound during the period September 6–17, 1955.

artificially heated to 102°F. At dawn on October 25 the male appeared but did no digging; at 10 a.m. the egg chamber temperature was further increased to 109°F. This was done in the presence of the male, but he made no response. At 7 a.m.

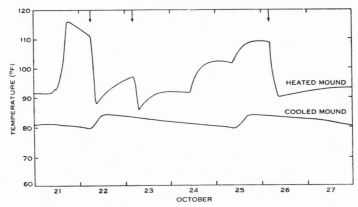

Fig. 2.—Temperature in the egg chamber of the heated mound and the cooled mound during the period October 21–28, 1955. The arrows indicate occasions on which the bird opened the heated mound.

on October 26 he appeared, opened the mound, and reduced the temperature to 90°F.

Several similar experiments were conducted during October and early November, with similar results. In all, the temperature in the egg chamber was raised on 13 occasions, but on only nine was it followed immediately by digging-out by the male. Clearly the bird could not detect unsatisfactory conditions without digging.

Insufficient data are available to generalize, but it was noted that when the interior was heated only moderately, i.e. to about 100°F. no alteration in behaviour followed. The mound was opened every two or three days following such moderate heating, as it had been before. However, on the five occasions on which the temperature was raised to 120°F or more, after the initial opening by the male bird, he invariably returned for several consecutive days and opened the egg chamber at least partly, until he had apparently satisfied himself that satisfactory conditions had been restored.

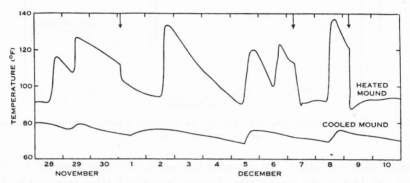

Fig. 3.—Temperature in the egg chamber of the heated mound and the cooled mound during the period November 28 to December 10, 1955.

Summer.—Figure 3 shows the temperature of the egg chamber of the heated mound during a typical experiment in the November–December period. At the end of November the organic matter was providing very little heat. The mound was heaped to a height of 24 in., and the bird visited it every fourth day to open it up, with no immediate effect on its temperature.

The bird did not appear at the mound on November 28 or 29, when the temperature was artificially raised to 126°F. As was expected, he visited the mound on the morning of December 1, beginning his normal opening-up procedure just before dawn and bringing the temperature down rapidly to 104°F. On reconstituting the mound, the bird increased its height by 4 in. The bird attended the mound all day on December 2, did not open it, but increased its height by another inch.

At midday on December 2 the heat was switched on and the temperature raised to 133°F. By December 5 it had cooled markedly, and the temperature was again increased—to 120°F. The male visited the mound at dawn on December 6 (i.e. still maintaining his four-day routine) and increased its height by 1 in., but did not open it up.

At midday on December 6, when the bird had gone, the mound temperature was raised to 124°F. On December 7 the male appeared at dawn, and after a little scratching and probing with his bill on the surface of the mound, he opened it rapidly. reducing the temperature to 92°F. On December 8 he reappeared but did no work on the mound: the temperature of the egg chamber that day was 93°F.

As soon as the bird had left. the heat was turned on and the temperature raised to 136°F. At dawn on December 9 he came again, and after walking over the mound he immediately opened it. reducing its temperature to 87°F (though it subsequently rose again by conduction from below to 92°F). He also increased the height a further 2 in. On December 10, when the internal temperature was 92°F. the bird did no work on the mound.

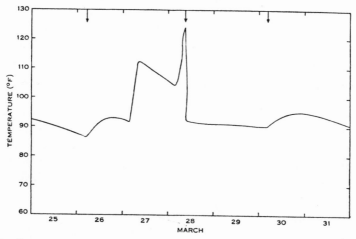

Fig. 4.—Temperature in the egg chamber of the heated mound during the period March 25–31, 1956.

It is impossible to say whether the opening of the mound on any of the occasions observed was done in response to the artificial heating. The raising of the temperature in summer, however, undoubtedly led to a rapid increase in the height of the mound.

Autumn.—Abnormally heavy rains interfered with mallee-fowl breeding at the end of the 1955–56 season. Fortunately, however, some observations had been carried out at the heated mound for a short period in March 1956. The temperature records for the week March 25–31 are shown in Figure 4. The birds, at this time, were opening the mound daily in the heat of the day. in order to increase the amount of solar heat reaching the egg chamber. One such warming occurred on March 26, and was concluded at 5.10 p.m.

At 4.20 a.m. on March 27. the heat was turned on and the mound temperature raised from 91°F to 112°F. At 9.30 a.m. the male arrived as usual and began to open the mound; but when he had dug down about 1 ft he was apparently satisfied, refilled the excavation and departed. This activity had no effect on the temperature.

During the night March 27–28, the mound temperature declined to 105°F. Before daylight on March 28 the heat was turned on and the temperature raised to 124°F. At 9.20 a.m. the male appeared and opened the mound. In a normal autumn-type opening, having reached the egg chamber, he would then have retired and allowed the sun to warm the eggs. On this occasion, however, after much probing with his bill, he refilled the excavation and repaired the mound immediately. His activity was thus altered to a typical spring-type opening, designed to cool the eggs, and resulted in the temperature being reduced to 92°F. The bird next visited the mound on March 30, when the internal temperature had been allowed to fall to 90°F, and carried out a normal autumn-type opening.

(iii) *Temperature Detection*

It is to be inferred from the experiments described in the preceding section that the mallee-fowl is capable of registering differences in the egg chamber temperature. It has not been possible to determine what organ is involved in the temperature measurement, but several observations strongly suggest that it is the tongue, or some part of the mouth cavity.

Whilst the mound is being opened the birds frequently peck the soil rapidly, and drive their bills into the sand as far as the eyes. For a long time it was thought that this action was designed to locate existing eggs, but as it was continued long after the last eggs were laid, and was frequently observed before there were any eggs in the mound, this theory was discounted. With one bird it was possible to watch the procedure very closely from a distance of 6 ft, and there is no doubt that it has nothing to do with locating eggs. Similar actions have been described (Fleay 1937) for *Alectura*.

The bill probing activity was studied from close range and part of it is shown in Plate 1, Figure 2. In carrying it out the bird first made several rapid pecks at the sand with an open bill; next it drove it, still partly open, deeply into the sand. Though the bill was closed when withdrawn, it was immediately opened and seen to be filled with sand which then trickled out.

On several occasions towards midsummer, when the fermentation heat was waning, the males were seen to carry out what amounted to detailed surveys of the temperature of the egg chamber. On opening it sufficiently to expose part of the side wall, the bird then scratched several holes in this and probed thoroughly before refilling that section of the egg chamber and exposing another. The process occupied up to one hour, exclusive of the time taken in the original opening of the mound, and was continued until every part of the egg chamber had been examined. Following these surveys the birds often changed their activity pattern, and a different phase of temperature control began.

IV. DISCUSSION

In a previous paper (Frith 1956) it was shown that the mound-temperature regulating activities of the mallee-fowl were efficient because the changing behaviour pattern of the birds was appropriate for the different seasonal conditions experienced during the incubation period, and also sufficiently adaptable to deal with variations in these conditions. The observations recorded here show that the adaptability of the birds' behaviour is such that they can respond to artificial manipulation of the mound temperature by changing, almost immediately, from one seasonal behaviour pattern to another.

The reaction of the male bird to the artificial heating of the mound in summer was actually inappropriate; for increasing the height of the mound would tend to conserve, rather than help to dissipate, heat reaching the egg chamber from below. This reaction, however, is of considerable interest, for it indicates that the birds have an appreciation of the air temperature and that this plays a part in determining the pattern of activities for a particular season.

For regulating the internal temperature of the mound, the bird has three variables, viz. the time of opening, the length of time the mound is allowed to remain open, and the thickness of the soil cover. In very hot weather and in very cold weather the mound is heaped high, in the former case to reduce the effect of the sun's rays, and in the latter to conserve the internal warmth. In moderate weather, however, it is unnecessary to heap the mound high; but it is opened regularly, and having opened it the bird is capable of determining the internal temperature and of varying its activity accordingly. If the temperature is found to be high, the mound is closed again after some heat has been allowed to escape; but if it is low the mound is allowed to remain open during the heat of the day. When the air temperature is high and the internal temperature is found to be high also, in addition to closing the mound the soil layer is increased in thickness.

Set out in these terms, the temperature regulating activities of the mallee-fowl can probably be reduced to a fairly simple system of responses. High air temperatures stimulate the response to increase the height of the mound; and moderate air temperatures stimulate the response to open the mound. The opening of the mound then stimulates the bill-probing activity for internal temperature detection: a high internal temperature leads to immediate closure of the mound, and a low internal temperature to delayed closure.

While it is probable that the mound-temperature regulating activities of the mallee-fowl are based on a limited number of responses, it is clear that other factors must have had a part in the development of the complete behaviour pattern. Some sort of appreciation of the rhythm of the seasons, as periods in which a certain pattern and time-table of activity can be stabilized, is suggested by the following observations—and it is possible that others with a similar implication could be cited. The frequency with which the birds visit their mounds varies with the season; at certain times of the year the male mallee-fowl seems to know that the mound can safely be left unattended for days on end. The type of opening of the mound characteristic of spring differs from the autumn opening not only in the time of day at which it is carried out and the period during which the mound is left opened up, but also in the way the digging-out is done: the spring opening takes the form of a relatively narrow conical pit, while in the autumn a broad saucer-shaped depression is excavated, and the birds start their work by scratching out the sand on the margin, not digging straight down at the centre as they do for the spring opening. The birds in fact approach the task of opening the mounds in the autumn with a preconception of the nature of the temperature adjustment that is required.

V. Acknowledgment

Mr. G. M. Storr assisted in the field throughout the 1955–56 breeding season and at some of the experiments reported.

VI. References

Fleay, D. H. (1937).—Nesting habits of the brush turkey. *Emu* **36**: 153–63.

Frith, H. J. (1956).—Temperature regulation in the nesting mounds of the mallee-fowl, *Leipoa ocellata* Gould. *C.S.I.R.O. Wildl. Res.* **1**: 79–95.

Explanation of Plate 1

Fig. 1.—Male mallee-fowl, much disturbed by the interference with the internal temperature of the mound by man, attempts to restore it despite his continued presence. (Photograph by E. Slater.)

Fig. 2.—Male mallee-fowl, watched by the female, probes the interior of the mound with his bill to test the temperature. (Photograph by author.)

EXPERIMENTS ON MALLEE FOWL MOUNDS

211

17

Reprinted from *Auk* 45:33–38 (1928)

THE ORIGIN OF HOST SPECIFICITY IN THE PARASITIC HABIT IN THE CUCULIDAE.

BY HERBERT FRIEDMANN.

ONE of the most puzzling features of the parasitic habit as exhibited by the Cuckoos (Cuculidae) is that of host-specificity. In the classic case of the European Cuckoo (*Cuculus canorus canorus*) it is now well established that generally each female deposits all her eggs in nests of a single species. That is, one Cuckoo may parasitize only Meadow-Pipits, another may lay its eggs only in nests of Hedge-Sparrows, while still another may victimize Reed-Warblers exclusively. Each individual has its own particular species of victim to which it generally limits its attention. The species *Cuculus canorus canorus* lays its eggs in the nests of a great number of different kinds of birds, but each individual tends to use the nest of but one kind. The parasitic habit in *Cuculus canorus canorus* may therefore be said to be characterized by *individual* host-specificity. In the Indo-Malayan region there are a great many genera and species of parasitic Cuckoos some of which have carried this specificity to an extreme with the result that the great majority, if not all, of the eggs are laid in nests of a single species or group of allied species. Thus the Indian Koel (*Eudynamis honorata*) lays its eggs wholly in nests of Crows and Jays. In British India it victimizes the Indian Crow (*Corvus splendens*) and the Jungle Crow (*Corvus macrorhynchus*); in Burma it foists its eggs upon the Burmese Crow, (*Corvus insolens*) and the Burmese Jay (*Pica sericea*); in southern China the victim is another Jay (*Graculipica nigricollis*). In large districts in its range practically all the individual Koels victimize the same species of bird. In other words, within each of these districts the individual host-specificity of each individual Koel is the same as that of every other one, and taking into consideration the entire range of the species the number of host species is so small and the species so closely related that the individual host-specificities of all the Koels are very similar. The parasitic habit in *Eudynamis honorata* may therefore be said to be characterized by *specific* host-specificity.

The development of specific from individual host-specificity may readily be accounted for by natural selection operating under conditions which would tend to emphasize the value of small differences. Thus, in the case of *Eudynamis honorata* the bird (and its egg) is too large to be successful with small fosterers. The Crows are everywhere common and their nests open and plainly visible and the birds (and their eggs) fairly close in size to the Koel's. An abundant, accessible group of species being everywhere available, the individual Koels having Crows as their individually specific hosts would rapidly increase and gradually eliminate their less successful fellows that depended on more precarious and more uncertain specific hosts. In time the entire membership of the species *Eudynamis honorata* would be composed of individuals parasitic on Crows.

The real problems, then, are those dealing with the origin and inheritance of individual host-specificities. This paper has to do only with the origin and not with the mode of transmission of the individual host-specificity.

During 1924 and 1925 I had the good fortune to be in Africa on behalf of the National Research Council on a special mission to study the parasitic habits of the Cuckoos, Honey-guides, and Weaver-birds with the object of comparing the parasitic habit in these groups with the similar habit in the Cowbirds of the New World. The question of host-specificity early entered into the study in-as-much as it is one of the most noticeable differences between the reproductive habits of the Cuckoos and the Cowbirds, and many and protracted observations were made in many parts of Africa in the hope of obtaining a clue as to the significance and origin of this peculiarity.

My experience with the parasitic Cuckoos of Africa is limited to the species of the following genera:—*Cuculus, Clamator, Chrysococcyx* and *Lampromorpha.* There are two other genera that are probably parasitic—*Pachycoccyx* which is extremely rare and known from only a few specimens, and *Cercococcyx,* which is also rare and local but not as rare as the other. Chapin indeed has found a young *Cercoccocyx mechowi* being cared for by an adult *Turdinus fulvescens.* Of the four genera studied in the field practically all the component species were observed. No two genera were found to have originally occupied the same ecological niche and therefore no competition for host-species existed between parasitic Cuckoos of different genera.[1] *Cuculus* is essentially a forest-inhabiting genus although one species *C. clamosus* is a bird of the bushveldt, and is found in the same type of country as some species of *Clamator.* The Crested Cuckoos forming the genus *Clamator* are birds of the thorny thickets and the *Acacia-Mimosa* association of the dry savannas. Just as in these larger Cuckoos we find one genus primarily sylvan and the other chiefly in more open country, so too in the smaller metallic Cuckoos the same thing holds true. The Emerald Cuckoos (*Chrysococcyx*) are entirely restricted to dense forests while the Golden Cuckoos (*Lampromorpha*), are birds of the bushveldt and of open farming country that is not entirely devoid of trees. The ecological restrictions are more fully emphasized within the individual species, than in the genera. The ecological factors affecting the ranges and habitats of the various parasitic Cuckoos necessarily limit the number of host-species available to each species of Cuckoo. In the tropics the number of species and of individuals of birds is. very large and the resulting struggle for existence more intense than in the more lenient regions to the north and south. As a result of the keenness of the competition we find that similarity in habits survives side by side only (c̅r̅ at least chiefly) where those habits do not affect the same species. That is, a habit such as the parasitic one, could survive far more easily in many species in the same region if they did not conflict with each other than if all were parasitic on the same group of species. So then, in the bushveldt of Africa we find that the little Golden Cuckoos (*Lampromorpha*) victimize Weaver-birds, Grass-Warblers, and a few other types of birds, chiefly limiting their attention to the Weavers and *Cisticolas.* Most (almost all) of their victims build domed or covered nests, some of them on the ground. In the same districts we find that the Crested Cuckoos (*Clamator*) confine their visitations to open, arboreal nests, such as the Golden Cuckoos never molest. The Cuckoos of the *Clamator* group parasitize Bulbuls (*Pycnonotus, Andropadus,* etc.) almost exclusively although other similar

---

[1] This is true in a general way but exceptions do occur. However such exceptions as are found are not typical and are never very numerous.

nests are occasionally used. The largest species of the genus, *Clamator glandarius,* like the Indian Koel, has taken to laying in the nests of Crows. Bulbuls are undoubtedly too small for a large bird like *C. glandarius* and Crows suit its purpose better. Probably the evolution of its corvine specificity followed much the same lines as suggested above for *Eudynamis honorata.* Environmental influences restrict and limit the number of ecologically available host-species for each particular species of parasite. However, with a fair number of species to choose from there is no environmental reason why a certain individual parasite should further limit its range of activities by tending towards extreme host-specificity. It is not of obvious benefit to the parasite to be still further restricted in this way. It is often and erroneously stated that the degree of specificity attained is an index of the evolutionary excellence of the parasitic habit in any given species. Every field biologist knows that the more generalized, and hence more adaptable, a species remains without suffering evolutionary stagnation, the more successful that species is apt to be. Conversely it is true that the more specialized an organism tends to become the less are its chances for survival in case of environmental change. No parasitic Cuckoo in the world is as successful a species in the full biological sense as is the common Cowbird of South America (*Molothrus bonariensis*). Probably no other parasitic bird has the parasitic habit less well developed, probably none is so eminently an amateurish generalist in its habits as is this Cowbird, yet there is no Cuckoo that is able to maintain its race over so wide an area of the earth's surface with anything like the numerical status of this species. Let those who doubt this journey to Argentina and they will find, as I did, that *Molothrι ɩ bonariensis,* with all its imperfections of reproductive instincts, is second to no other bird in ubiquitousness or general abundance.

The only way to arrive at a proper understanding of the way in which host-specificities might have begun is to study individual birds as well as species. In working on the reproductive habits . birds one of the first things to be determined is the extent and definiteness of the individual breeding territories. Chance and others have done this for the European Cuckoo (*Cuculus canorus canorus*) with splendid results. In the case of the African species of parasitic Cuckoos I found that all of them establish definite breeding territories in which they remain during the egg laying season. The males of some species, such as *Lampromorpha caprius, Chrysococcyx cupreus,* and *Cuculus solitarius,* are very faithful to their territories. The breeding territory in the case of a parasitic bird is based not upon a sufficiency of food for the young but upon an adequacy of nests for the eggs. As stated above the small Golden Cuckoos parasitize Weaver-birds[1] very frequently. A great many species of these Weavers are arboreal and build their nests in large colonies, often as many as a hundred or more nests in a single tree. I found that in several cases a pair of Didric Cuckoos (*Lampromorpha caprius*) had established their territories around trees containing colonies of Weavers and in at least four cases the territories were entirely restricted to single trees. These Weaver colonies very seldom contain more than a single species of Weaver, at least in my experience. In such cases the individual Cuckoos, by restricting their territories to single trees, automatically limit their parasitism to single species. These Weaver colonies are very common all over the African continent south of the Sahara and the Didric Cuckoos are also common and wide-spread. Therefore it seems very likely that individual host specificities are being formed in many individual Cuckoos in the way just mentioned. It is unthinkable to imagine any Cuckoo as originally going around the country-side, inspecting various kinds of nests, making notes of the dietetics of the different species, and then repairing to its favorite perch to cogitate upon its researches and finally decide to limit itself to any one of them. Specificities must have originated unpremeditated and survived because they were convenient. The fact that all parasitic Cuckoos are not specific indicates that some never went through any such experience as the Didric Cuckoo is subject to. Host-specificity is decidedly convenient to Didric Cuckoos fortunate enough to have within their territories whole colonies of suitable nests. Their territorial instincts of defence, like those of most parasitic species are faulty, and if they had to wander far afield in their search for nests the chances are they would not be able to keep any territory for themselves. That is

---

[1] *Ploceus, Hyphantornis, Otyphantes,* etc.

what seems to have taken place in the Indian Koel (*Eudynamis honorata*). In this species individual territories as such seem to exist no longer. Baker (Bull. B.O.C., xlii, March 13, 1922, p. 106) writes that the Koel, " . . . sets all Cuckoo laws in defiance; many birds breed in the same area and even in the same tree; and as many as eleven have been taken together."

It is quite easy to see how individual host-specificity could originate in a Cuckoo in the way that is taking place in the Didric at present, but so far we have absolutely no information as to how (and if) this specificity is inherited. In the case of *Cuculus canorus canorus*, many English workers seem to incline to the idea that it is inherited through the female and that the male has little hereditary influence in this regard. Without wishing to commit myself to any theory on this subject I may say that in the case of the African Cuckoos with which I have had personal experience the fact that the male is far more faithful to tł e territory than is the female, coupled with the fact that the territories in many cases are the apparent predetermining causes of host specificity seems to indicate that the male may have as much influence as the female on the maintenance of the host-specificity in each case. Whether this influence is genetic or merely euthenic I cannot say.

Department of Biology, Amherst College,
    Amherst, Mass.

The Advantage of being Parasitized

by
NEAL GRIFFITH SMITH

Smithsonian Tropical Research Institute,
Box 2072, Balboa, Canal Zone

Host–brood parasite interactions may not always be as they seem.

IT is usually assumed that avian brood parasitism is disadvantageous to the host species. Certainly parasites like cuckoos, some icterids, honey-guides and viduine finches, which place their eggs in the nests of other birds so that the hosts incubate their eggs and raise the parasites, do so often to the detriment of the hosts' offspring[1]. The parasites may destroy the hosts' eggs and/or chicks, or simply outcompete the other chicks for food brought by the hosts. The elaborate mimicry in egg colour[2] and other features[3,4] evolved by many brood parasites seems to be the result of intensive selection by the adult hosts against eggs or chicks unlike their own.

But data presented here indicate that brood parasitism is not always deleterious to the hosts. In certain conditions it is actually advantageous for the host species to accept the parasite's offspring. More host offspring reach breeding age from parasitized nests than from non-parasitized nests.

Scaphidura oryzivora, the giant cowbird, is a parasitic icterid which breeds from south-eastern Mexico to northern Argentina. Its hosts are almost exclusively the colonial-nesting icterids, the oropendolas and caciques. In the Republic of Panama, *Scaphidura* parasitizes at least three species of oropendolas—*Zarhynchus wagleri*, *Psarocolius decumanus*, *Gymnostinops montezuma*—and one species of cacique, *Cacicus cela*.

The normal clutch size in oropendolas and caciques was two, although one and three egg clutches occurred. There were significant differences among the eggs of the four host species in Panama in size, colour and pattern. Within Wagler's oropendola (*Z. wagleri*) and the yellow-rumped cacique (*C. cela*), the species that I studied most, significant variation in egg colour, shape and pattern occurred between colonies. Intracolonial variation was usually slight but important exceptions occurred (discussed later).

The eggs of the parasite *Scaphidura* were also quite variable but, when associated with the colony in which they were laid, it was clear that *Scaphidura* had evolved a rather elaborate egg mimicry (Fig. 1). There were five types of *Scaphidura* females with respect to their eggs:

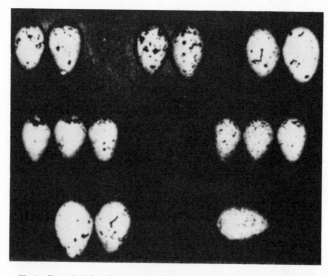

Fig. 1. Eggs of the brood parasite *Scaphidura oryzivora* and those of its hosts. The parasite's egg is on the right of each set which were complete clutches from single nests. Top: eggs from different colonies of *Zarhynchus wagleri* and their mimics; middle: eggs from different colonies of *Cacicus cela* and their mimics; bottom left: *Psarocolius decumanus* and the *Scaphidura* mimic; bottom right: the non-mimetic *Scaphidura* egg laid by a "dumper-type" female. All from the Republic of Panama and the Canal Zone.

three oropendola mimics (*Zarhynchus, Psarocolius, Gymnostinops*), a cacique mimic (*Cacicus*), and a type, which I designated "the dumper", which laid a non-mimetic, generalized icterid egg. The eggs of the first four varied in colour and size depending not only on the host species but also on the particular type of egg associated with any one colony (Fig. 1). Colonies were composed of one species of oropendola and *C. cela*. Thus, only three types of *Scaphidura* females visited any particular colony, the two mimics and the females which laid non-mimetic eggs.

Females who laid mimetic eggs usually placed only one egg in any nest, and then in nests containing only one host egg. Occasionally these females placed their egg in nests with two eggs but never in nests with more. Females of the fifth type ("dumpers" with non-mimetic eggs) laid as many as five eggs in one nest but two or three was the normal clutch. *Scaphidura* females of this group laid in empty nests as well as in nests with full host clutches. They often placed their eggs in nests previously parasitized by mimics but never in nests parasitized by other dumpers. Dumper females were aggressive, moved about a colony in groups, and often drove host females from their nests to parasitize them. Females who laid mimetic eggs were shy, did not move about in groups but skulked around the colony and waited until the host females had departed before attempting to enter a nest (Fig. 2).

Many colony sites were historical, the oropendolas and caciques returned every year in the dry season (January to May). Several colony sites in the Canal Zone were known to have been used for at least 20 yr. During the past 2 yr. two of these sites which averaged 173 nests of *Z. wagleri* and *C. cela* raised an average of 111 *Scaphidura* to fledging each year—an extraordinarily high rate of success for the parasites. I was informed by local people that "black oropendolas" (= *Scaphidura*) have

been produced in such numbers as long as the colony sites had been in existence. From the beginning it was obvious that certain colonies of oropendolas and caciques produced many more *Scaphidura* than other colonies of equal size, and that this was not related to the number of female *Scaphidura* around the colony when the hosts were laying eggs.

I placed model eggs of various types, and eggs of the hosts and those of other birds in the nests at night when the hosts were absent from the site, and observed their reactions to those eggs when they returned in the morning. (Although the bag-like nests of these species were often clustered high in trees, by using ladders and long, telescoping aluminium poles, I was able to remove the nests, examine and manipulate their contents, and to replace with sticky tape or contact cement, more than 1,750 nests in 4 yr.) The results showed that oropendolas and caciques were either "discriminators" or "non-discriminators" regarding eggs unlike their own. Discriminators tossed out non-mimetic eggs, usually within 5 min of returning. Non-discriminators accepted a wide variety of eggs in size, colour and pattern. This was a colony phenomenon, all of the individuals of both oropendolas and caciques reacted one way or the other. Colonies which produced large numbers of *Scaphidura* were composed entirely of non-discriminators. Egg colour variation within non-discriminator colonies was noticeably greater than in discriminator colonies. Discriminator females usually accepted eggs the colour and pattern of which fell within the narrow variation peculiar to their particular colony. Another difference between the two types of colonies was that males from discriminator colonies chased off *Scaphidura* while those from non-discriminator colonies did not.

The mechanism for this dichotomy in the hosts' behaviour will be discussed elsewhere (my yet unpublished experimental studies of adaptations for and against brood parasitism. Pt. 1. *Scaphidura oryzivora* versus the oropendolas and caciques). But the answer to the riddle of why in the first place some colonies discriminated against cowbirds and others did not is to be found by examining the sources of mortality in the hosts.

Oropendolas and caciques often clustered their nests near or even around the huge nests of wasps (*Protopolybia* and *Stelopolybia*) and stingless, but biting, bees (*Trigona*). This was probably very effective protection against vertebrate predators like opossums. Merely pulling on a nest caused the wasps or bees to storm out. At night, a disturbance caused the bees to buzz loudly, and the wasps to make an ominous grating noise in unison against the walls of their nest. Instead of sites with wasps or bees, the oropendolas and caciques chose parasol-like trees and placed their nests at the tips of the branches, making it difficult for a vertebrate predator to reach the nests.

The chief source of mortality is probably not from a vertebrate predator but from an invertebrate parasite—botflies of the genus *Philornis*. (The taxonomy of this group of botflies is poorly known and no specific determinations were made.) These conspicuous, noisy flies place their eggs or living larvae on the chicks. The larvae burrow into the chick's body, eventually crawling out to pupate on the bottom of the nest. Chicks with more than seven larvae usually die.

The pressure exerted by *Philornis* sp. on forty-seven colonies during 4 yr was measured by examining 2,139 chicks of *Zarhynchus, Psarocolius* and *Cacicus* for botlarvae, 3,011 nests for pupae, and by hanging "flypaper" at the colony sites to snare adult flies. In colony sites lacking bees or wasps, host chicks were heavily parasitized and mortality was high. Many adults of *Philornis* sp. were caught on the sticky paper. In colonies with large nests of bees or wasps, host chicks were rarely parasitized (exceptions noted later). No adult flies were caught at these sites.

Fig. 2. A female cowbird about to parasitize an oropendola nest in a "non-discriminator" colony. The female oropendola (right) made no attempt to drive the parasite away.

Large nests of stingless bees offered more "protection" than large or small nests of wasps. Large numbers of bees often hovered in a sphere of space 22 m in diameter around the nest, which was sufficient to cover the entire crown of a big tree (Fig. 3). The nests of oropendolas and caciques were spaced widely. Wasps tended to be either at their nest or away. The birds usually clustered their nests around the wasp colonies. Botflies did occasionally parasitize chicks whose nests were more than 7 m from the wasps (nine out of sixty-seven cases). The exact nature of the anti-*Philornis* protection offered by stingless bees and wasps is not known. I recovered ninety-four fly wings whose veination patterns matched those of some *Philornis* from traps set under three *Stelopolybia* nests for 2 yr. Playback of recorded *Philornis* "buzzing" failed to elicit any exceptional reactions from either bees or wasps. Other factors, perhaps pheromones, may be involved.

But in oropendola and cacique colonies lacking bees or wasps—thus opened to *Philornis*—host chicks from nests containing *Scaphidura* were almost never parasitized (fifty-seven out of 676 individuals, discussed later), while those from nests lacking *Scaphidura* were heavily parasitized (382 out of 424 individuals) and suffered high mortality. Observation of nests, in experimental and natural conditions, containing one or two host chicks and one or two *Scaphidura* chicks showed that the *Scaphidura* chicks actually preened their nest mates, removing any eggs and larvae of *Philornis*. Host chicks did not reciprocate. Stomachs of *Scaphidura* chicks 5 or more days old usually contained the remains of eggs, larvae and even adults of *Philornis*. Stomachs of the host chicks did not; they contained only what the female brought. Adults of *Cacicus* and *Zarhynchus* rarely ate *Philornis* at any stage of its life cycle (three out of eighty-nine stomachs examined).

All fifty-seven cases where host chicks had bot larvae although they had come from nests containing *Scaphidura*

Fig. 3.   Stingless bees swarming amid a colony of oropendolas (female upper right) and caciques.

could be accounted for by two factors. Often there wer more than two chicks in a nest, and sometimes the cow bird could not preen one of the host chicks (the mor distant one). Second, when *Scaphidura* placed its egg(s in a nest containing well incubated eggs, the cowbird(s then hatched much later than the host chicks and wer unable to provide protection until it was too late, or were outcompeted for food by the older host chicks and died.

The incubation period of *Scaphidura* was 5–7 days less than that of its hosts. (The incubation period in *Zarhyn chus*, *Psarocolius* and *Gymnostinops* was 17–18 days and in *Cacicus* was 16 days.) Chicks of *Scaphidura* wer covered with down when they hatched while the host chicks were naked. The eyes of *Scaphidura* chicks opened within 48 h of hatching while those of the host chick required 6–9 days. That *Scaphidura* chicks were hatched with protective down and were so aggressive towards moving objects (when the chicks were at least five days old) probably accounts for the relative scarcity of bot larvae on themselves. In nests containing two *Scaphidura*, the cowbirds preened one another, if they were together. This was probably one of the factors maintaining the dumper-type females in the population—complete freedom from *Philornis* parasitism. This preening behaviour and the active, directed snapping, rather than passive gaping, of *Scaphidura* chicks 5 days old or older was unique for an altricial, passerine bird.

Table 1 shows the most significant points. Nests from oropendola and cacique colonies, lacking bees or wasps, and containing one or two host chicks and one or even two *Scaphidura* chicks, produced in 4 yr more host individuals to fledging age than nests in the same conditions (attack by botflies) containing one, two or even three host chicks but no *Scaphidura*. Thus the probability of bringing off at least one host chick was about three times greater if the other chick(s) in the nest were cowbirds rather than siblings. This was the advantage of having *Scaphidura*, and why, lacking the protection of bees or wasps against botflies, the hosts did not discriminate against the cowbirds. But given protection against botflies by wasps or bees, it was then disadvantageous to have *Scaphidura* in a nest. Nests from colonies with bees or wasps and containing *Scaphidura* produced fewer host chicks to fledging than those lacking *Scaphidura*.

The fledging success figures in Table 1 were directly correlated with the percentage of the host individuals which reached breeding age. Oropendolas and caciques did not usually breed until their second year. Banding data showed that the post-fledging mortality was 63 per cent for the 1964 fledglings and 61 per cent for the 1965 group. There were no significant differences between any of the groups in Table 1 in their post-fledging success. The groups that produced the most fledglings produced the most individuals to breeding age. The necessary circle for selection to work on was thus completed.

While the nature of the advantage of having parasite chicks with host chicks in "non-wasp or bee" situations (thirteen colonies) was fairly clear, the disadvantage of this in "wasp or bee" situations (thirty-four colonies) was less clear but some points may be mentioned. At the beginning of egg laying, the ovary of a typical *Z. wagleri* or *C. cela* individual showed three follicles capable of being ovulated. Usually, after the second egg was laid, the third follicle became atretic. The average clutch size for the two host species where they were not parasitized by *Scaphidura* was 1·80 (1,768 nests) but in parasitized nests was 1·27 (1,502 nests). In situations where cowbirds parasitized nests less than 24 h after the host had laid her first egg, the hosts rarely laid a second, and the two remaining follicles were absorbed. Thus the presence of a parasite's egg often tended to depress the host's clutch size. The aggressive *Scaphidura* nestlings might have outcompeted host nestlings for food brought by the

Table 1. SUCCESS OF HOST BROODS FROM INDIVIDUAL NESTS WITH AND WITHOUT CHICKS OF THE PARASITIC COWBIRD *Scaphidura oryzivora* AND FROM COLONY SITES WITH AND WITHOUT WASPS AND STINGLESS BEES

Degrees of success for hosts	Colonies	
	With bees or wasps no botflies Hosts discriminators	Without bees or wasps botflies present Hosts non-discriminators
Most successful brood composition		
No. 1	3 Hosts, —— fledging success per nest 0·55 No. (%) of such clutches—239 (7%)	2 Hosts, 1 *Scaphidura* fledging success per nest 0·53 No. (%) of such clutches—117 (9%)
No. 2	2 Hosts, —— fledging success per nest 0·53 No. (%) of such clutches—870 (44%)	1 Host, 1 *Scaphidura* fledging success per nest 0·52 No. (%) of such clutches—284 (22%)
No. 3	2 Hosts, 1 *Scaphidura* fledging success per nest 0·28 No. (%) of such clutches—189 (9%)	1 Host, 2 *Scaphidura* fledging success per nest 0·43 No. (%) of such clutches—227 (18%)
No. 4	1 Host, —— fledging success per nest 0·28 No. (%) of such clutches—417 (36%)	2 Hosts, 2 *Scaphidura* fledging success per nest 0·43 No. (%) of such clutches—68 (5%)
No. 5	1 Host, 1 *Scaphidura* fledging success per nest 0·24 No. (%) of such clutches—279 (14%)	1 Host, 3 *Scaphidura* fledging success per nest 0·24 No. (%) of such clutches—239 (19%)
No. 6	1 Host, 2 *Scaphidura* fledging success per nest 0·20 No. (%) of such clutches—59 (3%)	3 Hosts, —— fledging success per nest 0·19 No. (%) of such clutches—19 (2%)
No. 7	2 Hosts, 2 *Scaphidura* fledging success per nest 0·20 No. (%) of such clutches—40 (2%)	2 Hosts, —— fledging success per nest 0·19 No. (%) of such clutches—236 (18%)
Least successful brood composition		
No. 8		1 Host, —— fledging success per nest 0·17 No. (%) of such clutches—87 (7%)
Total No. of colonies	34	13
Total No. of nests with eggs (4 yr)	3,309	1,498
Total No. of nests with clutch data (4 yr)	1,993	1,277
Total No. of nests containing host chicks at late brood stage (4 yr)	1,433	906
Total No. of host chicks fledged (4 yr)	308	289
Average No. of host chicks fledged per nest	0·39 (782 nests)	0·43 (665 nests)
Total No. (%) of nests containing *Scaphidura* with host chicks	567 (28%)	935 (73%)

Hosts were the combined total of *Zarhynchus wagleri* and *Cacicus cela* individuals which nested together in these colonies. Data are for 4 yr in the Republic of Panama and the Canal Zone.

female* Direct evidence was lacking but in some cases the indirect data suggested that this occurred, for example, Table 1, left, No. 5 versus No. 2. I do not understand why this was less marked in the "non-wasp or bee" colonies, for example, Table 1, right, No. 4 versus No. 3. Lack's hypothesis[5] that clutch size is adjusted to the greatest number of offspring that the parent(s) can raise (feed) is pertinent here. Female oropendolas and caciques were capable of feeding and raising three or even four nestlings, but nests containing two nestlings were the most successful and two egg clutches were the commonest. The situation was complicated because body weight of male oropendolas (not caciques) was almost twice that of the females and this weight was attained before leaving the nest. Adult sex ratio in oropendolas was skewed in favour of the females, the normal ratios were between 5 : 1 and 10 : 1. (Adult sex ratio in *Cacicus cela* was usually 2 : 1 in favour of the females.)

* In the oropendolas and caciques studied, only the female built the nest, incubated the eggs and brought food to the nestlings. The less abundant males were promiscuous and copulation occurred normally away from the colony site. The few males at each site moved about the colony defending against other males the particular portion of space that they were in at any one time.

These were the same ratios that occurred at fledging. If sex ratio at hatching was 50 : 50 (I do not know), then the disparity was due to differential mortality in males between hatching and fledging. (There was no evidence which suggested that *Philornis* parasitized males more than females.) If we suppose that the difference in the amount of food necessary to raise a male rather than a female was in proportion to the weight difference, and if food was limiting, then the success of nestlings, host and parasite alike, would have depended heavily on the presence and number of male nestlings in each nest. This aspect remains to be investigated.

The overall picture for 4 yr suggested that the oropendolas and caciques preferred sites with bees or wasps to those without (thirty-four against thirteen). They had protection against botflies and vertebrate predators, and discriminated against *Scaphidura*. But as Table 1 shows, these colonies with bees or wasps were no more, in fact slightly less, successful in bringing up young than those in sites lacking bees or wasps where the colonies accepted *Scaphidura* as their protection against botflies. Twenty-two per cent of the clutches in non-discriminator colonies produced fledglings while 15 per cent survived in discriminator colonies. Thirty-two per cent of the nestlings survived in the former, but only 21 per cent in the latter colonies.

Put another way, why bother to get involved with *Scaphidura* when the hosts could avoid the botfly problem by nesting close to bees or wasps? There were at least three answers. Colonies of wasps or stingless bees often deserted their nest sites for no obvious reason, leaving the discriminating hosts without protection against *Philornis*. Second, the tendency of the hosts to pack their nests around wasp colonies often caused the limb of the tree to snap with a total loss of all nests. Finally, the wasps and bees tended to construct their nests rather late in the dry season. Colonies of oropendolas and caciques historically associated with wasps or bees did not begin to breed until the insects had arrived. The result was that these colonies had a breeding season almost 2 months shorter than those at sites lacking bees or wasps. The first rains were early in 1964 and in 1967, catching many nestlings from such colonies in their nests. Nine colonies were wiped out in 1964 and six in 1967 with a total loss of 187 chicks.

In view of these disadvantages, why did not the hosts switch away from sites with wasps and bees and develop complete commensalism with *Scaphidura*? There were at least several reasons. There was a scarcity of trees which lacked bees or wasps but the physical characteristics of which made the sites relatively predator proof. But the relatively recent planting of royal palms (*Roystonea regia*) and similar palm species in rows for ornamental purposes did provide many potentially "safe" sites. It was very difficult for even flying predators like toucans and owls to grasp the bag-like nests hung from the tips of palm fronds. The disadvantage of palms as colony sites was that the "life expectancy" of an individual leaf was about the same as the breeding time for an oropendola. In very dry periods, palm leaves often dried up earlier and fell, destroying the attached nests. Another factor was that dumper-type females occasionally laid their eggs in nests before the host had laid her first egg. The host female then absorbed her three follicles and incubated the cowbird's eggs. Dumpers preferred the non-discriminator (no wasps or bees) colonies. Thus the balance between hosts and *Scaphidura* in these colonies was very delicate, for there, the probability of a host raising only cowbirds was much greater and the protective advantage in these cases was, of course, nullified.

Tables 1 and 2 show that *Scaphidura* of all types preferred non-discriminator colonies to discriminators (73 per cent versus 28 per cent). This was expected, but why did *Scaphidura* bother at all to penetrate discriminator colonies when "assured" of acceptance in non-dis-

Table 2. SUCCESS OF THE PARASITIC COWBIRD *Scaphidura oryzivora* IN TWO TYPES OF COLONIES OF OROPENDOLAS AND CACIQUES IN THE REPUBLIC OF PANAMA AND THE CANAL ZONE

	Colonies with bees or wasps Botflies absent Hosts discriminate against *Scaphidura*	Colonies without bees or wasps Botflies present Hosts accept *Scaphidura*
Total No. (%) of host nests parasitized	547 (28%)	935 (73%)
Total No. of eggs laid	666	1,708
Average No. of eggs placed in one nest	1·17	1·82
Total No. (%) hatching	559 (84%)	1,263 (74%)
Total No. (%) reaching late brood stage	508 (91%)	795 (63%)
Total No. of *Scaphidura* that fledged	433 (0·76 fledglings per nest)	682 (0·73 fledglings per nest)

Data are for 4 yr.

criminator colonies ? That *Scaphidura* had evolved elaborate egg mimicry indicated that there was a strong advantage to getting into colonies associated with wasps or stingless bees. Once the parasites' eggs had been accepted, the probability of a young *Scaphidura* reaching fledging age was slightly greater in the discriminator colonies (Table 2, 0·76 versus 0·73). Predation from vertebrates and the uncertainty associated with nests attached to palm leaves were certainly principal factors. Also, *Philornis* parasitism did cause some mortality in *Scaphidura*. If a chick was attacked by botflies before it was old enough to protect itself, then it often died. Wasps and stingless bees were useful also to *Scaphidura*. This was counterbalanced in non-discriminator colonies, for in these, nests containing more than one cowbird were frequent (progeny of dumpers), and as already noted, the cowbirds could preen one another.

This complicated association and interaction between several species of oropendolas and caciques, wasps and stingless bees, botflies, predator-proof breeding sites seasonality and parasitic cowbirds probably did not come about all together simultaneously. Without guessing at the steps leading to the present arrangement, it seems clear that the situation is analogous to one of balanced polymorphism. The picture of *Scaphidura* versus its hosts was not one of simple commensalism. The association was distinctly advantageous to the hosts in some conditions and quite disadvantageous to the hosts in others. Advantages and disadvantages for the hosts and parasites alike accrued to each situation. It appears to be the first case, at least in the vertebrates, where parasitism often confers an advantage to the host species.

Scaphidura had a remarkably high breeding success. But no part of the world, not even Panama, is overrun with giant cowbirds. They are rarely seen outside the breeding season and their numbers seem rather stable. They apparently lose badly between leaving the nests and returning for the next season.

The oropendolas and caciques, like many tropical birds, have a low breeding success but apparently long adult lives. Some banded females are known to have attempted breeding every year for the past 9 yr. The number of chances that a female oropendola or cacique has at least to replace herself in the population counterbalances the very low breeding success. The association with *Scaphidura* increases the probability of this success.

Received May 23, 1968.

[1] Hamilton, W. J., and Orians, G. H., *Condor*, **67**, 361 (1965).
[2] Southern, H. N., in *Evolution as a Process* (edit. by Huxley, J.), 219 (Allen and Unwin Ltd. London, 1954).
[3] Nicolai, J., *Zeit. für Tierpsychol.*, **21**, 129 (1964).
[4] Payne, R. B., *Amer. Naturalist*, **101**, 363 (1967).
[5] Lack, D., *Population Studies of Birds* (Oxford University Press, 1966).

19

Reprinted from *J. Embryol. Exp. Morph.* 5(2):293–299 (1957)

A Critical Period for the Turning of Hens' Eggs

by D. A. T. N E W [1]

From the Department of Anatomy and Embryology, University College London

INTRODUCTION

THE artificial incubation of hens' eggs involves four factors—temperature, humidity, air supply, and at intervals a rotation or 'turning' of the eggs. This last factor is perhaps the most curious and unexpected of the four, yet there is no doubt that it is necessary for development as shown by good hatchability, and in the natural state it is carried out by the sitting hen. Eycleshymer (1906), Chattock (1925), and Olsen (1930) have all concluded from observations on the hen's nest that the hen frequently rotates the eggs during the incubation period; Olsen considers it occurs as often as 96 times in 24 hours.

Various abnormalities have been recorded in eggs incubated without turning. Dareste (1891) stated that absence of turning causes the allantois to adhere to the yolk sac; Eycleshymer (1906) confirmed this, and added that during the first week of incubation, absence of turning may also cause the embryo to adhere to the shell membranes. Randles & Romanoff (1950) have observed further effects: in unturned eggs the formation of the albumen sac is delayed and the physical properties of the amniotic and allantoic fluids differ from normal.

Although the relation between frequency of turning and hatchability has been intensively studied (see Landauer, 1951, pp. 54–55 for references), little appears to be known as to the mechanism by which insufficient turning induces abnormalities. A useful line of approach to this question would be to find the relative importance of turning eggs at different stages of the incubation period. If it were known that at a particular stage of development the embryo was exceptionally sensitive to the absence of turning, it might be possible to deduce what the function of turning was from a consideration of the condition of the egg contents at that time; or at least the range of possible explanations might be greatly narrowed. Such few indications as have been obtained previously suggest that turning is more important in early than in late incubation. For example, Byerly & Olsen (1936) conclude that turning in the third week probably has little effect on hatchability, and Card (1926) observed that eggs turned during the first 6 days hatch nearly as well as those turned throughout incubation, but he gives no figures.

[1] *Author's address*: Department of Anatomy and Embryology, University College London, Gower Street, W.C. 1, U.K.

Consideration of the arrangement of the egg contents suggests that a particularly critical time for turning the eggs might be the latter part of the first week of incubation. At this time a large area of chorion lies close to the shell membranes and the layer of albumen between the two has been greatly reduced by a loss of fluid from the albumen to the yolk. Abnormal adhesions between the chorion and shell membranes, therefore, seem a possibility at this stage unless the shell and its membranes are periodically moved relative to the egg contents, i.e. unless the eggs are turned. The present work consists of evidence in support of this explanation. It will be described in two sections, the first indicating that the most important time for turning the eggs is between the 4th and 7th days of incubation, and the second suggesting that turning during this period exerts its marked effect on hatchability by preventing harmful adhesions arising between the chorion and shell membranes.

METHOD AND RESULTS

(a) Hatchability of eggs turned only between the 4th and 7th day

Table 1 summarizes the results of four experiments designed to test the effect on hatchability of turning the eggs only between the 4th and 7th day (inclusive). The eggs were incubated at 39° C. (heated from above), at about 60 per cent. relative humidity, and turned on the days indicated. Turning was carried out once every 12 hours and the eggs rotated by hand approximately 120° about the long axis; they were rotated alternately right and left.

TABLE 1

Hatchability of eggs after various turning treatments during incubation

Experiment No.	Turned 4th–7th day		Turned 8th–11th day		Turned throughout		Unturned	
	Failed	Hatched	Failed	Hatched	Failed	Hatched	Failed	Hatched
1 . .	2	6	5	3	2	6	4	4
2 . .	2	7	7	2	5	4	8	1
3 . .	6	3	9	0	2	7	6	3
4 . .	4	5	8	1	2	7	8	1
TOTALS .	14	21	29	6	11	24	26	9

In each experiment the eggs were divided into four equal groups. The first group was turned between the 4th and 7th day, the second between the 8th and 11th day, the third throughout incubation up to the 18th day, and the fourth was not turned at all. The eggs of the four groups were interspersed to avoid differences arising from microclimates within the incubator. The purpose of including a group turned only between the 8th and 11th day was to find whether any improvement in hatchability of the 4–7 day group was restricted to eggs turned specifically during this period, or whether it resulted from turning them during any 3-day period at about the middle of incubation. Turning of the 'turned-throughout' group was stopped at the 18th day to conform with the normal practice in poultry rearing.

The results of the four experiments added together show clearly that turning the eggs between the 4th and 7th day is similar in its effects on hatchability to turning them throughout the whole incubation period, and turning the eggs only between the 8th and 11th day gives similar results to leaving them completely unturned. Calculation of χ squared values for these two pairs of results also fails to detect significant differences. But χ squared calculated for a comparison of the results of turning 4th to 7th day with those of turning 8th–11th gives a P value of less than 0·001. It seems unquestionable, therefore, that turning the eggs between the 4th and 7th day of incubation gives an improvement in hatchability similar to that obtained by turning the eggs throughout incubation, and the results further suggest that this improvement is limited to eggs which are turned specifically at this stage.

In adding the results of separate experiments together the question of homogeneity arises. The numbers involved in each experiment are here too small to test for homogeneity by the χ squared method. However, if for each treatment the two results are selected which show the greatest difference, it can be shown that the probability of their belonging to different populations is insignificant; e.g. the extreme cases resulting from turning between the 4th and 7th day were 2 failed, 7 hatched; and 6 failed, 3 hatched. The probability of these belonging to the same population is 0·0767.

TABLE 2

Distribution of mortality during incubation

	Infertile	Died 1st week	Died 2nd week	Died 3rd week	Hatched
Turned 4th–7th day .	4	1	2	7	21
Turned 8th–11th day .	7	7	7	8	6
Turned throughout . .	4	5	0	2	24
Unturned . .	6	3	2	15	9

The eggs that did not hatch were opened on the 25th day and the time of death estimated from the extent of embryonic development. Table 2 shows the totals from the four experiments. The division between 'infertile' and 'died 1st week' can only be considered as roughly established since in those cases where the blastoderm died very young it is not always possible to distinguish the egg contents from the infertile condition, particularly after 3 weeks of incubation. The table shows that by far the heaviest mortality of unturned eggs occurs during the third week and it is this mortality which is most affected by turning throughout incubation or from the 4th to 7th day. The results also suggest that turning from the 8th to 11th day reduces the third-week mortality but correspondingly increases mortality in the second week; however, the figures are not large enough to constitute a convincing proof of this.

In those eggs which had died after the 18th day of incubation, malpositions of the chick were fairly common. Of the 26 dead chicks in this group, 6 were found to have the head in the small end of the egg, 4 to have the beak away from the air sac, 3 to have the head under the left wing, and 1 to have the head between the thighs.

(b) Adhesion between the chorion and shell membranes

Statements that insufficient turning leads to adhesions between the chorion (or blastoderm) and shell membranes have frequently been made (e.g. Eycleshymer, 1906) but it is not quite clear what evidence they are based on. The chorion is very close to the shell membranes whether the eggs are turned or unturned, and in the second week of incubation the chorio-allantois adheres to the shell membranes as a normal feature of development. Furthermore, a small proportion of unturned eggs develop successfully and produce normal chicks at hatching. It would need, therefore, fairly carefully controlled observations to be certain that abnormal adhesions were occurring.

With this in mind the following experiment was devised. Two groups of fertile eggs were incubated for 7 days. The first group (11 eggs) was left unturned during this period; the second group (8 eggs) was turned every 12 hours between the 4th and 7th day (inclusive). On the 7th day all the eggs were candled and the position of the embryonic membranes marked by drawing a pencil line on the shell over some prominent blood-vessel (usually the sinus terminalis). Each egg was then rotated through 90° and left in this position for a few minutes. On candling for the second time the embryo was seen to have begun to move back towards its former position at the top of the egg, and the embryonic membranes had correspondingly moved relative to the pencil mark on the shell. By measuring with calipers the distance moved by the membranes and also the circumference of the egg, sufficient data were provided for calculating the angle through which the egg contents had turned. The results are plotted in the accompanying graph (Text-fig. 1), which shows the angle through which the egg contents have rotated towards their former position at a given time after the egg was turned through 90°. It can clearly be seen that in the eggs that were unturned during the period of incubation, movement of the embryonic membranes relative to the shell is very much slower than in eggs that were turned between the 4th and 7th days; in fact two of the 'unturned' group have not moved at all.

The eggs were returned to the incubator and examined again 3 hours later. In all those of the 'turned' group the contents had moved through more than 80°, whilst in over half those of the 'unturned' group the contents had rotated less than 60°. The only explanation that seems reasonable is that turning the eggs between the 4th and 7th day of incubation lowers the resistence to relative movement of the chorion and shell membrane; or in other words, absence of turning encourages adherence between the two.

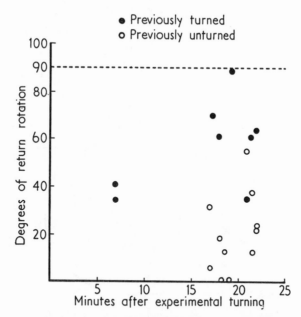

TEXT-FIG. 1. Each egg was turned through 90°. The subsequent internal rotation by which the egg contents returned towards their former position is plotted (in degrees) against time after the whole egg was turned. Of the two groups of eggs shown, one had been systematically turned during incubation, the other had not.

After these observations had been made, incubation of the eggs was continued without any further turning. Of the unturned group (actually turned once through 90° during the experiment on the 7th day) 2 out of 11 eventually hatched. Of the group turned between the 4th and 7th day 6 out of 8 hatched. This lends further support to conclusions reached in the previous section.

DISCUSSION

It is clear that the turning of hens' eggs is particularly important during the latter half of the first week of incubation. If the eggs are turned at this time good hatchability is obtained even if they are left unturned during the remainder of the incubation period. If they are not turned at all during incubation, however, hatchability is low. It has been shown that absence of turning results in an abnormal degree of adhesion between the chorion and shell membranes during the first week of incubation; the probable inference is, therefore, that this abnormal adhesion is closely connected with the causes of the high mortality found in unturned eggs.

Yet there remains the curious fact that, during the second week of incubation, firm adhesion between the chorion (chorio-allantois) and the shell membranes occurs as a normal feature of development. This normal adhesion is evidently primarily related to the respiratory needs of the embryo. It is important that respiratory surfaces should be applied as closely as possible to the shell; in fact it has been shown (Danchakoff, 1917) that the allantoic capillaries actually move to a position outside the chorionic ectoderm.

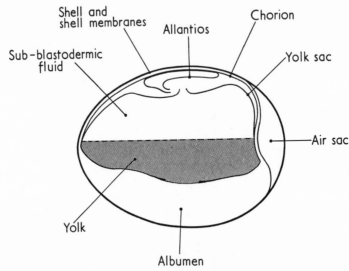

TEXT-FIG. 2. Distribution of some of the egg contents at the 6th day of incubation. Based on a sagittal section of an egg fixed by boiling. (For the sake of clarity the distance between the chorion and shell membrane has been exaggerated.)

It seems, therefore, that turning is not a device to prevent adherence altogether, but to prevent it occurring too soon. Why is this so important? The answer may be in the state of development of the chorion and yolk sac during the first and second weeks. Although these two membranes have practically surrounded the yolk by the 4th day, they continue to expand rapidly until the end of the 6th day (Grodzinski, 1934) owing to the increase of the yolk-sac contents as a result of sub-blastodermic fluid formation. Probably, therefore, any adhesion of the chorion to the shell membrane before the 6th day would hinder its normal development, but after this time would not be so serious.

Part of the mechanism involved in premature adhesion is likely to be the rapid water loss from the albumen which is occurring at this time. Until the end of the first week the albumen is constantly losing water (a) by evaporation from the shell surface, and (b) via the blastoderm to form the sub-blastodermic fluid (New, 1956). Since the albumen layer between much of the chorion and shell membranes has become extremely thin by the end of the first week of incubation and the area of chorion close to the shell membrane has become very large (Text-fig. 2), desiccation of this albumen layer would seem greatly to increase the chances of adhesion. Turning the egg at this stage might well prevent such desiccation by supplying a fresh layer of watery albumen over the dry regions of the membranes.

As mentioned previously a variety of abnormalities have been observed in unturned eggs. The present work suggests, however, that many of these abnormalities are likely to be secondary effects of an initial interference with development occurring between the 4th and 7th day.

SUMMARY

1. Turning hens' eggs between the 4th and 7th day of incubation gives a hatchability similar to that of eggs turned throughout incubation.

2. Turning the eggs between the 8th and 11th day gives a hatchability similar to that of unturned eggs.

3. New evidence is presented that absence of turning leads to abnormal adhesion between the chorion and shell membranes.

4. The mechanism by which turning effects development is discussed.

ACKNOWLEDGEMENTS

I should like to thank Mr. M. Abercrombie for much useful advice and helpful criticism. I am also indebted to Dr. D. E. Barton for examining the statistics and to the Agricultural Research Council for financial support.

REFERENCES

BYERLY, T. C., & OLSEN, M. W. (1936). Certain factors affecting the incidence of malpositions among embryos of the domestic fowl. *Poult. Sci.* **15**, 163–8.

CARD, L. E. (1926). Incubator eggs turned first six days hatch well. *Thirty-ninth Annual Report Illinois Agricultural Experiment Station.*

CHATTOCK, A. P. (1925). On the physics of incubation. *Phil. Trans.* B. **213**, 397–450.

DANCHAKOFF, V. (1917). The position of the respiratory vascular net in the allantois of the chick. *Amer. J. Anat.* **21**, 407–19.

DARESTE, C. (1891). *Recherches sur la production artificielle des monstruosités ou essais de tératogénie expérimentale.* (2nd edn.) Paris (quoted from Eycleshymer, 1906).

EYCLESHYMER, A. C. (1906). Some observations and experiments on the natural and artificial incubation of the egg of the common fowl. *Biol. Bull. Wood's Hole*, **12**, 360–74.

GRODZINSKI, Z. (1934). Zur Kenntnis der Wachstumvorgänge der Area vasculosa beim Hühnchen. *Bull. int. Acad. pol. Sci.* Ser. B, 415–27.

LANDAUER, W. (1951). The hatchability of chicken eggs as influenced by environment and heredity. *Bull. Storrs agric. Exp. Sta.* **262.**

NEW, D. A. T. (1956). The formation of sub-blastodermic fluid in hens' eggs. *J. Embryol. exp. Morph.* **4**, 221–7.

OLSEN, M. (1930). Influence of turning and other factors on the hatching power of hens' eggs. Thesis. Iowa State College (quoted from Landauer, 1951).

RANDLES, C. A., & ROMANOFF, A. L. (1950). Some physical aspects of the amnion and allantois of the developing chick embryo. *J. exp. Zool.* **114**, 87–101.

20

Reprinted from *Compt. Rend. Soc. Biol.* 97:1296–1297 (1927)

CONTRIBUTION A L'ÉTUDE DE L'INCUBATION

R. Lienhart

Habituellement les Oiseaux ne couvent qu'après la ponte en série d'un nombre plus ou moins grands d'œufs ; aussi paraît-il raisonnable de penser que le phénomène d'incubation est déclanché par la fonction ovarienne. Par une série d'expériences, dont je donne ici le résultat, j'ai tenté de mettre en évidence la probabilité de ce fait.

1° Du sérum sanguin de Poule en état d'incubation, obtenu par sacrifice de la Poule et filtration du sang recueilli sur papier filtre stérilisé, est injecté à raison de 5 c.c., par voie sous-cutanée, à une Poule de race commune. Celle-ci entre en incubation dans les quarante-huit heures qui suivent l'injection, et elle accepte les œufs qui lui sont donnés, mais elle les abandonne au bout de huit jours pour reprendre la vie normale.

2° Une deuxième Poule, traitée de la même manière, mais recevant une dose double de sérum, soit 10 c.c., couve pendant vingt et un jours les œufs qui lui sont confiés et élève parfaitement les poussins qui éclosent après cette incubation provoquée artificiellement.

3° 10 c.c. de ce sérum sont injectés a une Poule de race Bresse noire, race qui a la réputation justifiée de ne jamais couver ; ils provoquent l'incubation après vingt-quatre heures. La Poule accepte des œufs, mais elle ne les couve que pendant onze jours.

4° Enfin 10 c.c. de ce même sérum sont injectés à un Coq de race Wyandotte, race très bonne couveuse ; après trente heures, le Coq tombe en incubation et après avoir accepté des œufs, reste en cet état pendant quatre jours, après quoi il reprend sa vie normale (1*).

Cette première série d'expériences paraît bien prouver qu'une substance hormonique ou autre, sécrétée sans doute par l'ovaire et certainement présente dans le sérum sanguin, est la cause déterminante de l'incubation chez la Poule et probablement chez tous les Oiseaux. Il y aurait donc parallélisme entre le phénomène d'incubation chez les Oiseaux et le phénomène de lactation chez les Mammifères.

Mais comment admettre l'action d'une sécrétion ovarienne

(1*) Pour bien m'assurer qu'il ne s'agissait pas d'une pseudo-incubation, conséquence possible du choc provoqué par l'injection de sérum, j'ai injecté à trois Poules différentes 10 c.c. de sérum sanguin de trois Poules distinctes qui ne couvaient pas. Dans les trois cas l'injection n'a aucunement déterminé l'incubation chez les sujets en expérience.

comme déterminant l'incubation chez les Oiseaux dont le mâle partage avec la femelle le soin de couver les œufs, comme c'est le cas pour les Pigeons, les Passereaux et nombre d'autres Oiseaux ? L'observation montre, fait au moins curieux, que chez ces Oiseaux, l'incubation est toujours précédée du phénomène d'abecquement, c'est-à-dire que mâle et femelle, bec à bec, échangent leurs sucs digestifs en des gestes que les poètes ont pris pour de tendres embrassements. Chez les Oiseaux dont la femelle seule couve, comme c'est le cas pour la Poule domestique par exemple, l'abecquement n'a pas lieu. La constance de l'abecquement chez les espèces d'Oiseaux où l'incubation est partagée par le mâle, ne permet-elle pas de penser que l'agent déterminant l'incubation qui existe, comme je l'ai montré, dans les sérums sanguins de certaines couveuses, passe grâce au torrent circulatoire dans toutes les sécrétions de l'organisme ; l'abecquement devenant ainsi un acte physiologique nécessaire chez les Oiseaux dont le mâle couve lui aussi. L'expérience, difficile à réaliser, le montrera peut-être un jour. Ce qui est certain, et tente à prouver que mon hypothèse n'est pas trop hardie, c'est que chez les Pigeons, les Passereaux, et probablement aussi chez d'autres Oiseaux, il arrive parfois, pour des raisons difficiles à préciser, que le mâle acceptant de féconder sa femelle se refuse à tout abecquement avant et après la ponte. Dans ce cas, le mâle ne couve pas et une fois les œufs pondus, contrairement aux habitudes de son espèce, il ne se soucie plus de sa femelle.

(Laboratoire de zoologie de la Faculté des sciences.)

20

CONTRIBUTION TO
THE STUDY OF INCUBATION

R. Lienhart

*This article was translated expressly for this Benchmark volume
by Rae Silver, from* Compt. Rend. Soc. Biol. *97:1296-1297
(1927)*

Usually birds don't incubate until they lay a number of eggs in succession; thus it seems reasonable to think that the phenomenon of incubation is initiated by ovarian activity. In a series of experiments presented here, I tried to provide evidence for this fact.

1. Five cc of the blood serum of an incubating chicken, obtained by sacrificing the chicken and filtering the blood through sterilized filter paper, was injected subcutaneously into a chicken of the same breed. The latter began incubating within 48 hours after the injection, accepted eggs which were given to her, but abandoned them after 8 days.

2. A second chicken, treated with a double dose of serum, that is 10 cc, incubated eggs which were given to her for 21 days and reared the chicks which hatched.

3. 10 cc of the serum was injected into a chicken of the breed "black Bresse", a breed which has the justified reputation of never incubating; incubation was induced after 24 hours. The chicken accepted the eggs but only incubated for 11 days.

4. Finally 10 cc of the serum was injected into a rooster of the breed "Wyandotte", which is very broody; after 30 hours, the rooster began to incubate and once having accepted the eggs, remained in this state for 4 days.*

The first series of experiments seemed to prove that a hormone, secreted without a doubt by the ovary, and present in the blood, is the cause of incubation in the chicken and probably in all birds. Thus there is a parallel between the phenomenon of incubation in birds and the phenomenon of lactation in mammals.

But how can one claim the action of an ovarian secretion as a determinant of incubation in birds in which the male participates with the female in incubating eggs, as is the case for pigeons, passerines and a number of other birds? Observations indicate, curiously enough, that in birds, incubation is always preceeded by billing. That is to say male and female, beak to beak, exchange digestive juices in movements which the poets took for tender embraces. In birds in which the female incubates alone, as is the case for the domestic chicken, billing does not occur. Doesn't the constancy of billing in species of birds where incubation is shared by the male permit one to think that the determining factor for incubation exists in the blood of the incubating individuals and passes through the circulatory system

*To assure myself that it wasn't a case of pseudoincubation as a consequence of the stress provoked by the injection of serum, I injected 3 different chickens with 10 cc of blood serum from 3 chickens which were clearly not incubating. In these 3 cases the injection did not induce incubation.

into all the secretions of the organism; billing thus becomes a physiological necessity in birds in which the male incubates. The relevant experiment may be done one day. What is certain, and tends to show that my hypothesis is not too rash, is that among pigeons, passerines, and probably also in other birds, the male sometimes fertilizes the female but refuses to bill before or after the laying of the clutch. In this case, the male doesn't incubate the eggs nor does he any longer bother his female.

21

Reprinted from *J. Comp. Physiol. Psychol.* **64**:219–224 (1967)

EFFECTS OF LOCAL BRAIN IMPLANTS OF PROGESTERONE ON REPRODUCTIVE BEHAVIOR IN RING DOVES[1]

BARRY R. KOMISARUK

Institute of Animal Behavior, Rutgers University[2]

Progesterone crystals chronically implanted into brains of reproductively experienced ring doves affected 2 components of reproductive behavior differentially, depending on the site of implant. Incubation, a component of parental behavior, in both sexes was induced by implants in the preoptic nuclei and lateral forebrain system, and male courtship, having sexual and aggressive components, was suppressed by implants in these regions and also in the anterior hypothalamus. These behavioral effects were also induced by systemic injection of progesterone. It is proposed that widespread, and partially overlapping, regions of the brain are involved in the control of these behavior patterns, and that these regions may be directly sensitive to progesterone.

In the normal reproductive cycle of the ring dove, courtship behavior of the male stimulates the female to lay eggs, and both then take turns incubating the eggs. Male courtship behavior, dependent on the testes (Carpenter, 1933a, 1933b; Erickson & Lehrman, 1964), declines sharply toward the onset of incubation (Fabricius & Jansson, 1963), and virtually disappears by the time incubation is established (Riddle & Lahr, 1944). Progesterone administration induces both the behavioral changes: suppression of courtship in the male (Riddle & Lahr, 1944) and induction of incubation in both sexes (Lehrman, 1963; Riddle & Lahr, 1944). The purpose of the present study was to determine the loci in the brain at which progesterone exerts these behavioral effects.

METHOD

Subjects

A total of 206 ring doves, each of which had been raised in this laboratory and had reared 1–3 broods of young, were maintained as described by Lehrman and Brody (1964).

Experimental Techniques

Stereotactic methods. Implantations were performed using Equithesin anesthesia (Jensen Salsbery Company, Kansas City, Missouri; 0.25 ml. per 100 gm. bw) supplemented with ether. The head was oriented by first placing the ear pins (short-tipped compass points mounted in brass rods) into the center of the small protuberance of bone just rostral to the posterior angle of the external ear openings, then pivoting the head until the center of the pineal body, made visible by an injection of sesame oil into the porous bones of the skull, was exactly vertical to the center of the ear pins. A pointed rod mounted in the median sagittal plane and in the same horizontal plane as the ear pin centers was then pushed into the skull and tightened in place, providing a firm support of the head with minimal torque. This head holder was used on a modified Hoebel stereotactic instrument (Scientific Prototypes Inc., New York City). A firm gripping surface for the dental cement (Kadon, L. D. Caulk Company, Milford, Delaware) was provided by shaving off the superficial layer of bone above the eyes.

Each cannula (30-g. stainless steel hypodermic tubing, inside diameter 0.009 in.) was filled by tapping the polished tip into solidified progesterone which had been previously heated once in a small crucible until just melted. A small pin vise was used to hold the cannula to be implanted. The cannula was wiped several times with cotton pellets soaked in 70% ethanol, and was then lowered into the brain. Dental cement was allowed to flow around its shaft and into the exposed spongy bone above the eyes. The vise was loosened and removed after the cement hardened, and a second portion of cement was allowed to flow over the top of the shaft and onto the remaining exposed skull. The scalp was then sutured over

[1] Contribution No. 38 from the Institute of Animal Behavior; based on a doctoral dissertation submitted to Rutgers University, 1965. This work was supported by a National Defense Education Act Graduate Teaching Fellowship, by a predoctoral fellowship from the National Institute of Mental Health, and by Research Grant MH-02271 (D. S. Lehrman, principal investigator) from the National Institute of Mental Health, for all of which I express my grateful appreciation. The advice of D. S. Lehrman is gratefully acknowledged.

[2] Newark, New Jersey 07102.

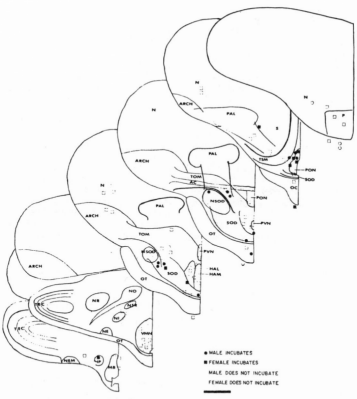

FIG. 1. Effects of progesterone implants on incubation behavior. (Each point represents the tip of one implant. Each of the frontal sections is a projection of points and structures in a 1-mm. thickness of tissue. Abbreviations for the nomenclature, after Huber and Crosby, 1929, and Crosby and Woodburne, 1940, are as follows: AC—anterior commissure; ARCH—archistriatum; HAL—anterior hypothalamus, lateral nucleus; HAM—anterior hypothalamus, medial nucleus; HP—posterior hypothalamic nucleus; MB—mammillary bodies; N—neostriatum; NE—entopeduncular nucleus; NEM—ectomammillary nucleus; NI—nucleus intercalatus; NO—nucleus ovoidalis; NR—nucleus rotundus; NSOD—nucleus of the supraoptic decussation; NSR—nucleus subrotundus; OC—optic chiasm; OT—optic tract; PAL—paleostriatum; PON—preoptic nuclei; PVN—paraventricular nucleus; S—septum; SOD—supraoptic decussation; TEC—optic tectum; TSM—septo-mesencephalic tract; VMN—ventromedial hypothalamic nucleus. The solid line in the legend represents approximately 1 mm.)

most of the assembly.[3] The effect of this procedure was to place into contact with brain tissue a disk of progesterone with a diameter of 0.009 in.

At autopsy, the brains were fixed in 10% formalin without perfusion for several days with the cannula left in place, and were then dissected free, dehydrated for 2 days in methanol, cleared in benzene, and embedded in Tissuemat (Fisher Scientific Company, New York City; melting point of 52.5° C.), sectioned at 25 μ, and stained with

thionin. This procedure permitted histological verification of the implantation sites.

Hormone injections. Progesterone was injected systemically into the pectoral muscle (100 μg. in sesame oil per day) on alternate sides on alternate days. Progesterone used for implantation and for injection was prepared from the same supply.[4]

Behavioral observations. Placed into a cage, 32 × 18 × 14 in., with a bird of opposite sex and a nest containing eggs, each S was observed continuously for the first 10 min.; then, brief obser-

---

[3] I am indebted to Robert Lisk of Princeton University for teaching me the basic stereotactic technique, and to Elizabeth C. Crosby of the University of Michigan for advice on the identification of brain regions.

[4] Progesterone (Lot No. 353) was generously supplied by the CIBA Pharmaceutical Company, Summit, New Jersey, through the courtesy of R. Gaunt.

vations were made at least once per hour between 9:30 A.M. and 6:00 P.M. for a 2-day period.

The criterion for incubation was that on at least one occasion S was fully settled down on the eggs with the wings at the sides of the body, and the ventral skin in contact with the eggs. Courtship behavior in males was measured by the number of bow-coos performed in the first 10 min. of observation. In each bow-coo, the male stands up erect, then "bows" to a horizontal position, emitting a cooing sound.

Experimental design. A total of 104 male and 102 female intact ring doves received progesterone implanted into the brain. After an initial group of 21 males and 17 females showed apparent localization of responsive brain regions, the testing procedure was refined to include control pretests for the remaining 168 Ss. This modification consisted of pretesting each S with systemically administered progesterone to determine which individuals were capable of showing courtship

suppression or incubation in response to injection of this hormone. All Ss were then removed to individual cages and kept in visual isolation for 2½–8 wk. without any treatment and then retested to determine whether the effects of the previously injected progesterone had worn off. Progesterone was then implanted into the brain and the implant test was performed 7 days later.

Excluded from the implant-site analysis were 30 females and 16 males which did not incubate after receiving progesterone systemically, 7 males which incubated without treatment, and 2 males and 8 females whose implants loosened. The implant sites for 79 males and 64 females are presented in Figure 1. In the case of courtship suppression, the implant sites of 86 males are plotted in Figure 2; of the 18 males not included, 7 showed no courtship suppression effect under the influence of systemically injected progesterone, and 11 failed to show courtship after no treatment, when normal courtship was expected.

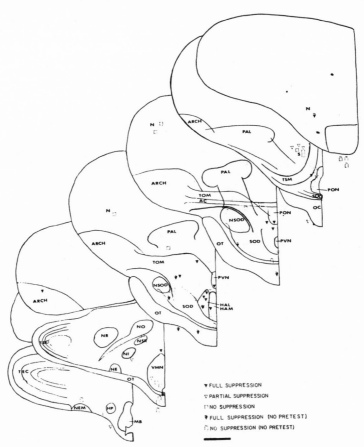

FIG. 2. Effects of progesterone implants on courtship behavior.

236

RESULTS

Effects of Progesterone Injected Systemically

Incubation behavior. When systemically injected with progesterone, 122 Ss (72.2%) sat on the eggs, but when the same individuals were retested without progesterone, only 7 (4.2%) incubated ($\chi^2 = 43.0$, $p < .001$: McNemar's test for correlated observations, Siegel, 1956). The incubation response to progesterone was significantly more frequent in males than in females (67 of 83 males and 55 of 85 females; $\chi^2 = 5.41, p < .02$).

Male courtship behavior. When the males received systemic injections of progesterone, the median number of bow-coos during the 10-min. test was zero (interquartile range: 0–1), but when the same birds were tested without progesterone, the median number of bow-coos was 21 (interquartile range: 8–39). Since 89% of the birds performed 5 or fewer bow-coos after progesterone injection, while 81% performed more than 5 bow-coos without progesterone, we have selected 5 or fewer bow-coos as the criterion for courtship suppression by the progesterone implant. Partial suppression of courtship is defined empirically as a reduction by at

least 40% from the control (no-treatment) level; in these cases, the median reduction was from 55 bow-coos in the control test to 19 bow-coos in the implant test.

Effects of Progesterone Implanted into the Brain

Figures 1 and 2 and Table 1 show a marked difference in the brain sites from which courtship and incubation were affected. In the hypothalamus, courtship was suppressed fully or partially in 10 of 15 males, while in only one female of 24 birds (11 males) with similarly placed implants was incubation induced. Incubation was elicited in 7 (3 males) of 16 birds (9 males) implanted in the preoptic area, however, while with similarly placed implants, 4 of 11 males showed full or partial courtship suppression. These results suggest that regional differentiation exists for the control of incubation behavior, whereas the sites for courtship suppression are more diffuse. The proportion of implants which induced incubation was significantly greater in the preoptic area and lateral forebrain tract region than in all remaining sites combined ($\chi^2 = 20.1$, $p < .001$). The proportion of implants which suppressed courtship was significantly greater in the hypothalamus, preoptic region, and lateral forebrain tract region than in all remaining sites combined ($\chi^2 = 5:3$, $p < .05$). In each part of the brain from which incubation could be elicited, neither sex showed a significantly greater incidence of incubation than the other.

Both behavioral effects were obtained in Ss with implants whose tips ended in the lateral forebrain tracts. These tracts also contain cell bodies scattered sparsely among them. Incubation occurred in birds bearing implants in the supraoptic decussation but not in the hypothalamus just above these. The decussation thus may be part of a common system which includes the lateral forebrain tracts and extends into the forebrain hemispheres. In support of this, courtship was suppressed by an implant in the archistriatum (homologue of the amygdala) and incubation was

TABLE 1
PERCENTAGE OF PROGESTERONE IMPLANTS IN VARIOUS REGIONS OF THE BRAIN INDUCING INCUBATION BEHAVIOR IN BOTH SEXES OR SUPPRESSING MALE COURTSHIP BEHAVIOR

Brain region	Incubation induced		Male courtship suppressed		
	N	Percentage responding	N	Percentage responding	
				Fully	Partially
Preoptic area	16	44	11	27	9
Lateral forebrain tracts	41	29	24	29	33
Hypothalamus	24	4	15	40	27
Mammillary bodies	5	0	2	0	0
Cerebral gray	25	0	18	11	17
Other areas[a]	32	6	16	19	19

[a] Thalamus, optic tract, mesencephalon, and below the base of the brain.

induced by an implant near the paleostriatum (homologue of the globus pallidus).

The two behavior patterns were affected separately in all but 4 implanted males, which showed both effects; 9 incubated even though courtship was not suppressed, and in 22 courtship was suppressed even though incubation was not induced. Thus it is not necessary for bow-cooing to be suppressed before incubation can occur. This is also the case when progesterone is injected systemically: of the same 35 males, 26 showed both effects, 5 incubated even though courtship was not suppressed, and in 4 courtship was suppressed but incubation did not follow.

Diffusion of hormone through the ventricles does not seem to account for widespread effects, for although many hypothalamic implants were situated in or near the third ventricle, none of these induced incubation. The effectiveness of the two implants which passed through the base of the brain may be due to hormone diffusing upward along the implant track, for each of these passed through or very near effective sites.

DISCUSSION

There are several interpretations of the apparent widespread progesterone effect. A single neural system could be diffuse, having multiple excitatory or inhibitory controlling pathways of varying threshold in different brain regions. Other possibilities are that a final common inhibitory path could be blocked by progesterone or a lesion produced by the implant tube at many different sites, thus releasing the same behavior pattern from widespread regions; differential diffusion through the brain or into the blood from different brain regions could occur; pituitary function could be altered, indirectly changing behavior.

While none of these possibilities can be definitely excluded on the basis of this experiment, there is reason to believe that widespread neural systems may in fact be involved. Components of courtship behavior have been elicited from implants of testosterone propionate in the preoptic

and lateral forebrain regions in essentially the same regions where progesterone implants blocked courtship in this experiment (Barfield, 1965). It is of particular significance that bow-cooing has been elicited by electrical stimulation in pigeons from all the widespread regions from which progesterone implants blocked the behavior in this experiment (Åkerman, 1966). This suggests that progesterone may act in diverse regions rather than diffusing to a single "center." Further evidence of diffuse hormone-sensitive regions in the brain is presented by Davidson (1966), for mating in castrated rats was elicited by testosterone propionate implants in and between the preoptic and posterior hypothalamic regions. Behavioral sleep has been elicited from widespread regions of the brain by implantation of crystalline acetylcholine in cats (Hernández-Peón, Chavez-Ibarra, Morgane, & Timo-Iaria, 1963). It is not yet possible to state whether a unitary mechanism underlies the present effects, but a widespread neural substrate is indicated.

The finding that testosterone propionate can induce a courtship pattern (Barfield, 1965) in the regions where progesterone implants can block it suggests that an antagonism between these two hormones may exist at the neural level. Recent findings in ring doves (Erickson, Bruder, Komisaruk, & Lehrman, in press) support this. Castrated doves injected with testosterone propionate showed full courtship behavior, but when progesterone was added to the same testosterone propionate dosage, courtship was fully suppressed. Antagonism between these two hormones has also been shown in male sexual behavior (Diamond, 1966) and sterility (Kincl & Maqueo, 1965).

Direct effects of progestins on the brain have been demonstrated in the induction of premature ovulation in chickens by injection of progesterone into the preoptic region (Ralph & Fraps, 1960), in the blockage of ovulation and sexual receptivity in rabbits by hypothalamic implants of norethindrone (Kanematsu & Sawyer, 1965), and by the effects of sys-

temically injected progesterone on the
(a) differential alteration of the lateral
and ventromedial hypothalamic EEG in
rats (Ishizuka, Kurachi, Sugita, & Yoshii,
1954), (b) induction of sleep spindles in
cortical EEG (Kobayashi, Kobayashi,
Takezawa, Oshima, & Kawamura, 1962),
and (c) suppression of neuronal responsiveness to afferent input (Komisaruk,
McDonald, & Sawyer, 1966). Kawakami
and Sawyer (1959) have shown that progesterone, administered intravenously to
estrogen-primed rabbits, increases the
threshold of electrically induced EEG
arousal after an initial decrease in
threshold. This later phase may be related to the anesthetic effect of relatively
large doses of certain steroids (Atkinson,
Davis, Pratt, Sharpe, & Tomich, 1965).
These depressant-like effects of progesterone may be related to the behavioral
quiescence characteristic of incubation behavior in birds.

REFERENCES

ÅKERMAN, B. Behavioural effects of electrical
stimulation in the forebrain of the pigeon. I.
Reproductive behaviour. Behaviour, 1966, 26,
323–338.

ATKINSON, R. M., DAVIS, B., PRATT, M. A., SHARPE,
H. M., & TOMICH, E. G. Action of some
steroids on the central nervous system of the
mouse. II. Pharmacology. J. med. Chem., 1965,
8, 426–432.

BARFIELD, R. J. Induction of aggressive and courtship behavior by intracerebral implants of
androgen in capons. Amer. Zoologist, 1965, 5,
203.

CARPENTER, C. R. Psychobiological studies of social
behavior in aves. I. The effect of complete
and incomplete gonadectomy on the primary
sexual activity of the male pigeon. J. comp.
Psychol., 1933, 16, 25–57. (a)

CARPENTER, C. R. Psychobiological studies of social
behavior in aves. II. The effect of complete
and incomplete gonadectomy on secondary
sexual activity with histological studies. J.
comp. Psychol., 1933, 16, 59–97. (b)

CROSBY, E. C., & WOODBURNE, R. T. Comparative
anatomy of the hypothalamus. Proc. Assn.
Res. Nerv. Ment. Dis., 1940, 204, 134–169.

DAVIDSON, J. M. Activation of the male rat's sexual
behavior by intracerebral implantation of
androgen. Endocrinology, 1966, 79, 783–794.

DIAMOND, M. Progestagen inhibition of normal
sexual behaviour in the male guinea-pig.
Nature, 1966, 209, 1322–1324.

ERICKSON, C. J., BRUDER, R. H., KOMISARUK, B. R.,
& LEHRMAN, D. S. Selective blocking by progesterone of behavioral effects of androgen in
the ring dove. Endocrinology, 1967, in press.

ERICKSON, C. J., & LEHRMAN, D. S. Effect of
castration of male ring doves upon ovarian
activity of females. J. comp. physiol. Psychol.,
1964, 58, 164–166.

FABRICIUS, E., & JANSSON, S. Laboratory observations on the reproductive behavior of the
pigeon (Columba livia) during the pre-incu-
bation phase of the breeding cycle. Anim.
Behav., 1963, 11, 534–547.

HERNÁNDEZ-PEÓN, R., CHAVEZ-IBARRA, G., MORGANE,
P. J., & TIMO-IARIA, C. Limbic cholinergic
pathways involved in sleep and emotional
behavior. Exp. Neurol., 1963, 8, 93–111.

HUBER, G. C., & CROSBY, E. C. The nuclei and
fiber paths of the avian diencephalon with
consideration of telencephalic and certain
mesencephalic centers and connections. J.
comp. Neurol., 1929, 48, 1–207.

ISHIZUKA, N., KURACHI, K., SUGITA, N., & YOSHII,
N. Studies on the relationship between EEG
of the hypothalamus and sexual function.
Med. J. Osaka Univ., 1954, 5, 729–740.

KANEMATSU, S., & SAWYER, C. H. Blockade of
ovulation in rabbits by hypothalamic implants
of norethindrone. Endocrinology, 1965, 76,
691–699.

KAWAKAMI, M., & SAWYER, C. H. Neuroendocrine
correlates of changes in brain activity thresholds by sex steroids and pituitary hormones.
Endocrinology, 1959, 65, 652–668.

KINCL, F. A., & MAQUEO, M. Prevention by progesterone of steroid-induced sterility in
neonatal male and female rats. Endocrinology,
1965, 77, 859–862.

KOBAYASHI, T., KOBAYASHI, T., TAKEZAWA, S.,
OSHIMA, K., & KAWAMURA, H. Electrophysio-
logical studies on the feed-back mechanism of
progesterone. Endocrinol. Jap., 1962, 9, 302–
320.

KOMISARUK, B. R., MCDONALD, P. G., & SAWYER,
C. H. Non-specific effects of progesterone and
sensory stimuli on brain activity. Amer.
Zoologist, 1966, 6, 571.

LEHRMAN, D. S. On the initiation of incubation
behaviour in doves. Anim. Behav., 1963, 11,
433–438.

LEHRMAN, D. S., & BRODY, P. N. Effect of prolactin
on established incubation behavior in the ring
dove. J. comp. physiol. Psychol., 1964, 57,
161–165.

RALPH, C. L., & FRAPS, R. N. Induction of ovulation in the hen by injection of progesterone
into the brain. Endocrinology, 1960, 66, 269–
272.

RIDDLE, O., & LAHR, E. L. On broodiness of ring
doves following implants of certain steroid
hormones. Endocrinology, 1944, 35, 255–260.

SIEGEL, S. Nonparametric statistics for the behavioral sciences. New York: McGraw-Hill, 1956.

(Received September 21, 1966)

22

Reprinted from *J. Endocrinol.* 30:355–359 (1964)

EFFECT OF EXOGENOUS HORMONES ON THE TACTILE SENSITIVITY OF THE CANARY BROOD PATCH

R. A. HINDE AND ELIZABETH STEEL

Sub-department of Animal Behaviour, High Street, Madingley, Cambridge

(*Received* 3 *April* 1964)

SUMMARY

0·05 mg. oestradiol benzoate, injected thrice weekly alone or in combination with either 0·25 mg. progesterone or 75 i.u. prolactin, induced an increase in tactile sensitivity in the female canary's brood patch. Progesterone and prolactin alone did not produce this effect; 0·3 mg. of the oestrogen was also ineffective alone and in combination with the other hormones.

INTRODUCTION

During the period immediately preceding egg-laying the female canary develops a brood patch—that is, the ventral surface becomes defeathered, vascular and, later, oedematous (Hinde, 1962). This brood patch is probably important for heat transfer during incubation. It may also have another role, for stimuli from the nest cup accelerate reproductive development and influence reproductive (e.g. nest-building) behaviour (Warren & Hinde, 1961; Hinde & Steel, 1962). In conformity with this, the ventral surface becomes more sensitive to tactile stimulation as the reproductive season advances (Hinde, Bell & Steel, 1963). Vascularization of the brood patch can be induced by exogenous oestradiol benzoate and defeathering by this oestrogen in combination with either progesterone or prolactin (Steel & Hinde, 1963). The experiments reported here were intended to assess the effects of these three hormones on the tactile sensitivity of the brood patch.

MATERIAL AND METHODS

Animals. Domesticated canaries of the Border variety were used. They were obtained from dealers as first-winter females. In the earlier experiments some males may have been included, but later all birds were sexed by laparotomy before injection.

The birds were kept in standard metal cages and, except when specified, were without nest pans or nest material. Some groups were given nest pans which were either 'standard' canary nest pans, 10·5 cm. in diameter and 4·5 cm. deep, or 'small' pans, 7·5 cm. in diameter and 3·5 cm. deep. When nest pans were provided nest material (grass and feathers) was also present.

Methods. All birds were injected thrice weekly into the pectoral musculature. The procedure has been described in detail by Steel & Hinde (1963). The hormones used were: oestradiol benzoate in aqueous suspension (Oestroform, British Drug Houses Ltd.), progesterone B.P. in aqueous suspension (Lutoform, British Drug Houses Ltd.), prolactin in aqueous solution (supplied by National Institutes of Health, Bethesda, Md.). Tyrode solution was used for the control series. Table 1 shows the number of birds in the groups and the treatment each group received. The experiments were performed during a period of 2 years. The months in which each group was studied are indicated on the Table.

Table 1. *Treatment of birds and the times of year at which the experiments were performed. a and b refer to groups used in the same months, but in different years*

	Group										
Treatment	I	II	III	IV	V	VI	VII	VIII	IX	X	XI
Tyrode	×	.	.	.	.	.	.	.	.	.	.
Oestrogen, 0·05 mg.	.	×	.	.	.	×	×	.	.	.	.
Oestrogen, 0·30 mg.	.	.	×	.	.	.	.	×	×	×	×
Progesterone, 0·25 mg.	.	.	.	×	.	×	.	×	.	.	.
Prolactin, 75 i.u.	.	.	.	.	×	.	×	.	×	.	.
Small nest provided	.	.	.	.	.	.	.	.	.	×	.
Standard nest provided	.	.	.	.	.	.	.	.	.	.	×
No. of birds used in:											
Jan.–Feb.	8	$a = 11$ $b = 12$	11	3	4	17	.	8	9	9	10
Oct.–Nov.	11	.	.	5	5	14	8	.	.	.	.

Tactile sensitivity was tested with a Semmes–Weinstein aesthesiometer, consisting of a series of nylon filaments of increasing diameter. During the test, the bird was held in the hand and its breast touched lightly with one of the filaments. A response consisted of a visible movement of one of its legs or toes. The filaments were used in sequence of descending or ascending diameter, each sequence being continued until two successive filaments had failed to evoke (descending series) or evoked (ascending series) a response. The threshold on any one day was taken as the mean of two thresholds obtained on descending runs and of two thresholds on ascending runs. The nylon filaments were graduated in terms of the common logarithm of the force in milligrams required to bend the filament: the range employed was 5·07 to 1·65 units. Further details of the procedure are given by Hinde et al. (1963).

Thresholds were obtained on 4 days before the injections started. The first estimation was rejected, and the initial threshold level for each bird was taken as the mean of those obtained in the second, third and fourth tests. The fourth test was made immediately before the first injection, on a Monday (day 0). Testing and injections were continued on successive Wednesdays, Fridays and Mondays until nine injections had been given. A final test was made 3 days after the last injection (day 21).

For analysis, four experimental periods were defined, to include: (i) the mean of the second, third and fourth pre-injection readings (see above), (ii) the mean of post-injection readings, days 2 and 4 (week 1), (iii) the mean of post-injection readings on days 7, 9 and 11 (week 2), and (iv) the mean of post-injection readings on days 14, 16, 18 and 21 (week 3). The data were evaluated by analysis of variance.

RESULTS AND DISCUSSION

Before discussing the main results, two preliminary points must be made.

(1) *Seasonal effects.* Several groups of birds shown in Table 1 contain sub-groups tested in different months. Before combining the sub-groups it was necessary to look for seasonal effects. Using a Mann–Whitney U-test to compare means, no significant differences between sub-groups given identical treatments at different times of the year were found. However, the difference between the two groups which had received 0·05 mg. oestradiol benzoate, though not significant, was considerable. For the analysis of variance Group IIa (February 1962) was used; as discussed later, use of Group IIb (January 1964) would not have affected the conclusions.

(2) *Effect of initial threshold level.* Although there was considerable individual variation in the initial thresholds, the group means, except that of the October–November sub-group of the controls, were all very similar. The January–February control sub-group alone was therefore used in the analysis of variance, since its mean initial level fell within the range of that of the experimental groups. Neither of the control sub-groups showed a change in threshold with time. The mean initial thresholds of the groups used for the analysis of variance were not considered as a separate variable, since after the omission of the October–November control sub-group they fell within a narrow range.

With the two exceptions described in 1 and 2 above all the results for each treatment were combined, irrespective of the time of year.

Figures 1, 2 and 3 show the mean sensitivity thresholds of the four experimental periods defined above for Groups I–IX. Two separate analyses of variance were carried out to evaluate the effects of (a) oestradiol benzoate alone and in combination with prolactin and (b) oestradiol benzoate alone and in combination with progesterone. In both analyses there were two significant ($P < 0·01$) factors, days ((a) $F = 6·17$, D.F. 3/150; (b) $F = 7·05$, D.F. 3/210), and days × oestrogen interaction ((a) $F = 6·88$, D.F. 6/150; (b) $F = 4·07$, D.F. 6/210). Figures 1–3 show that the latter was due to the effect of the dose (0·05 mg.) which produced a marked fall in threshold. The 0·03 mg. group differed little from the controls. Although the curves for Groups VI and VII (0·05 mg. oestradiol benzoate with either progesterone or prolactin) fall more steeply than the curve for the Group IIa (0·05 mg. of the oestrogen alone) the differences were not significant. Thus these data do not show a synergistic effect of progesterone or prolactin with oestrogen on the tactile threshold.

If Group IIb (injected with 0·05 mg. oestradiol benzoate) had been used in this analysis, the conclusions would have been the same (see Fig. 3). The influence of this hormone alone would have shown even more strongly and the suggestion of synergism would have been less marked.

During the natural breeding season the threshold of the canary brood patch to tactile stimulation falls gradually over a period of 2–3 months. Birds building early in this period have a higher threshold than those building later. The fall in threshold therefore does not necessarily depend on nest-building and feedback from the nest (via nervous and/or hormonal pathways) is unlikely to have a large effect on it. However, it is still possible that stimulation from the nest cup augments the action of an otherwise ineffective oestrogen dosage. To test this hypothesis two groups of

birds (X and XI) injected with 0·03 mg. oestradiol benzoate were supplied with
either standard or small nest pans and building material. Although it is difficult to
induce nest-building in canaries in winter with oestrogen (Warren & Hinde, 1961),
this high dose did induce complete nest-building behaviour in some birds. However,
there was no significant difference between the sensitivity changes shown by these
birds and those without nest cups.

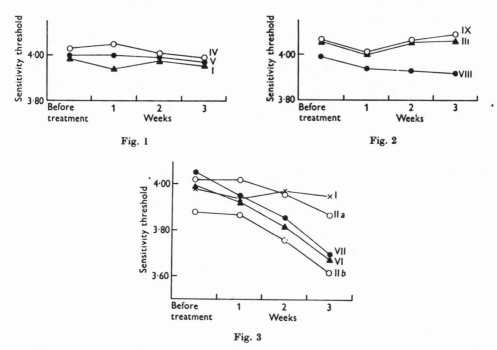

Fig. 1. Changes in the threshold of responsiveness to tactile stimulation of the brood patch in
birds injected thrice weekly with: 0·05 ml. Tyrode solution (I), 0·25 mg. progesterone (IV), and
75 i.u. prolactin (V).

Fig. 2. Changes in the threshold of responsiveness to tactile stimulation of the brood patch in
birds injected thrice weekly with: 0·3 mg. oestradiol benzoate (III), 0·3 mg. oestradiol benzoate
and 0·25 mg. progesterone (VIII), 0·3 mg. oestradiol benzoate and 75 i.u. prolactin (IX).

Fig. 3. Changes in the threshold of responsiveness to tactile stimulation of the brood patch in
birds injected thrice weekly with: 0·05 mg. oestradiol benzoate (IIa and b), 0·05 mg. oestradiol
benzoate and 0·25 mg. progesterone (VI), 0·05 mg. oestradiol benzoate and 75 i.u. prolactin (VII),
0·05 ml. Tyrode solution (I).

These experiments show that 0·05 mg. oestradiol benzoate administered thrice
weekly alone or in combination with either 0·25 mg. progesterone or 75 i.u. prolactin
produces a fall in the tactile sensitivity threshold of the brood patch. The fall is comparable in magnitude to that which occurs during the natural breeding season but it
occurs more rapidly. 0·05 mg. of the oestrogen with either prolactin or progesterone
in the proportions given above produces defeathering of the brood patch at a rate
similar to that which also occurs in the breeding season. During that time the sensitivity changes occur mainly after defeathering is completed and during the period of

increasing vascularity. Even when replacement oestrogen is given, the response of ovariectomized females to tactile stimulation never increases as much as that of intact controls injected with the same dose (Steele & Hinde, 1964) which suggests that an ovarian hormone in addition to oestrogen is needed for the full response.

The ineffectiveness of large doses of oestrogen may be due to the gross changes in the mechanical properties of the skin which they induce. At such dose levels the increase in vascularity is marked (Steel & Hinde, 1963); the absence of changes in sensitivity is therefore additional evidence for the view that sensitivity changes are not solely a function of alterations in blood supply (Hinde *et al.* 1963).

We are grateful to Professor V. H. Denenberg for advice on statistical methods. The pituitary hormones were kindly supplied by the National Institutes of Health, Bethesda, Maryland, U.S.A. This work was supported by a grant from the U.S. Air Force, Biosciences Division.

REFERENCES

Hinde, R. A. (1962). Temporal relations of brood patch development in domesticated canaries. *Ibis*, **104**, 90.

Hinde, R. A., Bell, R. Q. & Steel, E. (1963). Changes in sensitivity of the canary brood patch during the natural breeding season. *Anim. Behav.* **11**, 553.

Hinde, R. A. & Steel, E. A. (1962). Selection of nest material by female canaries. *Anim. Behav.* **10**, 67.

Steel, E. A. & Hinde, R. A. (1963). Hormonal control of brood patch and oviduct development in domesticated canaries. *J. Endocrin.* **26**, 11.

Steel, E. & Hinde, R. A. (1964). Effect of exogenous oestrogen on brood patch development of intact and ovariectomized canaries. *Nature (Lond.)*, **202**, 718.

Warren, R. P. & Hinde, R. A. (1961). Roles of the male and of the nest-cup in controlling the reproduction of female canaries. *Anim. Behav.* **9**, 64.

23

Reprinted from *Nature* 184:2030 (1959)

PROLACTIN IN
THE COWBIRD'S PITUITARY IN RELATION
TO AVIAN BROOD PARASITISM

E. O. Höhn

University of Alberta

THE European cuckoo provides a familiar example of complete brood parasitism, that is, failure to build nests, coupled with the habit of laying eggs in the nest of other birds which generally take over incubation and hatching of the foreign eggs. This habit is also shown by the North American brown-headed cowbird (*Molothrus ater*) as well as other species. Birds showing brood parasitism do not develop brood patches. Bailey[1] has shown that brood patch development in passerines requires the action of an œstrogen and of prolactin. Prolactin can also induce broodiness in pigeons and poultry.

It has therefore occurred to me, and independently to others, that failure to produce prolactin might be a basic factor in the development of brood parasitism. Hence a comparison of the prolactin content of the pituitaries of female cowbirds in breeding condition, with that of pituitaries of breeding females of the closely related red-winged blackbird (*Agelaius phoeniceus*) which has normal breeding habits, was made. Birds of both species were shot in the field and their heads were frozen within a few minutes of death by placing them in a vacuum flask containing dry ice. On return to the laboratory the heads were kept at $-20°$C. in a deep-freeze refrigerator. The bodies of the birds were dissected for examination of the ovaries and oviducts. The pituitaries were then dissected out and implanted on the external surface of one side of the crop sac of anæsthetized homing pigeons. After 5 days the pigeons were killed and the implanted crop sac area, aswell as a control section from the opposite side of the sac were removed. Stretched-out pieces of crop sac were first examined macroscopically for evidence of

epithelial proliferation and then fixed in 10 per cent formol saline. Paraffin sections were made and stained with hæmatoxylon eosin and with Giemsa. The results obtained are shown in Table 1.

All the red-winged blackbirds had well-developed brood patches; data on the ovaries and oviducts of all birds used are given in Table 2.

These results demonstrate the presence of prolactin in the pituitaries of ovulating cowbirds. Moreover, judging from the number of pituitaries of the two species needed for a definite crop sac response, the prolactin content of cowbirds and red-winged blackbirds pituitaries is fairly similar. It may be noted that the two species are of fairly similar body-weight; averages for three females of each species were, red-winged blackbirds 47 gm. and cowbirds 43 gm. The above experiments suggest that failure to respond to prolactin rather than lack of prolactin production is involved in the failure of female cowbirds to form brood patches. Dr. R. K. Selander, University of Texas, has informed me of some unpublished findings which confirm this view; he found that cowbirds of either sex treated with œstrogen and prolactin did not develope brood patches.

The prolactin used in this study was a gift from the Endocrinology Study Section, U.S. National Institutes of Health, and the work was supported by a grant from the Committee for Research in Problems of Sex, U.S. National Academy of Sciences—National Research Council.

[1] Bailey, R. E., *Condor*, **54**, 121 (1952).

Table 1. Epithelial Proliferation f Pigeon Crop Sacs Produced by 5-Day Implants of Pituitaries of Female Red-Winged Blackbirds and Female Brown-Headed Cowbirds

	Red-winged Blackbirds					Cowbirds			
		Epithelial proliferation					Epithelial proliferation		
Exp. No.	Implant	Macroscopic	Microscopic		Exp. No.	Implant	Macroscopic	Microscopic	
R.W. 1	8 pituitaries (June 8)	+	1 1/2		C.B. 1	12 pituitaries (June 5, 9, 12)	—	1/2	
	Control : muscle	—	—			Control : no implant	—	—	
R.W. 2	10 pituitaries (June 9, 11)	+	2		C.B. 2	12 pituitaries (June 14, 16)	—	—	
	Control : no implant	—	—			Control : muscle	—	—	
R.W. 3	9 pituitaries (June 11, 16, 18)	—	—		C.B. 3	14 pituitaries (June 16, 18)	+	2	
	Control : no implant	—	—			Control : no implant	—	—	
R.W. 4	14 pituitaries (June 29, July 2)	—	1/2		C.B. 4	15 pituitaries (June 29, July 2)	—	1/2	
	Control : muscle	—	1/2			Control : liver	—	—	

Dates in brackets indicate dates of collection of the birds used. Local crop sac response to total dose of 4 μgm. prolactin, 15 units/mgm. injected intradermally daily for four days rated as: macroscopic: + ; microscopic: 2

Table 2. Data on the Ovaries and Oviducts of Red-Winged Blackbirds and Cowbirds, Pituitaries of which were used for the Experiments shown in Table 1

	Red-winged Blackbirds				Cowbirds		
Exp. No.	No. with eggs in oviduct	Diam. of largest ovarian follicle, mean and range (mm.)	Oviducts	Exp. No.	No. with eggs in oviduct	Diam. of largest Ovarian follicle, mean and range (mm.)	Oviducts
R.W. 1	—	3·9, 1·5–10	All large	C.B. 1	7/12	7·3, 2–10	All large
R.W. 2	—	2·7, 1–8	All large	C.B. 2	3/12	7·0, 3–8	All large
R.W. 3	1	2·7, 1–8	All large	C.B. 3	8/14	5·5, 2–8	All large
R.W. 4	—	1·3, ·5–3	Small in 7/14	C.B. 4	4/15	3·8, 1–8	All large

24

Reprinted by permission of the authors and publishers from pp. 316–350, 379 of
Studies in Animal and Human Behavior, Vol. 1, by Konrad Lorenz
(R. Martin, transl.), London: Methuen & Co., and Cambridge, Mass.:
Harvard University Press, 1970

TAXIS AND INSTINCTIVE BEHAVIOUR PATTERN IN EGG-ROLLING BY THE GREYLAG GOOSE

K. Lorenz and N. Tinbergen

Introduction

The experiments which we conducted with greylag geese last season, to investigate the pattern of rolling eggs back into the nest-cup, are far from complete. In addition, our experimental animals were to a large extent related to domestic stock. Although some control experiments have already been carried out on pure wild-type Greylags, giving completely consistent results, we intend to carry out series of experiments on pure-blooded geese next year, placing particular emphasis on a detailed film-analysis of the observational material reported here. Nevertheless, we believe that there are two good reasons for publication of this paper in its present form. In the first place, discussion of two papers recently published by Lorenz (1937a and b) has uncovered a number of misunderstandings which can best be eradicated by discussion of our findings on the egg-rolling response of the Greylag goose. For this reason, we wish to see our findings published as soon as possible, as a sequel to the two papers mentioned. Secondly, we hope that discussion of our present results will provide us with additional knowledge, which would benefit the experiments to be conducted in the years to come. At this juncture, gratitude is expressed for permission from the *Deutsche Reichstelle für den Unterrichtsfilm* to make use of photographs from a film made in their commission by cand. zool. A. Seitz (Vienna) in Altenberg.

I Theoretical considerations of the taxis and the instinctive behaviour pattern

Whatever concept one chooses to associate with the word instinct. it will always be necessary to consider the existence of specific 'motor formulae' which are innately determined in an invariable form in the individual. These 'formulae' can represent characters of great taxonomic value, typifying a species, a genus or even an entire phylum. Since these motor patterns fairly closely represent the nucleus of what was interpreted in earlier descriptions of animal behaviour as the operational effect of the instinct, we have employed (or, more exactly, retained) the term *instinctive behaviour pattern*[112] to refer to them. The form of an instinctive behaviour pattern exhibits remarkable independence from all receptor process – not just independence from 'experience' in the broadest sense of the term but also from stimuli which impinge upon the organism *during* performance of the pattern. In this feature, instinctive patterns are quite distinctly demarcated from *taxes* or *orientation responses*, with which they share a purposive character (in terms of survival of the species) and independence from individual learning effects, and with which they are frequently bundled together under the concept of 'instinct'. In Drisech's definition of the instinct as 'a response which is complete from the outset', the conceptual delimitation covers both the motor pattern of innately-determined form *and* every innate orientation process, without taking account of the basic and far-reaching difference between the two. In order to clarify the conceptual framework of our work, it is necessary to briefly discuss the relationships and differences between the taxis and the instinctive behaviour pattern.

I. THE TOPOTAXIS

Most of the processes which we refer to as *taxes*, and particularly *topotaxes* (after A. Kühn) include – *in addition to* orienting responses – fixed instinctive behaviour patterns and 'employ' specific motor patterns, particularly those of locomotion. Nevertheless, the main motor component of the topotaxis – and *its only unique characteristic* – remains the *turning motion* of the animal's whole body or one of its parts (e.g. the eyes or the head), *which is quantitatively determined by the external stimuli*. This is not the case with Kühn's *phobic response* or phototaxis, and the relationship exhibited between this type of response and our instinctive behaviour pattern in terms of the fixity of the motor formula and its independence of the nature and direction of impinging stimulus permits us to distinguish it quite clearly from the topotaxis. Even if the analogies between the phobic response and the instinctive behaviour pattern are only external in kind and the two are little related in causal terms (e.g. as regards the dependence of the instinctive behaviour pattern upon the central nervous system), we do not wish to distinguish the phobic response as markedly as the topotaxis from the instinctive behaviour pattern in establishing our working hypothesis.

The turning motion (in its widest sense) leading to the spatial orientation of the animal has been termed the 'oriented turning response' by K. Bühler. This extremely broad concept includes every alteration of the motor status of the organism relating it to the spatial parameters of the environment, thus covering both the simplest 'towards' or 'away' turning motion and spatial orientation to external stimuli controlled by the highest forms of intelligence. Since the 'away' turning movement of the phobic response, which is *not* quantitatively controlled, is also covered by this concept, we must conclude that in this paper we are interpreting under the concept of the orienting turning motion only a fraction of the oriented turning responses categorized by Bühler. We are exclusively concerned with the quantitatively stimulus-controlled turning movement which is uniquely characteristic of the *topic* response. Since we are exclusively concerned with the topic response in the following comparison of taxis and instinctive behaviour pattern, we will confine the term taxis to this response for the sake of brevity. In the last analysis, it is only a question of convenience as to which concepts are defined, and the definitions selected here in order to permit the briefest possible presentation of the factual evidence are by no means intended as a critique of established concepts.

Even if some motor formulae which are themselves rigidly fixed play a part as elements of the overall movement pattern of the taxis, it is the *overall form* of the taxis which is dependent upon the quantitatively stimulus-controlled turning movement. The adaptation of the overall movement pattern to spatial environmental conditions, which is an important feature of the taxis, is itself ensured by the *norm of the response* to specific external stimuli. Without discussing the purely philosophical question of whether the animal subject steers towards these stimuli or is passively steered by them, it can be stated that we are here objectively concerned with *moulding of an adaptively purposive motor pattern by stimuli of external origin*. In the simplest, and best analysed case, this process is quite definitely *a reflex in the true sense of the word*. But even with orientational responses which are derived from the so-called higher psychological functions, the activity of the central nervous system is essentially a process of evaluation and reaction to external stimuli and is thus a reflex-like process, at least in functional terms.

2. THE INSTINCTIVE BEHAVIOUR PATTERN

The instinctive behaviour patterns also exhibit relationships to the reflexes, in that they are similarly elicited by particular external stimuli which are often very specific. However, closer investigation shows that it is *only the releasing mechanism* and not the subsequent performance of the motor formula which represents a genuine reflex. The overall form of the motor pattern, once elicited, is apparently independent not only of external stimuli but also of the animal's receptors generally. Instinctive behaviour patterns are not innate response norms like the taxes – they are *innate motor norms.*

However improbable it may at first sight appear that finely co-ordinated and extremely adaptive behavioural sequences of an animal can occur without interaction with the receptors, just like the movements resulting from spinal sclerosis (i.e. that they are *not* constructed from reflexes), there are several important reasons for assuming this to be the case. E. von Holst has been able to provide entirely convincing evidence, from deafferentiation studies of the central nervous systems of greatly different organisms, that *automatic, rhythmic stimulus production processes* occur in the central nervous system, *where the impulses are actually co-ordinated*, such that the impulse sequence transmitted to the animal's musculature arises in an adaptively complete form without interaction with the periphery and its receptors. The centrally co-ordinated motor sequences so far investigated are almost exclusively concerned with locomotion, and it is just these locomotor patterns which have been previously almost unanimously regarded as chain-reflexes, in which each movement was supposed to elicit the next only by a detour through the peripheral receptors.[113]

Without a trace of doubt, the motor patterns investigated by von Holst must be regarded as typical instinctive behaviour patterns in the sense of the concept defined here. This does not mean, however, that the converse must also apply – i.e. that all of the instinctive behaviour patterns which we have previously described must depend upon exactly similar processes in the central nervous system. The fact that the receptors (in the broadest sense of the term) do not play a part in the production of the adaptively-functioning motor pattern even in more highly-differentiated instinctive behaviour patterns is strongly indicated by a number of observational facts and experimental results which were described in detail in an earlier paper (1937). Even if it may seem premature at this time to equate the motor automatisms investigated by von Holst with the instinctive behaviour pattern, considerable emphasis should be given to the fact that two of the most important and most characteristic features of the instinctive behaviour pattern, which present virtually insurmountable obstacles to any other explanation, can be easily interpreted on the assumption that they are based on automatic, rhythmic production of the elicitatory stimuli and upon central co-ordination of the impulses. These two features are *lowering of the threshold* for the releasing stimuli and the consequent *vacuum performance* of the instinctive behaviour patterns in situations in which even the physical/mechanical prerequisites for adaptive function are absent.

Even in cases where the instinctive behaviour pattern is elicited by an unconditioned reflex, which normally responds only to fairly (or even extremely) specific combinations of external stimuli, it can under certain circumstances be observed that the *form of the elicited motor pattern is independent of that of the elicitatory stimuli*. If these stimuli are absent for a period of time longer than that to be expected under natural conditions, the instinctive behaviour pattern soon exhibits *reduced selectivity* of elicitation. The animal will then respond with this same motor pattern to other stimulus situations which are no more than generally similar to the adequate situation. The longer the period of 'damming' of the response, the smaller the requisite degree of similarity between the available stimuli and the normally specific elicitatory stimuli. In cases where the stimuli eliciting a particular instinctive behaviour pattern are to some extent open to quantitative determination, it is observed that the threshold value of the response to be elicited sinks continuously with the length of the inactive period. Since both of these phenomena are obviously based upon only *one* process in the central nervous system, I have included them under the concept of 'threshold-lowering' in the two papers already mentioned. Threshold-lowering can in very many instinctive behaviour patterns (perhaps basically in all) literally progress to the limiting zero value. Following a more-or-less extensive period of 'damming', the entire motor sequence can be performed *without* demonstrable operation of an external stimulus, and this has been referred to as the 'vacuum activity'. Threshold-lowering and vacuum activity present a double support for the assumption of automatic, rhythmic stimulus-production and for central co-ordination of the impulses involved.

The phenomenon of threshold-lowering indicates the existence of internal accumulation of response-specific arousal (von Holst assumes that stimulatory substances are involved), which is continuously produced by the central nervous system and reaches a level proportional to the period of absence of discharge resulting from performance of the instinctive behaviour pattern concerned. The higher the level of accumulation reached, the more intensive is the eventual eruptive performance of the instinctive behaviour pattern and the harder it is for the central nervous factors governing the instinctive behaviour pattern to maintain inhibition against such performance in the 'incorrect' situation. These considerations, which are reproduced here in grossly simplified form, were developed before von Holst's observations were known. We now know from von Holst's work that the operation of the continually-active automatic stimulus production processes is in actual fact persistently inhibited by the activity of higher (or more central) inhibitory components of the nervous system and that elicitation of the response signifies no more than removal of central inhibition. When left to themselves, the peripheral components (e.g. a spinal cord preparation in the case of von Holst's experiments) will continuously perform the motor pattern in accordance with the automatic rhythm. If the spinal cord preparation is then provided with a substitute for the inhibitory effects normally produced by the activity of higher centres – in the form of supplementary stimuli – one observes a phenomenon which Sherrington described under the term 'spinal contrast'. The greater the duration and intensity of operation of the inhibitory effect, the more violent the eruption of the inhibited motor pattern when inhibition is ceased. This correspondence between the automatic rhythms investigated by von Holst and the instinctive behaviour pattern (as it is here understood) goes much further and covers minor details which will not be discussed here.

To the same extent that the threshold-lowering and elevation of intensity which we can observe with virtually every instinctive behaviour pattern clearly supports the assumption of a stimulus-producing automatism, the phenomena which we observe during the eventual stimulus-independent eruption of the vacuum activity clearly indicate central co-ordination of the impulses which are transmitted by the automatism.[114] It is of undoubted importance that a motor sequence frequently performed virtually '*in vacuo*' resembles down to the finest detail the motor pattern observed during the normal, adaptive response. This is particularly obvious when the behaviour pattern performs mechanical work in the normal situation. In other words, the motor pattern is normally performed against an opposing force which is missing with the vacuum activity. (This will be considered in detail later on.) The complete independence of the form of the motor pattern from the conditions and stimulatory properties of the natural environment can only be explained on the assumption that the impulses for the individual muscular movements are transmitted in an already co-ordinated form and sequence from the centre, and that they (so to speak) 'experience nothing' of the prevailing environmental conditions. Otherwise, it is impossible to explain how a starling, for example, can perform the motor sequence of catching, killing and swallowing small insects in the complete absence of such animal prey; or to explain how a humming-bird can attach 'non-existent nest-material fibres' to a twig with a wonderfully co-ordinated weaving movement, as was recently observed by Lorenz at the zoological garden in Berlin.

Even though such vacuum performances of highly-differentiated instinctive behaviour patterns quite conclusively demonstrate the independence of the form of the motor pattern from all external receptor processes, we unfortunately do not possess an object on which we can exclude the co-operation of proprioceptor reflexes with the same degree of certainty as von Holst achieved with the spinal cord of the fish. For this reason, the part played by proprioceptive receptors in the perform-ance of highly-differentiated instinctive behaviour patterns is still ex-tremely obscure. Nevertheless, it can be observed that even extremely coarse mechanical effects which directly and forcibly cause temporary inhibition of the instinctive motor sequence or force the body of the responding animal into abnormal postures produce no alterations in the further performance of the response. Such coarse mechanical disruptive effects would automatically lead to alterations in the form of the sub-sequent parts of the motor pattern if the proprioceptive registration of each individual component movement were decisive for the form of the next in the manner postulated in the old chain-reflex theory of the instinctive behaviour pattern. We have so far never seen such alteration. After cessation of the mechanical pressure, the instinctive behaviour pat-tern – if it continues at all – continues in a completely typical manner. Examples of this will be provided later.

For all of these reasons, it would seem permissible to construct a working hypothesis on the assumption that the co-ordination of the individual movements of the instinctive behaviour pattern is indepen-dent of *all* receptors and is not influenced by them.[115] This may be a com-pletely gross simplification of the factual evidence and it may signify a too extensive restriction of the concept of the instinctive behaviour pattern, particularly since we know from experiments conducted by von Holst that an impulse emanating from a centrally co-ordinated automatism may be superimposed upon one emanating from a reflex to give *one* muscle contraction. Nevertheless, provisional disregard of this fact would seem to be justified in view of another fact which com-plicates the analysis of animal behaviour in general and of the instinctive behaviour pattern in particular. This complicating fact, which justifies an extensive simplification of the factual evidence as long as the provi-sional nature of this step is continually borne in mind, is based on the following circumstances:

3. INTERCALATION OF TAXIS AND INSTINCTIVE BEHAVIOUR PATTERN

An intact higher organism in its natural environment *almost never per-forms a centrally co-ordinated motor pattern in isolation.* Instead, the organism usually 'does several things at once', following (for example) an orienting response *during* the performance of an instinctive behaviour pattern. Nothing is farther from the truth than the consistently revived statement, usually employed in the distinction of the instinctive be-haviour pattern from the 'reflex', that the instinctive pattern represents a response of 'the organism as a whole', in contrast to the reflex, which involves the activity of only 'one part of the organism'. As far as one can really maintain that only a part of an organism is involved in a response, one can confidently state that this is the case with the instinctive behaviour pattern as well. The automatic rhythm and the central co-ordination of the instinctive behaviour pattern render it far more independent than the reflex, for the simple reason that the instinctive pattern is consequently independent of the receptors. The automatism of the instinctive behaviour pattern is a virtually closed system. It is subject to the systemic control of the central nervous system only with respect to the timing and degree of disinhibition. Despite the independence of the automatism of the instinctive behaviour pattern and its refractory nature with respect to the highest controlling factors of the central nervous system even in man (i.e. with respect to the 'ego'), its sphere of influence in the motor apparatus of the higher animal is always very limited. The automatism always transmits its impulses to quite specific groups of muscles and never (apart from a few exceptions) simultaneously to the entire body musculature. Therefore there is always the possibility that other muscles and groups of muscles can simultaneously perform movements which are brought about by quite different processes in the central nervous system.[116] Consequently, in the study of the instinctive behaviour pattern itself we shall have to select predominantly those processes in which the centrally co-ordinated impulse sequence of the instinctive behaviour pattern is transformed into visible and describable movements without disruptive side-effects derived from other nervous processes. Alternatively, we must select processes in which the operation of non-instinctive motor processes can be easily analysed and eliminated from considerations of the instinctive behaviour pattern proper.[117]

Certain non-instinctive movements, which can be easily analysed, accompany *every* instinctive behaviour pattern performed by higher animals: The tropotactically-controlled orienting response of equilibration proceeds continuously even when instinctive behaviour patterns are performed. With the sole exception of the copulation response of the rabbit buck, we know of no instinctive behaviour pattern in which a normally gravity-oriented organism loses its orientation and falls over! But even instinctive behaviour patterns for which the movements of tropotactic orientation to gravity are the only accompanying orienting responses represent definite exceptions. In the search for an instinctive behaviour pattern which could be clearly represented in curvilinear form with some degree of accuracy, for the purposes of a comparative and genetic investigation, it was not at all easy to find a centrally co-ordinated motor pattern which was not overlain by simultaneously active orienting mechanisms! The final choice fell upon the courtship motor patterns of certain duck species. These patterns are performed more or less in the sagittal plane by the duck whilst swimming in an upright posture, and it is possible to prepare a fairly complete reconstruction and graphic representation of the movement from a ciné-film taken from an exactly lateral position. From this projection, the individual, simultaneous orienting responses (the lateral movements of the equilibriating tropotaxis) can be completely ignored. A second orienting mechanism operates even with some similar instinctive behaviour patterns of the same birds, serving a biologically identical function. In such cases, the courting drake always orients his movement in the gaze direction of a conspecific female, such that the optically stimulatory markings displayed in the motor performance concerned are fully exposed to the courted duck.

The accompanying orienting responses naturally carry far greater significance for instinctive behaviour patterns whose adaptive value depends upon a *mechanical* interaction with certain environmental objects than for those which operate entirely optically (or partially acoustically) as stimulus transmitting patterns. Of course, only a few of these mechanically-operative motor patterns are functional *without* an orienting response to bring the organism into spatial relationship with the object of its instinctive behaviour pattern. The tactic movement 'towards' or 'away from' can rarely be dispensed with in the case of mechanically operative instinctive behaviour patterns, unless the animal is to interact mechanically with a homogeneous medium. With plankton-eating animals, even the food may represent a homogeneous medium under certain circumstances, and it is therefore among such inhabitants of open aquatic environments that we find the greatest independence from taxes existing among free-moving organisms. In the world of solid objects, however, no animal with the power of locomotion can maintain itself successfully without orienting responses, since even the most differentiated and finely-adapted system of centrally co-ordinated motor patterns cannot function unless the animal provides the correct spatial orientation of each individual instinctive behaviour pattern towards the object.

This alteration of position of the animal, which is always governed by a taxis, by an orienting response in the broadest sense, can be completed *before* the performance of the instinctive behaviour pattern, so that a purely tactic and a purely central co-ordinated motor sequence are seen in temporal progression. A typical case of such progression from a taxis to an instinctive behaviour pattern is provided by the snapping response of the sea-horse (*Hippocampus*). By means of an orienting response of undoubted complexity, this fish orients itself such that the prospective prey is situated in the sagittal plane of its head, obliquely above and in front of its mouth at a specific distance and in a specific direction. It can take some time before this is achieved. The sea-horse will often follow a small crustacean backwards and forwards for a period of minutes, twisting and turning as far as its rigid armour will allow. Only when the spatial orientation described has been properly established is the necessary stimulus situation for elicitation of the muscular response pattern or snapping presented. An orientation response of this type, preceding the performance of an instinctive behaviour pattern, provides the simplest case of what we call 'appetitive behaviour' following Wallace Craig.

Psychologically, it is doubtless the most important and remarkable property of the instinctive behaviour pattern (and at the same time the most difficult property to explain on a causal basis) that the purely automatic performance – which itself lacks any orientation towards a goal sought by the animal subject – may itself represent *in toto* a goal of this kind. The organism exhibits an 'appetite' for performing its own instinctive behaviour patterns. Disinhibition and performance of the instinctive behaviour pattern are accompanied by certain subjective phenomena, and both man and animals actively *strive* to enter stimulus situations in which these processes will occur.[118] We attribute such subjectively rewarding experience of the instinctive behaviour pattern to any organism possessing such patterns – not merely by way of analogy, but because we regard this subjective experience effect as one of the most important processes ensuring survival of the species. As has been particularly emphasized by Volkelt, subjective experience is not a chance side-effect (an 'epiphenomenon') of physiological processes! Without the 'sensual pleasure' which presumably represents the experiential aspect of every instinctive behaviour pattern, performance of the pattern would only take place when the organism entered the elicitatory stimulus situation purely by chance. It is undeniably the subjective experience effect which makes the instinctive pattern an attractive goal. The incorporation of subjective phenomena in the casual chain of adaptive, physiological process presents the greatest philosophical difficulties; in fact, it must be regarded as the central feature of the mind–body problem. It is particularly remarkable that these pleasurable sensations, which are largely proprioceptive, should accompany motor processes whose form and co-ordination characteristically arise without participation of the proprioceptors.[119]

The adaptively variable 'appetitive' behaviour directed towards elicitation of an instinctive behaviour pattern can, in the simplest case, be represented by one of the most common and best-analysed topic responses. However, there are all conceivable transitional stages between such responses and the highest known feats of learning and insight. In

the Lorenz publications already mentioned (1937a and b), detailed consideration was given to the remarkable fact that, from the standpoint
of objective ethology, no distinction can be made between the simplest
orienting response and the highest form of 'insight' behaviour.

Apart from taxes which precede the elicitation of the instinctive
behaviour pattern, *procuring* elicitation and thus clearly exhibiting an
appetitive character, there are also taxes which *continue to function during*
the performance of the instinctive motor pattern, as has already been
demonstrated with the example of topotactic orientation to gravity.
With instinctive behaviour patterns of mechanical function, they perform the task of maintaining the spatial orientation of the animal to the
object established by the appetitive orienting response. The fact that
this spatial relationship, in psychological terms, frequently corresponds
to a subjectively pleasurable stimulus situation automatically means that
one can very often interpret these continuing responses together with
the orienting mechanisms of the instinctive behaviour pattern as a kind
of appetitive behaviour. A grazing cow, for example, continuously holds
its head in a position dictated by orienting responses, such that the
necessary spatial relationship between the mouth and the grass is maintained. On the objective side, this relationship renders the purely instinctive motor patterns of the jaw, tongue and lip muscles mechanically
operative and ensures their adaptive function, while on the subjective
side the animal is presented with a pleasurable stimulus situation whose
preservation represents a rewarding goal.

Obviously, such simultaneous co-operation of taxes and instinctive
behaviour patterns is extremely common in unitary motor sequences of
higher organisms. Such 'simultaneous intercalation patterns', as we
shall call these composite behavioural units composed of simultaneously
operative heterogeneous factors, can equally obviously attain a virtually
unlimited degree of complexity. This can present almost insurmountable
difficulties in analysis employing the few methods of investigation
available. However, the comforting degree of optimism necessitated by
the size of the task involved is brought closer by one fact: Just as there
are cases in which we find pure instinctive behaviour patterns free of
taxes, or instinctive behaviour patterns and purposive behaviour in an
easily analysable sequence, there are also cases in which particularly
favourable circumstances permit us to separate the instinctive behaviour
pattern and the orienting response during a bout of stimulus activity.

II Aims of this study

The task of the research discussed in the following account was the investigation of the co-operation of an instinctive behaviour pattern and a simultaneously active orientation response, using a particularly simple case which was also especially suitable for analysis for supplementary technical reasons. The simplest imaginable case of such co-operation is provided when the centrally co-ordinated impulses of a motor pattern operate in *one* plane, while the movements produced by the taxis operate in a *perpendicular* plane, thus causing receptor-controlled deviations from the plane of the instinctive behaviour pattern. In order to illustrate a process of this kind, one may construct a simplified model. Let us consider a skeletal component which is rotatable around a ball-joint, something in the manner of a sea-urchin spine, and is moved by two perpendicularly-oriented pairs of antagonistic muscles, one pair operating vertically and one pair operating horizontally. Let us further assume that this skeletal component must oscillate through a particular spatial plane (e.g. vertical) to provide the necessary survival value. This oscillation of the skeletal organ can very well be operated by a centrally co-ordinated automatism, which transmits its impulses to the approximately vertically oriented pair of muscles. But in order to ensure that the movement is performed exactly in the vertical plane, a taxis controlled by a gravity receptor must be incorporated. Without affecting the spatial orientation of the animal's body as a whole, this taxis arrangement can ensure (through the orienting mechanism) that the second, horizontally contracting pair of muscles perform contractions which cause the skeletal component to be displaced from the plane of oscillation of the central automatism just enough to produce the exactly vertical movement necessary for survival of the species. This special case, in which an instinctive movement occurs exactly in one plane and is steered – like a horse between two reins – by perpendicularly arranged antagonistic muscles controlled by an orienting response, presumably occurs quite frequently. Similar relationships seem to be present in human chewing actions, the frictional movements of copulating male mammals and in other behavioural acts.

The mechanism outlined above appears to apply quite particularly to a motor pattern which can be observed with brooding birds of many different species. The adaptive function of this motor pattern is the return of eggs which have rolled out of the nest-cup. The behaviour pattern incorporates movements which can be immediately recognized as balancing actions, and can therefore scarcely be explained as anything other than products of an orienting response, and also movements which quite clearly exhibit all the characteristics of a true instinctive behaviour pattern. Since these movements also occur simultaneously and at a right angle to one another, we were convinced that we had found a particularly suitable object for the study, and attempted analysis, of taxis and instinctive behaviour pattern. When the Altenberg geese began to brood in the spring of 1937, we did not neglect the opportunity to conduct relevant observations and experiments.

The analytical approach governing the observations and the arrangement of experiments was based on a very simple consideration: If, as has been assumed, the taxis is a *response norm*, whilst the instinctive behaviour pattern is a *motor norm*, it must naturally follow that, in a case where the taxis and the instinctive behaviour pattern operate through separate muscles of the animal, absence of the steering stimuli must be accompanied by absence of the movements dependent upon the orienting response. Therefore, in the 'vacuum activity', the instinctive behaviour pattern must be performed *alone*, without the normally simultaneously functioning orienting response. In addition, alteration of the stimuli characteristic of the overall situation must lead to similar changes in the movements controlled by taxes, since the latter are dependent upon the former. These changes must take the form of adaptation to the new spatial conditions, whereas the instinctive behaviour pattern – because of its independence of receptor-control – must always be performed in exactly the same way, regardless of whether the performance occurs in the complete absence of an object, with a substitute object greatly deviating from the norm or with the biologically-adequate object. In addition, any alteration of the spatial relationships which requires even the smallest adaptation of the form of the motor pattern must lead to complete functional failure of the instinctive behaviour pattern. Finally, it was an evident step to exploit the action-specific fatigue effect characteristic of the instinctive behaviour pattern, in an attempt to cut out the central automatism through exhaustion, in order to see how far the appetite for the normally elicitatory stimulus situation and the associated orienting responses would be altered.

III Observations of the egg-rolling movement

The facts derived from pure observation without experiments will take first place, not only because they represent 'involuntary' experiments free of many of the sources of error of planned experiments, but also because they temporally preceded all experimental results in the collection of data.

The normal adaptively functional performance of the response takes roughly the following form: When the brooding goose sights an egg lying not too far away outside the nest-cup, the egg-rolling movement definitely does not emerge immediately as a 'decisive' action. Instead, the goose – on first sighting the stimulus-transmitting object – briefly gazes at the egg and then looks away. When the goose's gaze returns to the egg, it is fixated for a longer interval, and it is possible that a mild movement of the head towards the egg may be performed at this stage. After a brief series of intention movements of rapidly-increasing intensity, but of highly-variable length, the goose will stretch its neck

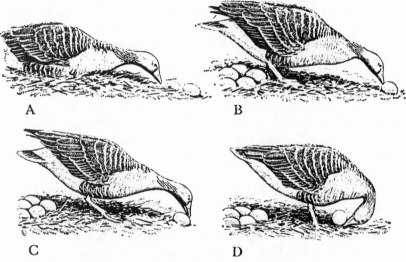

A   B

C   D

Fig. 1. *Normal performance of the egg-rolling movement*

towards the egg lying in front of its head, without moving the rest of its body (Fig. 1A). The goose frequently remains squatting in this position for a number of seconds, 'as if spell-bound', until rising slowly and hesitatingly from the nest (without altering the posture of the head and neck) and walking towards the egg with the peculiarly careful gait characteristic of all locomotion in the neighbourhood of the nest-cup. The premature stretching of the neck towards the egg gives the impression that the bird cannot localize the egg properly in space and 'hopes' to be able to reach it without having to get up from the nest. This impression, which is doubtless erroneous, is strengthened when the bird carefully approaches the egg, counting every step. This aversion to moving away from the nest-cup is based on the very good reason that the instinctive behaviour pattern is only really 'appropriate' and capable of transmitting the egg into the nest in a single performance when the goose does not move farther than the wall of the nest. When the goose has approached the egg sufficiently, the egg is first touched with the tip of the beak at the point on its surface nearest to the goose (Fig. 1B), and the slightly-opened beak is presumably pushed against the egg. The lower side of the beak is then passed slowly forwards over the upper surface of the egg, maintaining continuous contact, until it has extended beyond the egg and approaches the substrate on the side furthest from the goose (Fig. 1C). At this point, the bird's entire body is seized with a peculiar tension; the neck becomes taut and the head begins to quiver quite noticeably. This delicate quivering motion persists while the egg is slowly pushed or rolled towards the goose and into the nest-cup by means of a movement of the lower side of the beak accompanied by a strangely stiff and awkward-looking curved posture of the neck. The egg finally arrives on the goose's toes (Fig. 1D), and if it has by then passed the highest point of the nest wall it will roll into the nest-cup to join the other eggs. The tension and the delicate quivering motion is produced by a relatively strong accompanying innervation of the antagonists to the muscles which perform the actual work of the rolling movement. A similar phenomenon can be seen in voluntary human movements when these are opposed to an *unknown* resistance or to a resistance which can undergo unpredictable and sudden alterations in intensity. In such cases, in addition to the muscles performing the actual mechanical work of the movement concerned, *the antagonists* are activated. By means of a known and controllable intensification of the internal resistance to the movement, the unpredictable and sudden variation of the external resistance is rendered less effective, since the latter now represents only a minor contribution to the total resistance which must be overcome. This does indeed lead to a far greater consumption of muscular energy than is actually required for performance of the required external work, but stray movements are reduced to a fraction of their expected extent. This kind of control of muscular power, restrained by wasteful antagonistic activation, is the only means of stabilizing the form of the intended movement in the face of variations in the external resistance, which occur too rapidly for timely operation of receptor-controlled braking of the stray movement. On the basis of this necessity to do without receptor-control, we can explain the accompanying innervation of antagonists, the resultant tension of the part of the body employed and the frequently evident quivering movement observed with the instinctive behaviour pattern. The centrally co-ordinated motor pattern as such is not influenced by the receptors and therefore, in cases where mechanical work is performed against external resistances, the behavioural action *always* operates against a (so to speak) unknown resistance. In all cases where the resistance may vary, the pattern will require internal control of muscular power to a far greater extent than any voluntary pattern of movement. We do in fact find the described conditions of tension and antagonistic quivering in many mechanically-operative instinctive behaviour patterns. A particularly clear case is the lateral pushing movement with which herons and related birds insert twigs into the nest structure. A further indication of accompanying innervation of antagonists is provided by the fact that with instinctive behaviour patterns which have to overcome considerable resistances in their normal performance, the absence of resistance in occurrences of the vacuum response does not produce a noticeable alteration in the pattern or speed of movement.

Apart from the motor pattern described, which is performed more or less through the bird's sagittal plane, there are other motor acts which can be observed. These are *lateral* movements of the head and beak,

Fig. 2. *The hollowing movement*

which are normally restricted in extent and evidently perform the task of preventing the egg from slipping to the right or left and rolling back past the beak as it is rolled up the nest wall. *The egg is balanced on the underside of the beak.* This is particularly obvious when the path of the egg leads very sharply upwards so that a large part of the total weight is carried by the goose's bill. The extent of the movements immediately increases when the egg threatens to become unbalanced and can only be prevented from rolling away by application of a more extensive compensating movement. In such cases, the balancing nature of the lateral movements is immediately obvious.

After the conclusion of this egg-rolling movement, and always *before* the appearance of the settled brooding posture, the goose performs a specific instinctive behaviour pattern which will be referred to as the *hollowing movement*. The elbows are slightly extended such that the wing shoulder is lowered and pushed forward to press against the anterior edge of the nest. At the same time, the goose pushes itself forward with alternating backwards shuffling movements of the feet, so that the breast and the wing shoulders exert a strong pressure on the anterior edge of the nest-cup, whilst the feet themselves push nest-material towards the edge of the nest situated behind the bird (Fig. 2). This movement is employed alone for production of the nest-cup at the beginning of nest-building, and the same movement is used by the bird ιo remove the protective layer of down when returning from a brooding pause. Adoption of the normal brooding posture apparently never occurs without prior performance of this instinctive behaviour pattern.

From this normal, adaptive performance of the response itself, one can conclude fairly reliably that two fundamentally different motor processes are involved. An entire range of immediately evident characters of the instinctive behaviour pattern proper, such as the described tense condition accompanied by antagonistic quivering, the reliable photographic repetition of the movement, a number of comparative zoological features and other phenomena, permit the extremely safe assumption that centrally co-ordinated motor processes are present. On the other hand, the lateral movements of the beak which balance the egg are completely dependent in direction, extent and form upon the tactile stimuli which emanate from the egg in the course of its varying deviations to the right or left. These movements must therefore be regarded as the products of a taxis. A number of further facts, though derived from observations conducted on the basis of the analytical approach described rather than from experiments, support this assumption.

The most important of these observations, which strictly speaking includes all the results which later experiments were to provide, is the following: The egg rolled towards the nest with the motor pattern described will by no means always arrive at its goal. It frequently slips away from the beak as it is pushed up the nest wall, dropping back on more than half of the observed occasions when the wall was steeply constructed. But when there is a failure of this kind, the motor sequence is not always broken off – it is frequently continued in a peculiar manner. The tucking movement of the head and neck continues exactly as if the movement were really transporting an egg beneath the goose's abdomen. *But in this vacuous completion of the movement the balancing lateral movements of the head are absent.* The continuing movement, which proceeds exactly along the sagittal plane of the bird standing in the nest-cup, eventually leads to contact of the beak with the eggs remaining in the nest-cup. This tactile stimulus appears to elicit a new egg-rolling movement. At least, the contact almost always leads to particularly intensive and thorough rolling of the entire clutch, whereas this intensity is not usually observed when the egg rolled from outside really does reach the nest-cup. This gives the impression that the goose has been left 'unsatisfied' following the short stretch of the movement performed without the egg and consequently 'enjoys' with an enhanced appetite the tactile contact of the beak against the eggs. After rolling the clutch around, the goose settles onto the eggs, performs the hollowing response already described (Fig. 2), and – as long as there are at least a few eggs to be felt beneath the abdomen – will remain settled in completely contented fashion until the runaway egg is spotted afresh. To use an involuntary anthropomorphization: the goose seems to exhibit amazement at seeing 'yet another' egg outside the nest. At least there is no doubt that the goose is not aware whether the egg-rolling pattern has transported the egg into the nest-cup or not. Following an interval of varying length and renewed summation of the stimuli emanating from the egg, a new rolling response is performed.

The interpretation of the portion of the movement performed without the egg as a vacuum response may be countered with the supposition that the goose simply moves its beak along the shortest path to the nearest available egg after losing the egg it was rolling. For this reason, Lorenz intends next year to procure more extensive vacuum activities than those so far observed by artificially arranging for the rolled egg to disappear as rapidly as possible (e.g. by using a pit trap or by suddenly whisking away a light dummy with a thin thread). The responses will be recorded on film so that the form of the motor pattern in vacuum responses can be directly compared with that in the normal performance. We believe that a comparison of this kind will demonstrate the correctness of the assumption made here.

The fact that the balancing lateral movements disappear in the vacuum activity shows that these movements are dependent upon the tactile stimuli derived from the egg and therefore represent orienting responses. This does not necessarily mean, however, that the remaining part of the motor pattern lacks control through taxes. But if such additional, simultaneous responses are present they definitely play no more than a minor rôle. In any case, during the egg-rolling movement the goose's beak does *not* follow all of the minor vertical displacements which the egg undergoes because of unevenness of the substrate. The beak contacts the egg at its upper face when it is rolling through a small hollow and low down when the egg happens to pass over a prominence. Probably, the only process which orients the rolling movement roughly parallel to the substrate, to the pathway of the rolling egg, is the stretching of the neck towards the egg observed *before* the egg-rolling movement is observed. There is therefore no orienting response operating simultaneously with the instinctive behaviour pattern, but a preceding telotaxis belonging in the category of appetitive behaviour (in its narrowest sense). The motor pattern can also exhibit coarse irregularities in its parallel orientation to the substrate, under certain circumstances. The egg-rolling pattern is apparently quite specifically tailored for the situation in which the performing goose is standing exactly on the edge of the nest. Under natural conditions, this is doubtless usually the case. In the first place, the goose is subject to a considerable inhibition against leaving the nest (i.e. against proceeding further than the nest wall); in the second place, this is usually sufficient for the goose to reach an egg which has not been deliberately placed some distance away, but has been allowed to roll freely down from the edge of the nest, so that the spatial relationship between the egg and the goose is doubtless closest to that under natural conditions. Since a goose nest is always located in the middle of thick-stemmed vegetation, the egg always rolls just to the edge of the area covered with nest material, such that a fairly constant distance from the centre of the nest is established. Under these conditions, the motor pattern in the sagittal plane appears well adapted. The form of the pattern automatically determines that the vertically-operating component of the pressure from the beak is strongest at the point where it is most necessary, namely towards the end of the movement, where the egg has to pass the steepest gradient on the nest wall. The nest of our experimental goose was composed of pine needles (i.e. an extremely unnatural form of nest-material) and the wall was far lower and less steep than that of all the nests which we have seen under natural conditions. In Fig. 1D, one can almost see that the motor pattern is actually adapted for a somewhat steeper nest wall. Genuine disruption of the rolling movement actually occurred when the egg was *nearer* than expected under natural conditions. The goose, in this case, did not move forward but remained stationary over (or even behind) the eggs, so that the strongest lift exerted on the egg by the beak appeared at a point on the pathway where the substrate had already begun to decline again, because the peak of the nest wall had already been passed. The egg was then often raised into the air, only to fall from some height onto the rest of the clutch after several seconds of continuous juggling on the underside of the beak. This frequently led to cracking of the eggs.

According to these observations, there would definitely appear to be no taxis-controlled movement in the sagittal plane of the egg-rolling response, apart (of course) from the extension of the head towards the egg, which is to be regarded as appetitive behaviour. This latter movement only quite generally determines the angle of inclination of the neck and back of the goose and it occurs in temporal separation from the centrally co-ordinated motor sequence. Nevertheless, the possibility of finding other taxes which function simultaneously with the instinctive pattern must be examined in more detail. If it is at all possible, this will be attempted next year through preparation of ciné-film of the egg-rolling movement taken exactly from the side, with the profile of the path varied as much as possible in a number of successive shots. Such film material would permit graphic reconstruction of the curve of the movement, so that any influence of tactile stimuli from the egg upon the movement can be clearly demonstrated.

IV Experiments

This year's experiments were confined firstly to investigating the dependence of the balancing lateral movements upon tactile stimuli and secondly to demonstrating the fixity and receptor-independence of the sagittal movement which must be present if a pure instinctive behaviour pattern is involved. We at first looked for objects which would elicit the goose's rolling response and were nevertheless so different in form from a goose's egg that the stimuli necessary for the thigmotactic orienting response would be greatly modified or entirely absent. In addition, we looked for objects which would require receptor-controlled modifications of the sagittal movement and would literally cause a breakdown if the movement were independent of the receptors and purely centrally co-ordinated. This search for substitute objects had the result that we discovered a number of things concerning the conditions necessary for elicitation of the rolling response. This combination of hereditarily-recognized characters of the object – the so-called innate schema of the egg – will be briefly discussed.

1. THE INNATE RELEASING SCHEMA

The term innate releasing schema applies to the hereditarily-determined readiness of an animal to respond to a specific combination of environmental stimuli by performing a specific behaviour pattern. There is an innate receptor correlate for a stimulus combination, which (in spite of its relative simplicity) characterizes a biologically-important situation with sufficient exactitude to link the appropriate response firmly to the conditions of that situation and to prevent inappropriate elicitation by environmental stimuli exhibiting chance similarities. The elicitation of motor processes by the reaction of an innate schema corresponds in every detail to an *unconditioned reflex* and is allied to the simplest unconditioned reflexes by a continuous series of intermediate forms. However, this only applies to the releasing mechanism itself and it is not permissible to draw conclusions about the released or inhibited motor pattern. The latter can in fact be of an entirely different character. The

reaction of an innate schema can just as easily elicit a taxis-free instinctive behaviour pattern or appetitive behaviour leading to performance of
an instinctive pattern, and the appetitive behaviour can be represented
by a simple orienting response or by the highest forms of purposive
behaviour. Similarly, a pure taxis leading to the converse situation of a
refractory resting state, rather than to an elicitatory stimulus situation,
may be elicited by an innate schema.

In the case of the egg-rolling response, the innate schema of the
situation 'egg outside the nest' elicits appetitive behaviour in the form
of the previously-described orienting response of stretching the neck.
The stimulus parameters representing the characters of 'the egg' are so
simple and restricted in number that one is amazed that they characterize the egg to a biologically adequate extent. It is a well-established
fact (Koehler and Zagarus, Tinbergen, Goethe, Kirkman, etc.) that
many brooding birds, when performing the egg-rolling response, will
accept objects which bear very little similarity to the egg of the species
concerned. Herring-gulls will brood on polyhedrons and cylinders of
any colour and of virtually any size, and black-headed gulls appear to be
even less selective. At first, our Greylag geese behaved in a basically
similar way, but they later exhibited a remarkable change in their behaviour determined by individual learning. It is also possible that pure-
blooded individuals might react in a manner different from that of the
hybrid animals investigated. It is a particular feature of innate releasing
schemata that *loss of characters* can occur as effects of domestication, and
this leads to a considerable broadening of the object-schema. The observations must be repeated with pure-blooded geese in order to determine
whether such an effect operated with our experimental animals.

The purely optical elicitation of the appetitive orienting response of
stretching the neck is in fact linked only to the character of *surface
continuity* of the object. Any smooth object presenting an approximately
even contour will arouse the attention of the goose and will provoke
neck-stretching. However, a tactile character operates a moment later.
Even before the goose stretches its beak over the object, the degree of
hardness is determined by lightly pushing the beak against the object
(Fig. 1B). A rounded toy chicken made of white rubber will clearly
awaken appetitive behaviour for the rolling response. The goose rises
and stretches out its beak towards the object, but after gently tapping
with the beak the interest immediately disappears. A yellow children's
balloon blown up to the size of a goose's egg is treated in the same way.
A hard-boiled, peeled Muscovy duck egg is tested with the beak and
then greedily eaten. In our experience, the colour of the object plays
no part whatsoever in the goose's behaviour, and this provides a certain
contrast to gulls and plovers. Size has no effect over such a wide range

that one is scarcely able to decide whether failure of the response with too large or too small eggs is due to limits imposed by perceptive mechanisms or by purely mechanical factors. On the other hand, the requirements imposed with respect to surface continuity of the object at the beginning of the appetitive behaviour are considerably reinforced in the succeeding components of the response. The performance is particularly prone to immediate blockage by optical or tactile perception of any projecting edges or appendages on the object. A hollow wooden cube opening on one face is immediately rolled by the goose, but this persists only until the opening comes to point upwards. The rolling response then ceases as the goose begins to peck intensively at the exposed rim of the opening. A white toy chicken carved from a piece of wood the size of a goose's egg was not rolled but nibbled wherever projections (such as eyes, beak and feet) had been glued on to the egg-shaped body. Similarly, a large cardboard Easter egg was nibbled where the paper frills projected from the decorated seam. The right-angled edges of a cube or the edges of a plaster-case cylinder, on the other hand, are not noticed and do not disrupt the egg-rolling response.

This intensive response to all projecting appendages is not a chance phenomenon. The described form of the response is only elicited when the nibbled prominences are attached to an object *which otherwise agrees well with the egg schema* and arouses the appetite for rolling and brooding. This is doubtless a quite specific response which is presumably similarly characteristic of all Anatids and guarantees the adaptive function of *removal of broken eggs*. Experiments with various duck species and Egyptian geese have shown that such eggs are 'dismantled' from the edges of the crack and devoured, before the other eggs in the clutch are soiled by leaking egg content and consequently handicapped in respiration. Of course, this response must be inhibited prior to the onset of hatching, since the goose would otherwise kill the emerging chicks. It so happened at this time that pure-blooded Greylag goose eggs, which had been brooded by a hen, were just about to hatch. For technical reasons (regarding the requisite oiling of the down feathers of the goslings by the plumage of an adult goose), we intended to arrange for the actual hatching process to occur under our experimental goose. This proved to be utterly impossible, however, since the brooding goose immediately began to nibble greedily at the edges of the pipped areas on the eggs, responding at an evidently high intensity level. The goose pecked so uninhibitedly that the edge of the shell was at once pressed inwards, and blood began to flow from the torn egg membranes. The goslings would definitely have been killed, one and all, if the eggs had not been rapidly removed. When the goose's own offspring hatched only two days later, however, the response of nibbling edges and prominences had completely disappeared. We were not even able to elicit the response by directly presenting the goose with the pipped openings. The goose did in fact incline her head towards the peeping egg in an aroused and attentive fashion and the egg was even touched with the slightly opened beak, but no actual pecking resulted. The same movement of the head and beak is often exhibited by a guiding mother goose towards very young goslings, in which case the impression of an affectionate gesture is given. It only remains to provide experimental backing for our supposition that acoustic stimuli emanating from the hatching egg provoke response-reversal in the mother, so that even the slightly-pipped egg is already treated 'as a gosling'. This supposition is supported by the fact that pipped eggs are evidently no longer turned, since in the undisturbed clutches of hatching goose and duck eggs the pipped areas are virtually always uppermost and remain so. In addition, the mother no longer treads upon the eggs, as she does without hesitation in earlier stages of development (see also Fig. 5), and she does not even lower her full weight onto the clutch. Instead, there is already an indication of the squatting position which later serves to warm the young goslings. Unfortunately, we did not investigate whether a markedly pipped egg lying outside the nest-cup would still be rolled or not.

An attempt to elicit the rolling response with a freshly-dried, but still helpless, gosling gave an unusual result. Surprisingly, the goose initially responded in exactly the same way as at the onset of the normal egg-rolling response. She stood up, bent towards the gosling and touched it tentatively with her beak. On a few occasions, we even observed the mother reaching over the gosling with her beak. However, a retrieving, pushing movement never followed; instead, the goose always withdrew her head 'disappointedly'. This gave the definite impression that the goose *intended* to transport the gosling into the nest and attempted to achieve this with the rolling movement, but that the response was disrupted by the poor correspondence between the gosling and an egg. The goose was obviously aroused. A few seconds after the failure of the response, the mother once again stretched her neck towards the peeping gosling. However, since the relevant experiments with pure-blooded greylag geese have not yet been conducted, we will not discuss this behaviour at this juncture. It is possible that the continual backwards and forwards motion of the mother's head might eventually lead a scarcely mobile young gosling back into the nest, but it is equally possible that the experimental goose was subject to domestication-induced disruption of a specific response to a helpless chick situated outside the nest-cup.

Our experiments on the innate releasing schema of the egg-rolling response came to a remarkable close long before the end of the brooding phase of our experimental animals. After a pause of several days in the middle of the brooding phase, we attempted to induce the tamest of our geese (which had been used for almost all of the previous experiments) to roll a number of objects of different kinds, but we obtained no response. It soon emerged that the egg-rolling response, from this point onwards, *could only be elicited by genuine goose eggs*! All of the objects previously described as substitute objects were ignored and even an intact, small chicken's egg evoked no response. On the other hand, the egg-rolling response was performed with goose eggs immediately and without any particular tendency towards habituation; so our suspicion that the intensity of operation of the central mechanism had been reduced was shown to be unfounded.

2. EXPERIMENTS TO SEPARATE TAXIS AND INSTINCTIVE BEHAVIOUR PATTERN

The wide range of variation in substitute objects permitted by the crude and simple character-schema of the egg provided us with the opportunity of presenting appropriate objects of specific size and form in order to produce situations which facilitated the dissection of the entire behaviour pattern into tactic and instinctive components.

Our first task was experimental verification of the hypothesis that the lateral movements of the beak which balance the egg during the egg-rolling pattern are directly evoked by tactile guiding stimuli, as was deduced from the observations discussed on pp. 331–332. For this purpose, we needed to find a substitute object which – unlike the rolled egg – would not exhibit persistent lateral departures from the path and would thus fail to give tactile stimulation of the lateral margins of the underside of the beak. We first attempted to induce the goose to roll a plaster-cast cylinder along a smooth pathway by presenting the object on a broad plank resting on the nest wall. This experiment proved to be unsatisfactory, to the extent that the concomitant disturbance produced by the unfamiliar (and therefore fright-inducing) board weakened the goose's response and led to rapid habituation. In addition, the goose frequently failed to contact the cylinder in the middle (as would have been the case with egg-shaped objects). The cylinder was grasped so far to one side that pronounced lateral departures from the path occurred and elicited even more marked lateral head movements. These movements were nevertheless noticeably distinct from those usually observed. Since the straying movements of the cylinder were slower and less frequent than those of the egg, the relationship between each straying motion of the object and the compensating lateral movement of the beak was far more distinct than in the normal performance.

Exclusion of the balancing lateral movements proved to be much simpler and far more extensive when the goose was presented with an object which would slide rather than roll. A small, extremely light wooden cube, which remained stationary at any point on the nest perimeter and exhibited no tendency to slide back or to roll in a particular direction, proved to be the best object for our purposes. This was probably partially attributable to the purely mechanical fact that the cube usually tipped over so that one of the relatively broad, flat faces came to lie across the two rami of the underside of the beak (Fig. 4, p. 343). This naturally meant that the cube would tend to remain in the plane of movement, and it did in fact emerge that the cube was always pulled directly into the nest *without the slightest lateral movement.*

On the basis of these two experiments, we believe that it can be taken as established fact that the lateral movements of the beak are directly elicited by tactile stimuli produced by sidewards straying of the rolled object.

As a counterpart to this conclusion, it was necessary to establish whether the movement in the sagittal plane (which has been described as a pure, centrally co-ordinated instinctive pattern) is indeed free from modification by supplementary orienting responses serving to produce more detailed adaptation to the spatial relationships in any given individual situation. However improbable such modification may seem on the basis of the observations already discussed, it is nevertheless good practice to attempt to induce some slight modification of the innate co-ordination of the sagittal neck and head movement. The largest substitute object used in our experiments (the large cardboard Easter egg already mentioned) seemed to be suitable for this purpose. Since the diameter of this object required entirely different neck postures to those necessary for a genuine goose egg, it was obviously a promising step to observe the operation of the rolling movement under the resulting altered conditions. The size and curvature of the object determined a disruption right at the beginning of the movement. The action of reaching over the egg (as seen in Fig. 1C) is initially followed by a movement of the head alone, in which the head is inclined at an acute angle towards the neck, at first without altering the extended neck posture. This typical initiation of the sagittal movement will push a genuine goose egg towards the nest – but not the dummy egg used in the experiment. The beak is pressed against the surface of the substitute object at such an awkward angle that the head slips away, and we were forced to intervene slightly in order to permit the response to continue (Fig. 3A). All features of the motor pattern exhibited invariable adaptation to a much smaller object in other details of the response as well. Even before the pronounced flexure of the neck (Fig. 1D) could occur, the dummy egg *jammed* between the goose's beak and thorax so that the posture shown in Fig. 3B remained unaltered through 1½ metres of our film-strip. The response was then broken off and the goose adopted an upright posture. Characteristically, this event was never followed by the 'hollowing movement' typical of the 'satisfactory' performance of the response. Instead, the goose remained standing in the nest 'in discomfort'. Such forceful disruption of an instinctive behaviour pattern is apparently experienced as something extremely unpleasant, since the experimental goose exhibited 'disinclination' towards the large Easter egg (expressed

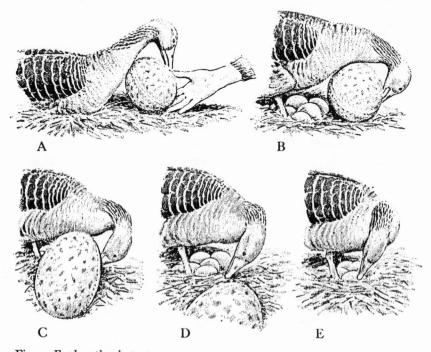

Fig. 3. *Explanation in text*

as lowered response motivation and rapid habituation) earlier and more intensively than towards other substitute objects. The goose never hit upon the idea of transporting the egg into the nest by walking backwards *without* the pronounced flexion of the neck.

In a number of our experiments, including that providing illustrations for Fig. 3A–E, the goose reached a subjectively far more satisfying solution: The dummy abruptly slipped out from between the beak and the thorax, like a cherry-stone pressed between two finger-tips. In Fig. 3C, it can be clearly seen how the beak follows the egg to the right 'in the attempt' to balance it. Four frames later in the film, the sagittal movement can be distinctly seen progressing *in vacuo*. In Fig. 3D, the head is still displaced somewhat to the right following the ultimate, extremely pronounced thigmotactic movement, but thereafter the movement of the beak is not related to the object, which is lying far out in the foreground. The movement is subsequently followed by the intensive rolling of the clutch described on p. 333 (Fig. 3D), and the goose eventually settles and performs the 'hollowing movement'. In all cases where this abrupt slipping effect of the large, but quite light, cardboard egg robbed the response of its object, the progression of the pattern *in vacuo* was particularly clearly observed, since the object disappeared instantaneously and without producing supplementary disruptive stimuli.

Although the *form* of the sagittal movement thus proved to be generally invariable in all observations and experiments, there remained the question as to whether the *power* of the response might be open to receptor-controlled adaptation to the obstructing resistance. Following the described observations on the associated innervation of the antagonists and on the detailed similarity between the vacuum activity and the normally-functioning mechanical performance, one might be inclined to think that an adaptation of the strength of the movement to meet the mechanical requirements might occur through the antagonists, for instance by reducing the resistance of the inhibitory muscles in the face of greater external resistance. According to our findings with substitute objects heavier than a goose's egg, this does not appear to be the case. These findings were initially unintentional, since the egg-rolling response failed in the face of resistances which we had assumed would be overcome as a matter of course. With the plaster-cast cylinder already mentioned, the power of the beak sweep was just sufficient to move the object over a smooth substrate, but the slightest obstacle immediately blocked the action. This, despite the fact that the power *applied* was only a fraction of the *potential* energy. Above all, the energy applied was definitely not confined to the direction in which the substitute object was rolled; the interplay of antagonistic muscles already described was just as obvious as ever before. This weak and unsuccessful wavering of the response with an object which is only a little overweight itself gives the most convincing impression of unintelligent, machine-like behaviour, at least for the initiate who is aware of the amazing power which a greylag goose's neck can otherwise exhibit. For example, playful performance of the grass-plucking pattern can produce a pull which will drag a heavy oak chair across a rough floor or whisk away the tablecloth from a table laid for several people. The fact that only an inordinately small fraction of this capacity can be employed in the rolling response for rolling a slightly overweight object supports the assumption that the energy employed in the sagittal movement is not accessible to any appreciable receptor-control.

With substitute objects which are lighter than a goose egg (e.g. the wooden cube mentioned several times previously), the movement in the sagittal plane showed virtually no difference from the vacuum activity. Towards the end of the movement, where the goose egg may normally be lifted and balanced in mid-air (see p. 334), the cube was regularly swept upwards and often pressed against the goose's thorax so that it was jammed in that position for a moment by the beak (Fig. 4). The ease and regularity with which the light cube was lifted created the definite impression that the tension of the muscles concerned in the sagittal movement had been calculated once and for all to cope with the normal average weight of a goose egg. On the other hand, this lifting of the cube itself awakens the suspicion that even in the muscles performing the

Fig. 4. *Retrieving a cube*

sagittal movement there are minor reflex processes operating alongside the centrally co-ordinated impulses. In the completely vacuous performance of the movement, in which the beak has nothing whatsoever to carry, one would expect the head to lift even further than in the carriage of the cube, and this is not the case. Instead, the beak is held noticeably low in the vacuum activity and even scrapes the substrate in some cases (Fig. 3D). The tactile stimulation from the rolled object presumably does not elicit an orienting response in the true sense of the term but produces a *tonus effect* in the muscles which incline the neck and (particularly) the head. A reflex tonus effect of this kind would of course be dependent upon receptors, but it could only be interpreted as an orienting response if it were to emerge that the intensity of the effect depended upon the weight of the rolled object. But this would appear to be improbable on present evidence. On the basis of the observations reported here, we incline more to the view that the power employed remains the same with both light and heavy objects, even if it may be slightly greater than the power involved in the vacuum activity.

3. APPETITIVE BEHAVIOUR FOLLOWING OVER-SATIATION OF THE INSTINCTIVE PATTERN

In conclusion, we should like to discuss two further observations which do in fact provide direct experimental evidence for the separation of taxis and instinctive pattern, but which are most important with respect to a different problem. From different quarters, compulsive elicitation of the 'instinctive' behaviour pattern (i.e. with the orienting responses often included under the same concept) has often been emphasized as an important characteristic. The instinctive behaviour pattern as defined here is, of course, not elicited with the same compulsive quality as a knee-jerk reflex or the like, since elicitation of the former always signifies *disinhibition* and since performance of the pattern is dependent upon central arousal states. Exhaustion of the central mechanism renders the normally elicitatory stimuli utterly inoperative. Such exhaustion appears in an action-specific form extremely rapidly with many instinctive behaviour patterns, long before the organism as a whole or the participating effectors are exhausted. Some instinctive behaviour patterns, such as the injury feigning performed by a number of birds to distract predators from the nest-site, can only be repeated a few times in succession. Even if the elicitatory stimulus situation persists, the behaviour pattern is not repeated. In the case of injury feigning, the bird presents an indifferent activity to the potentially elicitatory human observer and begins to search for food, to preen itself, etc. This is far from the case with the taxis; an orienting mechanism can usually be set in operation on a virtually limitless number of successive occasions. A beetle turned on to its back will right itself with a persistence greater than that of the experimenter. Similarly, an ant which is displaced from its path of locomotion will consistently re-orient itself with respect to the direction of the sun.

We must now ask whether this typical resistance to fatigue is also found with orienting responses which operate as appetitive behaviour, bringing the organism into the appropriate spatial relationship to the object for performance of the corresponding instinctive behaviour pattern. This is not generally thought to be the case; instead it is widely believed that the ease of elicitation of the appetitive behaviour, even where this is represented by a simple taxis, vanishes with satiation through the performance of the instinctive behaviour pattern. J. von Uexküll states: 'The effector signal always extinguishes the receptor signal – the behaviour pattern is thus concluded. The receptor signal is either objectively extinguished, for example when it is derived from a food object which is eaten, or it is subjectively wiped out when a state of satiation is reached – the sensory filter is closed.' But it is our impression that this extinction of elicitatory signals is not always complete, particularly when the elicited appetitive behaviour is a relatively simple orienting response. One can provide many examples in which an instinctive behaviour pattern can no longer be elicited after repeated performance, whilst the preceding orienting response can still be elicited. It is more or less the rule that an over-satiated organism will still attend to the stimuli normally eliciting a behaviour pattern (e.g. by orienting its gaze or performing some other orienting action), even when the instinctive behaviour pattern itself has been exhausted by repeated performance.

It is a remarkable fact that an orienting response which, by its very nature, is only functional as appetitive behaviour should be open to elicitation even in a case where the actual appetite (in terms of active purposive behaviour on the part of the subject) has been extinguished. Such elicitation has a distinctly compulsive, reflex-like appearance. It is also remarkable – and probably fundamentally important – that such reflex-like elicitation of such 'appetite-less appetitive behaviour' is evidently accompanied by subjective displeasure. Even we human beings, after satiation, will turn away in disgust from even the most delicious dishes and push the plate out of reach. In brief – we withdraw from the stimuli emanating from the object of the instinctive behaviour pattern just concluded. This itself is clear indication of a reflex-like, compulsive response to the stimuli concerned. It would not be necessary for us to withdraw from these stimuli if they had really been 'subjectively extinguished' by satiation (to use von Uexküll's terminology) and if we were not compelled by further operation of these stimuli, despite our lack of appetite, to an unpleasurable repetition of our response. Even if the above-mentioned quotation by von Uexküll is generally applicable, there are definitely a number of exceptions. It is possible to cite many cases in which an organism will not become completely indifferent to the stimulus situation eliciting appetitive behaviour at any level of satiation and will instead respond with a withdrawing response immediately after extinction of the approaching orienting response.

The described persistence of the introductory orienting response after exhaustion of the instinctive pattern was easily observable in the egg-rolling response of the greylag. The instinctive behaviour pattern concerned habituates rapidly. This is not surprising, since the tempo of central stimulus-production and the associated maximum frequency of performance of the pattern are adapted to the natural environmental conditions, under which the eggs are certainly not dislodged with the frequency pertaining in our experiments. It was completely unnecessary for us to conduct special fatiguing experiments; without our intervention we became more familiar with fatigue effects of the pattern than we actually desired. The differences in the behaviour elicited by a genuine goose egg and the oft-mentioned large Easter egg were conspicuous and probably significant. Since the observations concerned were conducted shortly before the point where the goose began to rigorously ignore all substitute objects, the paradoxical nature of the behaviour to be described is probably due to the fact that the goose was acquainted through personal experience with the goose egg as a satisfactory object for rolling and brooding, whereas the dummy had been experienced as a substitute object which did not provide full satisfaction.

Even after utter exhaustion of the instinctive pattern, the tactic

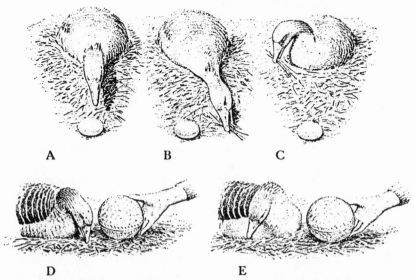

A   B   C

D   E

Fig. 5. A–C: *The 'gathering pattern'*. D and E: *Submissive gesture*

movement towards the egg persisted. The orienting stretching of the neck (Fig. 5A), or at least an attentive glance towards the egg, could be elicited virtually at any time. The stimuli emanating from the egg will apparently not allow the goose to settle down; the neck is repeatedly stretched towards the egg until the goose eventually performs a type of behaviour which we regard as being of great theoretical interest. In the place of the habituated, refractory egg-rolling pattern, the neck-stretching movement was abruptly followed by *a different instinctive behaviour pattern*. Immediately after stretching the neck towards the egg, the goose performed the *'gathering pattern'*, which is otherwise employed in nest-building. This motor pattern is exhibited by all Anatids and serves the function of collecting and consolidating nest-material at the nest-site (Fig. 5B, C).

Such substitution of a biologically inappropriate instinctive behaviour pattern for the expected pattern has been recognized in birds in a number of different situations. Birds which are greatly alarmed by the presence of a predator close to the nest (or offspring) will often interpolate, between the threatening movements and the well-known guiding-to-safety responses, behaviour patterns of food-seeking, preening and even settling down to sleep. Lapwings disturbed on the nest may rise and perform pecking movements; Corvids which are not quite confident enough to attack will peck with all their might at the branch on which they are sitting; and so on. Apparently, these 'erroneously' erupting instinctive patterns can be produced in *two* different ways. In some cases, such as that of nest-defending passerines which abruptly follow a small number of mock limping responses (or the like) with preening patterns, there would appear to be much justification for Howard's assumption that after exhaustion of the specific motor pattern appropriate to a given situation the receptor-controlled arousal state will flow over 'into other channels' and lead to the eruption of 'some other' instinctive behaviour pattern.[120] In other cases, such as that of a Corvid pecking at its perch, we are inclined to assume that two incompatible instinctive behaviour patterns (in this example, patterns of attack and escape) block one another so that the general arousal is channelled off into a third pattern. A further possibility for the production of similar phenomena would be the following: From some mammals we know of a functionless mode of performance of instinctive behaviour patterns which nevertheless exhibits some degree of 'insight'. Everyone is familiar with the 'begging' motion of scraping with the fore-hoof which Equids – from the carthorse to the wild stallion – will exhibit towards a human food-donor. Dogs which no longer dare to bite after a painful experience with a wasp or a hedgehog will regularly begin to dig in the direction of the arousing object. The individual scratching movements usually do not touch the object, so that a hedgehog rolled into a ball may eventually stand more or less on a pedestal, surrounded by a trough. In intelligence tests, monkeys will sometimes throw sand in the direction of the goal; and so on. In all of these cases, it is highly probable that an extremely primitive form of insight is involved, to the extent that 'something must be undertaken' in a particular direction.

It may be an overestimation of the powers of intelligence of the Greylag to assume that the behaviour described above is closest to that of the mammals mentioned, but this was our immediate impression when observing the process in action. It seemed as if the goose were 'unable to bear' the sight of the egg lying outside the nest-cup without doing something about it. This subjective description is probably less naïve than may appear at first sight, since human beings experience similar phenomena in just the same type of situation, where innate releasing schemata and instinctive behaviour patterns play a major part. We believe that the goose attempted to cope with the unsettling and unpleasantly intrusive stimulus situation following the failure of the egg-rolling response by the purposive 'application' of another accessible instinctive behaviour pattern. In fact, we feel justified in assuming that it was by no means a chance effect that the motor pattern which emerged happened to be one which similarly transported something towards the nest. A major feat of intelligence of this kind – the most impressive which we would attribute to a bird – itself demonstrates the animal's pronounced dependence on innate motor patterns.

Whereas a Greylag goose egg lying alongside the nest induced our experimental goose to perform the described behaviour after extreme exhaustion of the egg-rolling response, the cardboard egg had no effect at all once the response had suffered the least habituation. The goose then continued to brood in an entirely settled fashion in sight of the dummy egg. However, if the Easter egg was forced upon the goose (as shown in Fig. 5D), the beak was drawn in (as shown), as if to avoid any tactile contact with the egg. If the egg were pressed even closer, the goose turned her head right away (Fig. 5E). The head movement shown is not that of 'gathering' of nest-material but represents the so-called 'submissive gesture' which a brooding goose always exhibits when roughly harassed on the nest by a higher-ranking conspecific and does not quite dare to chase the intruder away. Even a highly-intrusive approach by a human being (e.g. in stroking or scratching the brooding goose) was not met with this gesture, so such an approach was obviously a less unpleasant stimulus for the goose than the forcible introduction of the literally 'unappetizing' dummy egg. Although the form of this movement is also innately determined, the motor pattern in this situation gave the observer an extremely strong recollection of the emotional state which well-bred children have often summarized with the following words in the face of cake-foisting aunts: 'Thanks awfully, but I feel sick already!' It is doubtless significant that this behaviour appears most prominently in response to an object which is inappropriate for the instinctive behaviour pattern.

Summary and conclusions

The subject of this study was the manner in which an instinctive behaviour pattern (i.e. a receptor-independent, centrally co-ordinated motor pattern) can co-operate with one or more receptor-controlled taxes and combine with them to produce a functionally unitary and adaptive behavioural sequence when performed by the animal concerned. The motor sequence which the Greylag goose (*Anser anser* L.) employs to return a stray egg to the nest-cup incorporates an instinctive behaviour pattern and both simultaneously and separately operating taxes. Oriented stretching of the neck represents an introductory orienting response, producing the stimulus situation in which the instinctive behaviour pattern is elicited and simultaneously providing the necessary spatial relationship which is a prerequisite for adaptive performance of the pattern. In other words, this is a typical case of 'appetitive behaviour' in Craig's terminology. The instinctive behaviour pattern which is then performed consists of ventrally-directed bending of the head and neck, such that the egg lies against the lower side of the beak and is pushed towards the nest. From the very beginning of this purely sagittal pattern of movement to the end, another orienting response is simultaneously operative. By means of thigmotactically-controlled lateral movements, the egg is balanced on the underside of the moving beak so that the major direction is maintained.

Our assumption that *the movement in the sagittal plane is a pure instinctive behaviour pattern* is based on the following findings:

1. The motor pattern exhibits the phenomenon of *vacuum performance*, which characterizes independence of the instinctive behaviour pattern from receptor-control (p. 333).

2. The form of the motor pattern is always the same. It proved to be impossible to produce receptor-controlled adaptations of the pattern to modified spatial relationships. Neither the characteristics of the pathway along which the egg rolls, nor the form of the rolled object produce recognizable modifications of the motor pattern. In cases where it was attempted to forcibly produce changes through mechanical factors (using a very large object – p. 340), the motor pattern *jammed* and was discontinued.

3. The power with which the sagittal movement is performed remains *constant* within extremely narrow limits. The tactile stimuli deriving from the object do have a minor tonus effect on the muscles performing the movement, but this effect would appear to remain at a constant level even if the weight of the object varies. Thus, as soon as the object is only a little overweight the motor pattern fails.

4. The motor pattern exhibits the property of response-specific *habituation* characteristic of the instinctive behaviour pattern as opposed to the taxis.

The assumption that the *lateral balancing movements* – in contrast to the motor processes occurring in the sagittal plane – are *orienting responses* controlled by tactile stimuli is based on the following facts:

1. The lateral movements do not appear in the object-less vacuum performance of the sagittal movement (p. 332).

2. The lateral movements are lacking during rolling of objects which do not exhibit lateral straying from the path of movement (p. 340).

3. With objects whose lateral departures differ from those exhibited by the goose egg, the lateral movements exhibit complete *adaptation to the manner of movement of the object* (p. 339).

In addition, the following supplementary findings should be mentioned:

1. The appetitive behaviour in the form of orienting stretching of the neck is elicited by any object which exhibits surface continuity and a more or less cohesive outline; the size of the object is immaterial over a wide range (Figs. 3, 4).

2. If the object possesses any protruding corners or appendages, the rolling movement is not continued. Instead, a different response is performed – the object is broken up and eaten. This response, which serves for the removal of broken eggs, is presumably extinguished by means of acoustic stimuli before the chicks are due to hatch.

3. In the course of the brooding phase, rough recognition of the object of the egg-rolling response, which is initially determined only by a few innate characters, is supplemented by acquired recognition of characters, producing enhanced selectivity in the performance of the response.

4. Following extinction of the instinctive behaviour pattern through the easily-procured habituation of the central automatism, *the introductory orienting response can still be elicited.*

5. Following exhaustion of the rolling response, the goose attempts to eliminate the stimulus situation presented by a goose's egg located outside the nest-cup by *employing a different instinctive behaviour pattern.*

6. The goose turns away to *withdraw* from the stimuli emanating from a potentially elicitatory object which has proved to be unsatisfactory.

Notes

[*Editor's Note:* Only those notes relevant to this paper are included here.]

112   *P. 317, l. 4.* The nomenclature conference at the ethologists' congress in 1949 decided on the term 'fixed motor pattern' for this concept.

113   *P. 319, l. 19.* At this point, the perspicacious reader who has suffered through all the inconsistencies of the author, ought to heave a sigh of relief.

114   *P. 321, l. 19.* These phenomena, including that of the 'spinal contrast' of Sherrington himself, are all the more convincing arguments for E. v. Holst's theory, since they are collected by people who were not yet aware of it.

115   *P. 322, l. 19.* Gray, Lissman and others have shown that, in certain cases, an influx of stimulation from one or two posterior roots is necessary to keep the automation going. Even in these cases, however, the afferent input has no influence on the co-ordination.

116   *P. 323, l. 20.* Later studies of mixed motivation have shown that this is actually the rule and that it is actually an exception when an animal's behaviour is motivated by a single automatic pattern.

117   *P. 323, l. 29.* Even central nervous mechanisms which are completely innate, in the sense of being entirely programmed in the phylogeny of the species, are here termed non-instinctive. This rather awkward (and later abandoned) terminology was motivated by my striving to emphasize as clearly as possible the unique properties of the fixed motor pattern.

118   *P. 325, l. 20.* What is not yet clearly stated here (though it ought to have been) is that the consummatory motor pattern, once it is set off, demonstrably acts as a reinforcement of the precedent appetitive behaviour.

119   *P. 325, l. 40.* The problem here raised is thoroughly discussed and, I think, solved in 'Innate Bases of Learning'.

120   *P. 347, l. 18.* This is the first, unclear description of what N. Tinbergen and A. Kortland called a *displacement activity* while G. F. Makkink called it *'sparking over'*.

Part V

POST-HATCHING BEHAVIOR

Editor's Comments
on Papers 25 Through 32

The newly hatched young of birds are by no means uniform in their requirements for food, heat, and protection. Based on the relative development at the time of hatching, birds are said to be either precocial or altricial. Precocial young are fully covered with down at hatching, and are soon able to move about. Altricial young are helpless when hatched. In both altricial and precocial species the behavior of the parent must be adjusted to the needs of the constantly growing young, and the behavior of the young influences the nature of parental care. Impekoven found that in the Laughing Gull (*Larus atricilla*) the vocalizations of the embryo produce behavioral changes in the parent (1970). She also demonstrated that the response of newly hatched chicks to the call of Laughing Gull parents differed in incubator-raised and in parent-raised chicks (1970). Thus, the embryo responds to the call of the incubating parent (1973). Schleidt, et al. (Paper 25) demonstrated the role of auditory stimuli in the mother-young interaction of turkeys (*Meleagris gallopavo*). They discovered that the following response was more marked in deafened poults than in controls. Deafened hens bred and incubated normally, but often killed their young upon hatching. The paper was originally published in German, and Dr. W. Schleidt has kindly reviewed and modified the English translation included here. Auditory cues from the young can be important in species with altricial young as well. Thus, Nottebohm and Nottebohm (1971) showed that deafened pairs of Ring Doves tended to underfeed their young and to abandon them prematurely.

Parental care of young includes brooding (i.e., behavior associated with the provision of heat by the parents to young), nest sanitation, feeding the young and leading them to food, and protecting them from predators with warning calls and distraction displays.

The role of hormones in care of young after hatching has been studied only in a few domesticated species. What are some of the functions which might be mediated by hormonal mechanisms and which remain to be explored? While they are incubating eggs and brooding young, the parent(s) at the nest do not feed; hormonal factors may influence feeding behavior, body temperature, and/or metabolic requirements at this time. During incubation and brooding, courtship and new egg laying do not occur, and nest building and nest defensiveness may increase. These effects may be hormone mediated. Although thorough and systematic analyses on the role of hormones in the care of the young are lacking (see Lehrman, 1961) there is evidence that hormonal factors play a role. One of the earliest demonstrations that sex hormones influence the display of parental behavior comes from the experiment of Goodale (Paper 26). Male chickens do not normally incubate eggs or brood young. Goodale showed that after castration, capons will show parental behavior. In general, the role of prenatal "organizing" hormones and adult "activating" hormones in the display of sex-typical parental behavior of birds has received little attention.

The work of Tinbergen and his colleagues (Paper 27) on egg shell removal by the Black-headed Gull (*Larus ribidundus*) is a classic in the study of the survival value of a behavior. The removal of the egg shell is a response that takes about half a minute a year and appears insignificant. The paper describes experiments on the function, and some of the stimuli eliciting this response. This paper is one of a number of studies by Tinbergen and his colleagues on the antipredator system of the Black-headed Gull. Beer (1960) described a wide variety of objects which are removed from the nest by gulls. Tinbergen et al. (1962) found that gulls which had been incubating black eggs responded more to black shells than did control animals incubating their own khaki-coloured eggs.

Shell removal is but one of the many mechanisms which serve to limit predation. For example, many birds respond to the appearance of intruders at the nest by performing behaviors which divert the intruder away from the eggs or young. These behaviors have been termed "diversionary display," and have long intrigued students of behavior, but much of the literature is anecdotal in nature (see Armstrong, 1965, for discussion). An important contribution to the understanding of antipredator behavior, including distraction display, has been made by Kruuk (1964). He describes interactions of gulls with several different predators and attempts to estimate the damage each of these predators could do if the gulls did not possess special antipredator responses.

Nest sanitation includes not only removal of egg shells but also the removal of other foreign objects. Many altricial species eat or carry away the fecal sacs of the young which, when voided, are enveloped in a clean, tough, mucous membrane. This behavior may have adaptive value by minimizing parasites in the nest, or by camouflaging the nest from potential predators. It is also possible that the feces of nestlings contain nourishment; such possibilities have not been examined. There is substantial variability within families in the occurrence of feces removal. Welty (1962) reports that the nest of the Wood Pigeon (*Columba palumbus*) is filthy, and that in one nest 6,573 vermin were found. On the other hand, the Ruddy Ground Dove (*Columbina talpacoti*) keeps its nest scrupulously clean by eating feces. In contrast to egg shell removal, the experimental analysis of survival value or adaptive significance of feces removal has not yet been done. Smith (Paper 28) addresses the question of whether or not feces removal can be called an instinctive behavior. That is, the problem of how the behavior develops is raised implicitly. The nature-nuture controversy has often arisen in the context of complex avian behaviors, and here, as in many other cases, conditions under which removal of objects from the nest can be elicited are studied. The next step in the analysis would involve clarification of the precise developmental requirements for the appearance of the behavior in adults.

Feeding the young is one of the major tasks of the parents. Over the years many studies have been directed to analyses of the type of food, the feeding frequency and duration, method of feeding, stimuli eliciting feeding, amount of feeding, work expended in feeding young, etc. Most of these studies deal with behaviors directed to young which are still in the nest. Paper 29 is one of a series of truly remarkable experiments done by Michael Norton-Griffiths (1969; Tinbergen and Norton-Griffiths, 1964) at the gullery in Ravenglass, England. Due to space limitations only one of these detailed and well-documented studies is presented here. Briefly, Norton-Griffiths found that individual adult Oystercatchers (*Haematopus ostralegus*) eat only prey of particular species (i.e., mussels, crabs, cockles, or limpets) and have one particular method of eating the prey. Among oystercatchers, there are two distinctive ways of opening mussels. One is by hammering the ventral surface of the mussels which are carried from the rocks to an area of suitable firm sand. The second involves stabbing the bill into the gape of the slightly open mussels which are still attached to rocks and covered by seawater. Breeding pairs share the same prey and the same technique of opening shells. Young Oystercatchers follow their parents down to the mussel beds and eat the same prey and acquire the same techniques as their parents, which they use exclusively thereafter. There is certainly the basis for cultural isolation between hammering and stabbing birds, but as Norton-Griffiths points out, Oystercatchers have many other foods besides mussels.

Norton-Griffiths' experiments were extended with cross-fostering studies (personal communication). If the eggs of mussel-stabbers are hatched out under crab specialists the young become crab specialists. Similarly, eggs of mussel stabbers hatched under mussel hammerers become mussel hammerers. In summary, the young acquire the same prey and techniques as their foster parents. With these cross-fostering experiments, Norton-Griffiths has proven experimentally, for the first time, the social transmission of acquired behavior.

In desert-dwelling birds, finding water for the young is an important aspect of parental care. One of the most bizarre adaptations to desert life is the manner in which male sandgrouse (*Pterocles namaqua*) carry water to the young. In an admirable study Cade and MacLean (Paper 30) tell the remarkable story of how sandgrouse soak their plumage, which can absorb more water than a sponge or paper towel, and return to the young, flying over 25 to 30 Km of the hot Kalahari desert. The laboratory and field techniques combined in this study confirm the original observations of field naturalists who previously described this fascinating behavior to an incredulous scientific community.

It is clear that behaviors which serve to foster the development of young are shown by individuals other than the parents. In some cases mutual benefit to young and caretaker is involved. Thomson (Paper 31) describes the Golden Shouldered Parrot (*Psephotus chrysopterygius*), which constructs its nest in a termitarium, or a "white ant" mound. The lepidopterous larvae in the nesting chamber eat the excreta of the nestlings, thus preventing feces from accumulating on the feet and feathers. The time required for the larvae to pupate and for the birds to fledge are about the same. Thus, the selection of the nest site by the parents determines the care the young will receive when they hatch.

As well as nesting in association with insects, birds associate with other avian species and even with conspecifics at the nest. In fact, eighty species in thirty-two different families are known to exhibit cooperative breeding (Woolfenden, 1975). The problem was first set by Skutch in 1935 (Paper 32) and again in 1961. In Paper 32, Skutch suggests a classification for these assistants at the nest. His classification provides useful framework which helps to keep in mind the types of nesting associations to be found among birds. When originally published, Skutch's paper, entitled "Helpers at the Nest," challenged the notion that birds necessarily establish exclusive territories which they defend from conspecifics. The occurrence of helpers at the nest is also disturbing for the synthetic theory of evolution (see Williams, 1966). How could natural selection favor genes that cause their bearers to expend resources which benefit their genetic competitors? The theory of kin selection explains the occurrence of "altruistic" behaviors by proposing that the individual facilitates survival of the kin by not reproducing. As the individual and its kin share many genes, "altruistic" behavior may have been selected because the individual's genes are best represented in the future through the kin.

REFERENCES

Armstrong, E. A. (1965). *The Ethology of Bird Display and Bird Behavior.* Dover Publications, Inc.: New York.

Beer, C. G. (1960). Incubation and nest-building by the Black-headed Gull. Ph.D. thesis, Oxford.

Impekoven, M. (1970). Prenatal experience of parental calls and pecking in the Laughing Gull (*Larus atricilla* L.). *Anim. Behav., 19,* 475–480.

Impekoven, M. (1973). The response of incubating Laughing Gulls (*Larus atricilla* L.) to calls of hatching chicks. *Behaviour* 46:94–113.

Kruuk, H. (1964). Predator and anti-predator behavior of the Black-headed Gull (*Larus ridibundus* L.). *Behaviour,* Suppl. XI.

Lehrman, D. S. (1961). Hormonal Regulation of Parental Behavior in Birds and Infrahuman Mammals. In *Sex and Internal Secretions,* W. C. Young, ed., Williams and Wilkins: Baltimore.

Norton-Griffiths, M. (1969). The organization, control and development of parental feeding in the oystercatcher (*Haematopus ostralegus*). *Behavior, 34,* 55–114.

Nottebohm, F., and Nottebohm, M. E. (1971). Vocalizations and breeding behavior of surgically deafened ring doves (*Streptopelia risoria*). *Anim. Behav., 19*, 313–327.

Tinbergen, N., Kruuk, H., and Paillette, M. 1962. Egg shell removal by the Black-headed Gull (Larus r. ridibundus L.) II. The effects of experience on the response to colour. *Bird Study* 9:123–131.

Tinbergen, N., and Norton-Griffiths, M. 1964. Oystercatchers and Mussels. *British Birds* 57:64–70.

Welty, J. C. (1962). *The Life of Birds.* W. B. Saunders Co.: Philadelphia and London.

Williams, G. C. (1966). *Adaptation and Natural Selection.* Princeton University Press: Princeton, N.J.

Woolfenden, G. E. (1975). Florida Scrub Jay helpers at the nest. *Auk, 92*, 1–15.

25

Reprinted from *Behaviour* 16:254–260 (1960)

STÖRUNG DER MUTTER-KIND-BEZIEHUNG BEI TRUTHÜHNERN DURCH GEHÖRVERLUST [1]

von

WOLFGANG M. SCHLEIDT, MARGRET SCHLEIDT UND MONICA MAGG

(Max-Planck-Institut für Verhaltensphysiologie, Seewiesen)

(Eingeg. 10-XII-1959)

Im Rahmen einer Untersuchung über das Verhaltensinventar der Trut-hühner *(Meleagris gallopavo)* wurde versucht, Einzelheiten über die Be-deutung a k u s t i s c h e r Auslöser für die Mutter-Kind-Beziehungen zu ermitteln. Für andere Experimente wurden taube Truthähne benötigt, und wir entschlossen uns, die Operation schon an Kücken auszuführen, obwohl wir in diesem Alter noch keine sicheren Angaben über das Geschlecht der Tiere machen konnten. Wir gewannen aber damit die Möglichkeit, 1. Beob-achtungen an ertaubten Kücken zu sammeln, und 2. mit ertaubten Hennen während des Brütens zu experimentieren.

VERSUCHSTIERE UND METHODIK

Die für die meisten Versuche verwendeten Bronzeputen stammen von einer Hausrasse, in die Wildblut eingekreuzt ist. Der Stamm wird seit 6 Jahren am Institut weitergezüchtet. 6 weitere Versuchstiere gehören anderen Zuchtrassen an (2 weiße Schneeputen, 1 Norfolk, 3 Bronzemischlinge); sie wurden mit gleichaltrigen Wildputenkücken (Subspezies unbestimmt) zusammen aufgezogen.

11 Kücken (5 Bronze aus eigener Zucht, 6 von den oben genannten Zuchtrassen) wurden zwischen dem 2. und 8. Lebenstag durch einseitige oder beidseitige Entfernung der Columella oder/und des Ductus cochlearis in ihrem Hörvermögen gestört oder ganz taub gemacht (siehe Tabelle I). 9 von ihnen erreichten die Geschlechtsreife, darunter 2 Hennen, die für die im zweiten Teil der Veröffentlichung beschriebenen Versuche verwen-det wurden. Die Operationstechnik lehnt sich an die von SCHWARTZKOPFF (1949) beschriebene an, sie wird an anderer Stelle ausführlicher behandelt (SCHLEIDT, 1961). Beidseitige Entfernung des Ductus cochlearis hat völlige Taubheit zur Folge, einseitige Entfernung muß zu Schwierigkeiten bei der

---

[1] Herrn Prof. Dr O. KOEHLER zum 70. Geburtstag gewidmet.

TABELLE I

Art der bei den einzelnen Tieren ausgeführten Operation.

Rasse	B	B	B	B	B	M	M	M	N	S	S
Geschlecht	♂	♂	o	♂	♂	♀	♂	o	♂	♀	♂
Ductus cochlearis											
links ⎫ entfernt	—	—	—	—	1	1?	1?	1	1	1	1
rechts ⎭	r	r	r	r	r?	r?	r?	r	r	r	r
Columella											
links ⎧ entfernt	—	—	—	—	—	1	1	1	—	1	—
rechts ⎭	r	r	r?	—	—	r	r	r	—	—	—

Erklärung der Abkürzungen: B = Bronzeputen, M = Mischlinge, N = Norfolk, S = Schneeputen. ? bedeutet: Organ nicht entfernt, aber wahrscheinlich zerstört. o bedeutet: als Kücken gestorben, Geschlecht unbestimmt.

Richtungslokalisation führen. Beidseitige Entfernung der Columella bedingt eine wesentliche Einschränkung des Hörvermögens, die sich im Verlauf von einigen Tagen etwas verbessern kann; es bleibt aber eine Erhöhung der Hörschwelle um ca. 40 db bestehen (SCHWARTZKOPFF).

> Die Kücken wurden in den ersten Tagen mit einem Weichfutter, bestehend aus gekochtem Ei, Quark, Haferflocken und Kückenaufzuchtmehl ernährt. Wir steigerten den Anteil an Aufzuchtmehl systematisch, und mit 4 Wochen wurden sie ausschließlich damit gefüttert. Um die Kücken besonders zahm zu halten, fütterten wir sie täglich einmal mit einigen Ameisenpuppen, die älteren Tiere bekamen Apfelstückchen als Leckerbissen. Die erwachsenen Tiere bekommen ein reichhaltiges Weichfutter abwechselnd mit Körnern (Weizen, Mais), außerdem steht ihnen eine Grasweide von rund 4 000 qm zur Verfügung.

BEOBACHTUNGEN AN TAUBEN KÜCKEN

Die führende Henne produziert eine ganze Reihe von Lautäußerungen — Führlaute, Warnlaute, etc. — die offensichtlich Auslöser für bestimmte Verhaltensweisen der Kücken sind. Es muß also erwartet werden, daß Kücken durch Einschränkung oder Verlust des Hörvermögens eine wesentliche Veränderung ihres Verhaltens zeigen. Leider führt die Entfernung des Ductus cochlearis auch zu Schädigungen angrenzender Labyrinthteile, sicher durch die Eröffnung des perilymphatischen und endolymphatischen Raumes, vielleicht auch durch Blutungen und Verwachsungen im Operationsgebiet. Wiewohl keines der Kücken während der Operation oder in den unmittelbar nachfolgenden Tagen gestorben ist, waren sie doch z.T. erheblich gestört (Veränderungen im Tonus der Halsmuskulatur; Drehtendenzen beim Versuch, zu laufen; seitliches Vorbeipicken an Futterbröckchen). Es erforderte viel Mühe, die Kücken zum Fressen anzuregen, z.T. mußten sie in den ersten Tagen auch gestopft werden. Die fünf nur einseitig taub gemachten Kücken waren (vielleicht wegen einer geringfügigen

Änderung der Operationstechnik) viel weniger gestört, so daß wir sie 24 Stunden nach der Operation bereits zu ihrer Mutter zurücksetzen konnten. Sie entwickelten sich wie ihre normalen Geschwister.

Leider wurde es versäumt, die Nachfolgereaktion der Kücken durch planmäßige Attrappenversuche zu kontrollieren, es fiel aber auf, daß die tauben Tiere dem Pfleger meist ganz dicht aufgeschlossen folgten, dichter als die hörenden und wie an seine Fersen geheftet schienen. Blieb der Pfleger stehen, so verstreuten sie sich wie die Vergleichskücken in der näheren Umgebung, rupften Grashalme ab und jagten nach Insekten; sie verschwanden im hohen Gras oder hinter Gebüschen. Merkte ein taubes Kücken dann plötzlich, daß es den (optischen) Kontakt verloren hatte, so blieb es stehen und fing mit hochgestrecktem Hals laut zu „weinen" an. Durch auffällige Bewegungen (winken mit den Armen, wackeln mit den Beinen) konnte man sie manchmal zurück„rufen", meist mußte man aber zu ihnen hingehen und sie regelrecht abholen.

Die Verstärkung des Nachfolgens bei den tauben Kücken verdient besondere Beachtung. Bei den bisher bekannten Attrappenversuchen über das Nachfolgen von Enten- und Gänsekücken findet man immer wieder, daß eine Muttertierattrappe mit Stimme wirksamer ist als eine stumme. Auch die vom Menschen aufgezogenen, normal hörenden Putenkücken schließen dichter auf, wenn man sie ruft. Wir hatten daher bei unserer tauben Kücken eher eine Verminderung der Nachfolgereaktion erwartet. Eine Reihe von Arbeitshypothesen sind denkbar, diese Diskrepanz in den Beobachtungen einstweilen zu überbrücken. So wäre z.B. denkbar, daß die aktive Fürsorge um die Kücken (das Füttern aus der Hand, das oftmals notwendige Zurückführen zur Herde) das Nachfolgen fördert. Oder, daß das generelle Ausfallen akustischer Eindrücke [1]) zu einer graduellen Überbewertung der optischen Informationen führt, und so den auslösenden Reiz wirksamer macht. Wir hoffen, in der nächsten Brut-Saison zur Klärung dieses Sachverhaltes beitragen zu können.

BEOBACHTUNGEN AN TAUBEN HENNEN

Im sozialen und sexuellen Verhalten unserer tauben Hennen konnten wir keine Unterschieden gegenüber den hörenden Tieren feststellen. Nachdem ihre Gelege vollständig waren, begannen sie zu brüten. Zwei normale, eben-

---

1) Es muß aber auch damit gerechnet werden, daß die operierten Tiere akustische Sensationen („Phantomeffekte") haben, die durch die Spontanerregung nachgeschalteter Hörzentren erklärt werden könnte. Es wäre denkbar, daß die Tiere genau so viel hören wie normale, nur stehen diese Wahrnehmungen mit denen aus anderen Sinnesgebieten in keinem Zusammenhang, und erscheinen in rein zufälliger Verteilung.

falls einjährige Hennen — eine war von uns, die andere von einer Pute aufgezogen worden — setzten sich zur selben Zeit auf ihr Nest, und wir benützten sie für die folgenden Versuche als Vergleichstiere. In diesem Jahr waren noch keine Kücken geschlüpft, keine der vier Hennen hatte also seit ihrer eigenen Kindheit ein Kücken gesehen. Wir tauschten die Gelege gegen unbefruchtete Eier aus, und nachdem die Hennen 4 Wochen lang gebrütet hatten, übersiedelten wir sie bei Nacht mit ihren Eiern in 4 Versuchskisten (Abstand von Kiste zu Kiste mindestens 8 m.; Maße der Kisten: 80 cm. hoch, 80 cm. breit, 120 cm. lang; die Vorderseite besteht aus einer Drahtgeflechttüre, die übrigen Seiten aus Brettern).

Nachdem wir festgestellt hatten, daß die Hennen in der neuen Umgebung unvermindert weiter brüteten, wollten wir ihre Gelege gegen je 4 Eintagskücken austauschen. Die Übersiedlung der Hennen im Finstern war so glatt verlaufen, daß wir für diesen Versuch ebenfalls die Nachtstunden (23^h-1^h) wählten. Wir hofften, dadurch optisch ausgelöste Verhaltensweisen unterbinden zu können. Die erste, normale Henne fauchte bei unserer Annäherung, sie unterbrach es aber, als eines unserer Kücken weinte und ließ einen singenden Ton hören. Beim Wegnehmen der Eier verstärkte sich ihr Fauchen, und sie hackte um sich. Als das erste Kücken piepsend im Nest saß, änderte sich ihr Verhalten, das Fauchen verebbte und Führlaute waren zu hören. Nachdem wir der Henne 4 Kücken untergeschoben hatten, beobachteten wir sie noch 20 Minuten lang im Scheine unserer Taschenlampe. Die Pute war durch unsere Anwesenheit nicht sonderlich beunruhigt. Sie saß in Huderstellung (auf den Fersen) und nur wenn wir versuchten, sie anzufassen, stellte sie die Rückenfedern auf und fauchte etwas. Kam ein Kücken hervor, so ließ sie wieder den Führlaut hören, gleichzeitig wurde sie vorne etwas hoch, so daß unter ihrer Brust und zwischen den gesenkten Flügeln eine Öffnung entstand („Tunnelstellung"). Schlüpfte das Kücken nicht unter, so zielte sie mit dem Schnabel ganz langsam nach ihm, und beförderte es schließlich mit einer, der Eirollbewegung (LORENZ u. TINBERGEN, 1938) ähnlichen Kopfbewegung unter ihr Brustgefieder.

> Durch ein auf den Nestrand gelegtes Ei war die Eirollbewegung sofort auszulösen, und der unmittelbare Vergleich ergab den folgenden Unterschied: Bei der Eirollbewegung wird das Ei durch feinschlägige Bewegungen des Kopfes in der Mediansagittalen (von seitlichen Balance-Bewegungen unterstützt) mit dem Unterschnabel ins Nest geschoben. Das Kücken wird genau so wie das Ei mit dem Unterschnabel berührt, dann schubst es aber die Henne mit einer einzigen Kopfbewegung bis zu 20 cm an sich heran, intendiert wieder nach dem Kücken, berührt es neuerlich mit dem Unterschnabel und manövriert es so mit zwei, drei Bewegungen unter ihren Bauch. Diese Unterschiede rechtfertigen es wohl, neben der Eirollbewegung eine eigene Kückeneinholbewegung als selbständige Bewegungskoordination abzugrenzen.

Bei der zweiten normal hörenden Henne verlief der Versuch Punkt für

Punkt ebenso. Da sie nicht so zahm gegen den Pfleger war, wunderte es uns nicht, daß sie etwas mehr fauchte, und — als wir schon Licht gemacht hatten — plötzlich das Rückengefieder sträubte und den Schwanz fächerte, wenn wir uns bewegten.

Dagegen waren die tauben Hennen besonders zahm. Als wir aber der ersten die Kücken untergeschoben hatten wurde sie auffallend unruhig und begann, wild um sich zu hacken. Wir hörten die Kücken aufschreien und als wir Licht machten, lag schon eines totgehackt im Nest, ein zweites blutete und es bedurfte großer Schnelligkeit, die drei Überlebenden zu retten. Nun setzten wir ein Kücken außer Reichweite ihres Schnabels an den Kisteneingang, und jeder Schritt näher zum Nest verstärkte das Sträuben des Rückengefieders und das Schwanzspreizen der Henne. Präsentierte man ihr dagegen an der selben Stelle ein Ei, so stand sie auf, oder rutschte vor, und rollte es in ihr Nest.

Bei der zweiten tauben Henne gelang es uns offenbar besser, die Kücken unter ihren Bauch zu stecken, sie verhielt sich ruhig, und keines der Kücken weinte. Wir machten Licht und konnten nichts Auffallendes bemerken, bis man eines der Kücken unter der Henne sich bewegen hörte. Da wurde die Henne zusehends unruhiger, sie rückte hin und her, und als das erste Kücken erschien, hackte auch sie sofort danach. Wir retteten die Kücken und gaben der Henne (wie der anderen tauben auch) ihre Eier zurück.

Am nächsten Morgen setzten wir zu den beiden normalen Hennen je 3 weitere Kücken, die sie ohne weiteres annahmen und unterkriechen ließen. Dagegen löste ein an den Nestrand gesetzter Goldhamster (als „Bodenfeind-Attrappe") sofort Fauchen, Knattern, Aufstellen des Rückengefieders und Schwanzspreitzen aus; kam er in Reichweite, so wurde er gehackt.

Bei den beiden tauben Hennen lösten Kücken die gleichen Reaktionen aus wie in der Nacht vorher, nur daß zum Fauchen noch Knattern als weitere Lautäußerung dazu kam. Der Goldhamster wurde völlig gleich behandelt wie ein Kücken, das sich dem Nest nähert.

Um den Einwand zu entkräften, die tauben Hennen könnten aus unbekannten Gründen durch die Versuche selbst gestört werden — wiewohl dies wegen ihrer Zahmheit unwahrscheinlich ist — schoben wir der einen Henne 3, der anderen 4 angebrütete Eier unter. Die Kücken waren unmittelbar vor dem Schlüpfen, die Eischale war aber noch nicht gepickt. Zum Vergleich ließen wir 7 weitere Eier im gleichen Bebrütungszustand im Brutapparat. Die Kücken im Brutapparat schlüpften normal, bei den Hennen fanden wir am entsprechenden Morgen die Leichen der geschlüpften Kücken um das Nest verstreut, mit blutigen Hackspuren an Kopf und Körper.

Nachdem die beiden tauben Hennen auch noch weiter brüteten, als die Kücken geschlüpft waren, sperrten wir sie aus den Versuchskisten aus und brachten sie unter Aufsicht mit Kücken zusammen, um so vielleicht ihre mütterlichen Verhaltensweisen zu stimulieren. Erfolg hatte es keinen: am ersten Tag gingen sie mit vorgestrecktem Hals und knatternd auf die Kücken los und versuchten, nach ihnen zu hacken. Dann gewöhnten sich die tauben Hennen an die Kücken, und nahmen überhaupt nicht mehr Bezug auf sie. Eine der tauben Hennen begann nach 3 Wochen wieder zu legen, und nachdem sie 4 Wochen lang gebrütet hatte, setzten wir ihr wieder ein Kücken ans Nest. Sie verhielt sich genau wie bei den früheren Versuchen, ihre Erfahrung mit den Kücken hatte also keinen Einfluß auf ihre Reaktion gegen das Kücken am Nest.

Sucht man aus diesen Versuchen herauszugliedern, durch welche Merkmale sich für eine brütende Pute ein Bodenfeind von einem eigenen Kücken unterscheidet, so kommt man zu der folgenden Charakterisierung:

Feind am Nest: etwas sich Bewegendes, (pelziges).
Kücken am Nest: etwas sich Bewegendes, (pelziges?), das in charakteristischer Weise piepst.

Wir haben die Absicht, mit Kückenattrappen die Richtigkeit der gebrachten Formulierung zu prüfen.

Wahrscheinlich kann das „Führen" der Putenhenne nur vom Nest aus in Gang gebracht werden, eben normalerweise durch die eigenen Kücken. Wir haben in den letzten Jahren mehr als 20 Bruten genauer überwacht, und nur zweimal ist es vorgekommen, daß eine Henne eine fremde Kückenschar adoptiert hat (daß es überhaupt zustande kam, liegt sicher an der unnatürlich hohen „Populationsdichte" durch die Haltung im Gehege). Beiden Fällen war gemeinsam, daß eine Henne, deren eigene Kücken kurz vor dem Schlüpfen waren[1]), an ihrem eigenen Nest von versprengten Kücken oder von den Jungen einer unmittelbar benachbart sitzenden Henne stimuliert wurde, ihre Eier zu verlassen. In beiden Fällen führten diese Hennen zuerst jeweils gemeinsam mit der richtigen Mutter, übernahmen aber nach einigen Tagen die Schar allein, während die beiden eigentlichen Mütter wieder zu legen begannen. Da die gehörlosen Hennen aber ein Kücken am Nest niemals als solches erkennen können, weil sie den akustischen Schlüsselreiz nicht wahrnehmen, bleibt der ganze Satz mütterlicher Verhaltensweisen unausgelöst.

---

1) Es wäre denkbar, dass schon das Piepsen des Kückens im Ei die Henne stimuliert. Dass dieser Reiz aber nicht unbedingt notwendig ist, zeigt ja der erfolgreiche Versuch mit den normalen Hennen, die Eier direkt gegen Kücken auszutauschen.

ZUSAMMENFASSUNG

11 Putenkücken wurden durch Entfernung der Columella oder des Ductus cochlearis in ihrem Hörvermögen beeinträchtigt oder ganz taub gemacht.

Die gehörlosen Kücken folgten dichter als normale Vergleichstiere. Die gehörlosen Hennen brüteten normal, verhielten sich aber am Nest so, als ob sie nicht zwischen Kücken und Bodenfeind unterscheiden könnten. Dies führte dazu, dass sie sogar selbst erbrütete Kücken unmittelbar nach dem Schlüpfen tothackten.

LITERATUR

LORENZ, K. und N. TINBERGEN (1938). Taxis und Instinkthandlung in der Eirollbewegung der Graugans. — Z.f. Tierpsychol. 2, p. 1-29.

SCHLEIDT, W. M. (1961). Exstirpation des Ductus cochlearis bei Truthahnkücken. — Im Druck.

SCHWARZKOPFF, J. (1949). Über Sitz und Leistung von Gehör und Vibrationssinn bei Vögeln. — Z. vergl. Physiol. 31, p. 527-608.

SUMMARY

Eleven turkey poults were either deafened, or their hearing ability reduced by removal of the Ductus cochlearis or the Columella. In deafened poults the following response was found to be more marked than in the normal controls.

Hens, who had been deafened as poults bred like normal hens, but behaved as if they could not differentiate poults and predators. Thus they even killed their own young immediately after hatching, as they would do with predators approaching the nest.

25

DISTURBANCES IN THE MOTHER-YOUNG RELATIONSHIP OF TURKEYS CAUSED BY LOSS OF HEARING

Wolfgang M. Schleidt, Margret Schleidt, and Monica Magg

This article was translated expressly for this Benchmark volume by Jonathan E. Fuller, from Behaviour *16:254–260 (1960)*

In the context of a more general study of the behavioral repertoire of turkeys (*Meleagris gallopavo*) we attempted to investigate the role of acoustical releasers in the mother-child relationship. Deaf male turkeys were needed for some other experiments, and even though we could not determine the sex of young poults, we decided to perform the operation at the earliest possible age. This provided the opportunity, however, to study the behavior of deaf poults, and to observe and perform experiments with deaf hens during their incubation period.

TEST ANIMALS AND METHODOLOGY

The birds used in most of the experiments were of a domestic breed which had been crossed with wild stock. This breed has been raised for six years at the Institute. Six other test birds belong to other breeds (two white "Schneeputen," one Norfolk, three Bronzehybrids): they were raised together with wild turkey poults of the same age.

Eleven poults (five Bronzehybrids from our own breed, six from the breeds mentioned above) were either deafened, or had their hearing impaired by a single or double removal of the Columella and/or the ductus cochlearis (see table 1). Nine of them reached sexual maturity, among them two hens which were used in the experiments described in the second part of this publication. The operational technique resembles that described by Schwartzkopff (1949), and is more completely described elsewhere (Schleidt, 1961). Double removal of the ductus cochlearis results in total deafness, single removal leads to difficulty in the localization of sound sources. Double removal of the columella causes a significant limitation in the hearing capacity, which can improve somewhat over the course of a few days; an increase in the threshold of hearing of approximately 40 dB remains however (Schwartzkopff).

The poults were fed in the first few days on a mixture consisting of hard-boiled egg, curds, oatflakes, and commercial turkey feed. We increased the amount of commercial feed systematically until, at four weeks, it was their sole source of nourishment. In order to keep the poults particularly tame, we fed them some ant pupae once a day. The older birds got pieces of apple as a treat. The adult birds were fed a varied soft feed, alternating with grain (Wheat, corn); besides that, the birds had a 4000 square meter grass pasture at their disposal.

Table 1 Type of operation performed on individual birds

Breed	B	B	B	B	B	M	M	M	N	S	S
sex	m	m	*	m	m	f	m	*	m	f	m
ductus cochlearis											
left ︶removed	-	-	-	-	l	l?	l?	l	l	l	l
right	r	r	r	r	r?	r?	r?	r	r	r	r
Columella											
left ︶removed	-	-	-	-	-	l	l	l	-	l	-
right	r	r	r?	-	-	r	r	r	-	-	-

Explanation of abbreviations: B=Bronze, M=Mixed breed, N=Norfolk, S=Schneepu-ten. ?=organ not removed, though probably destroyed. *=died as poult, sex undetermined.

OBSERVATIONS OF DEAF POULTS

The leading hen produces a gamut of calls—leading calls, warning calls, etc.— that are apparently releasers for certain kinds of behaviour in the poults. One must therefore expect that the poults would display a significant change in behaviour with the limitation or complete deprivation of their hearing capacity. Unfortunately, the removal of the ductus cochlearis also leads to damage of the neighboring labyrinth parts; certainly through the opening of the perilymphatic and endolymphatic cavities, and perhaps also through hemorrhaging or cicatrization in the operated area. Although none of the poults died during or shortly after their operations, they nevertheless experienced considerable trauma (changes in the tone of the neck musculature, a tendency towards inability to run in a straight line, a sideward tendency when pecking at morsels of food). It required a great deal of care to stimulate the poults to eat. In the first few days, they had to be force-fed. The five only partially deafened poults were (perhaps because of a minor change in the operational technique) much less disturbed; so much so that we could place the poults with their mothers only hours after the operation. They developed just as their normal siblings.

Unfortunately, we did not control the "following reaction" by planned experiments with dummies, but it struck us that the deaf birds followed their keeper closely—more closely than the normal birds—sticking right to his heels. If the keeper remained standing, the deaf birds, like the control birds, would disperse in the near vicinity, tearing up stalks of grass and chasing insects, and disappearing in the tall grass or behind the bushes. If a deaf poult noticed that it had lost (visual) contact, it remained standing and began to "cry" with its neck stretched skywards. With noticeable movements (waving the arms or legs), one could occasionally "call" them back; usually, though, one was obliged to go fetch them.

The greater tendency of the deaf poults to follow deserves special attention. With the hitherto familiar dummy experiments into the following of duck and goose poults, one finds again and again that the dummy hen is more effective with a voice than without. The normally hearing birds raised by humans also follow more closely when they are called. Therefore, we had expected to observe a reduction of the following reaction in the deaf poults. A range of working hypotheses which would explain this discrepancy is conceivable. For example, it is possible that the excessive care for the poults (hand feeding, herding the poults back into the flock as was often necessary) fosters following. Otherwise, perhaps the general lack of acoustical stimulation[1] leads to a general over-evaluation of the optical stimulation, making the releasing stimulus more effective. We hope, in the next breeding season, to be able to contribute to the illumination of this matter.

OBSERVATIONS OF DEAF HENS

In social and sexual behaviour, we could not find differences between deaf and normal birds. After completing their clutches, they began to incubate. Two normal, about one-year-old hens—one raised by us, one by a turkey hen—started incubation at the same time, and we used them for comparison in the following experiments. That year, no chicks had hatched, and so none of the four hens had seen any chick since their own youths. We exchanged their eggs for sterile ones, and after the hens had incubated for four weeks, we moved them at night with their eggs into four boxes (eight meters apart; the boxes themselves were 80 cm high by 80 cm wide by 120 cm long; the front side was a wire-mesh door, the remaining sides were boards).

After we had determined that the hens continued to incubate as before in their new surroundings, we exchanged each of their clutches for four one-day-old chicks. The move of the hens into their boxes had proceeded so smoothly in the darkness that we chose to perform this experiment in the night (from 11:00 PM to 1:00 AM). The first (normal) hen hissed upon our approach, but stopped as one of our little poults cried and emitted a singing tone. Upon the removal of the eggs, hissing increased, and it began pecking about itself. As the first poult sat peeping in its nest, the hen changed its behaviour, the hissing diminished, and yelping calls became audible. After we had placed four poults under the hen, we observed it for another twenty minutes in the light of our flashlights. The hen was not particularly unsettled by our presence. It sat on its heels and only when we attempted to touch it would it raise its backfeathers and hiss a bit. If a poult would appear from under its mother, the hen yelped and, at the same time, it raised itself in front so that an opening appeared between the wings and under the breast (tunnel posture). If the poult would not slip back underneath its mother, the hen would quite slowly reach for it with her beak and finally urge it back with a motion of the head similar to that of the "egg-rolling" movement (Lorenz and Tinbergen, 1938).

> The egg-rolling movement was released immediately by an egg placed on the border of the nest, and the direct comparison produced the following difference: In the egg-rolling movement, the egg is pushed into the nest by delicate back-and-forth movements of the head (supported by lateral movements for balance) with the lower beak. The poult is touched exactly as the egg with the lower beak, but then the hen push-

[1] One must however also take into account the possibility that the birds which were operated on experience acoustic sensations (phantom effects), which could be explained by the spontaneious stimulation of higher hearing centers. It is conceivable that these birds "hear" as much as they normally would, except that these perceptions do not bear any relation to those of the other senses, and occur at random.

es it with one single head movement to approximately twenty centimeters toward itself, reaches for the poult again, touches it afresh with the lower beak, and maneuvers it with two or three movements under its breast. These differences between poult retrieval movement and egg-rolling movement justify their separation as different behavioral patterns.

With the second normally hearing hen, the experiment ran point by point in the same way. Since this bird was not as tame with the keeper, it did not surprise us that it hissed somewhat more, and, as soon as we turned on the light, raised its backfeathers and fanned its tail when we moved.

On the other hand, the deaf birds were particularly tame. As we pushed the first of the poults under a hen, however, she became noticeably disturbed, and began pecking wildly about. We heard the poults scream, and as we turned on the light, one lay already dead in the nest, a second bled, and it required great speed to save the three survivors. The next time, we set a poult outside the range of its beak (our flashlights still on) at the entrance of the box, and each step nearer to the nest increased the bristling of the back feathers and the fanning of the tail. When the hen was presented an egg in the same place, it stood up or slid forward, and rolled the egg into its nest.

With the second deaf hen, we clearly had better success in placing the poult under its abdomen. It behaved itself, and none of the poults cried. We turned on the light, and could notice nothing until we heard one of the poults move under the hen. The hen became much less calm, it jerked back and forth, and as the poult appeared the hen pecked at it immediately. We rescued the poults and gave the hen back (like the other deaf hen) its eggs.

The next morning, we placed three more poults with each of the two normal hens, each of which accepted them, letting them crawl underneath them. On the other hand, a golden hamster placed on the edge of the nest (as a ground predator dummy) immediately triggered hissing, rattling, bristling of the back feathers, and faning of the tail. If it came within range, it was pecked at.

In the two deaf hens, the poults elicited the same reactions as during the previous night, except that rattling was added to hissing as an additional vocal expression. The hamster was treated just like a poult which approached the nest.

To forestall the objection that the deaf hens could have been disturbed for unknown reasons by the experiment itself—although this is improbable in view of their tameness—we placed the other four incubated eggs under the second deaf hen. The poults were on the verge of hatching though the eggshell had not yet been pecked. As a control, we left seven more eggs in the same developmental condition in an incubator. The poults in the incubator hatched normally; with the hens, we found the bodies of the poults strewn around the nest on the appointed morning, with bloody peckmarks on head and body.

Since the two deaf hens continued incubation, we locked them out of their boxes and brought them together with the newly hatched poults under close supervision in order to stimulate their maternal behaviour patterns. Without success: on the first day they chased after the poults, rattling with extended necks, attempting to peck them. Then the deaf hens got used to the poults and took no further notice of them. One of the deaf hens began to lay again after three weeks, and after it had brooded for four weeks, we gave her another poult. The hen behaved exactly as it had in the earlier experiments. Therefore, the hen's experience with the poult had no influence on its reaction to the poult in the nest.

If one attempts to analyze, based on these experiments, which features distinguish a ground predator from its own poult for the turkey hen, one can arrive at the following characterization:

> Predator near the nest: something moving (furry). Poult near the nest: something moving, (furry?), that peeps in a characteristic manner.

We plan to test the correctness of this formularization with dummy poults.

The "leading" of the turkey hens can probably be activated only from the nest, normally by its own chicks. We have carefully observed more than twenty broods in the past few years and only twice has a hen adopted a group of poults (that it happened at all occurred probably because of the unnaturally high population density caused by the penning of the turkeys). Both cases were alike in that a hen whose own chicks were about to hatch[2] was stimulated either by hatched poults in its own nest, or by the poults of a neighboring hen, to leave its own eggs. In both cases, these hens first led the flocks along with the real mother, and then took over alone after the two real mothers began laying again. Since the deaf hens could not recognize a poult in the nest as such, due to their inability to perceive the poult's acoustical key stimuli, the whole chain of maternal behaviour patterns remained unelicited.

SUMMARY

Eleven turkey poults were either deafened, or their hearing ability reduced by removal of the ductus cochlearis or the columella. In deafened poults the following response was found to be more marked than in the normal controls.

Hens, who had been deafened as poults bred like normal hens, but behaved as if they could not differentiate poults from predators. Thus they even killed their own young immediately after hatching, as they would do with predators approaching the nest.

BIBLIOGRAPHY

Lorenz, K. and N. Tinbergen. 1938. Taxis und Instinkthandlung in der Eirollbewegung der Graugans. *Z.f. Tierpsychol.* 2:1-29.

Schleidt, W. M. 1961. Operative Entfernung des Gehörorganes ohne Schädigung angrenzender Labyrinthteile bei Putenküken. *Experientia* 17:464.

Schwartzkopff, J. 1949. Uber Sitz and Leistung von Gehör und Vibrationssinn bei Vögeln. *Z. vergleichende Physiologie,* 31:527-608.

[2] It is conceivable that even the peeping of the poults in the eggs could stimulate the hen. That this stimulation is not unconditionally necessary is positively demonstrated by the successful experiments where eggs were exchanged for poults with the normal hens.

Editor's Note: Dr. Schleidt adds the following comments to the original manuscript:

(1) The general study referred to in the first sentence is: Hale, E. B., Schleidt, W. M., and Schein, M. W. 1969. The behaviour of Turkeys. In: The Behavior of Domestic Animals. (E. S. E. Hafez, ed.) London: Baillière, Tindall & Cassell, pp 554–592.

(2) The studies of deaf male turkeys, mentioned in the second sentence are described in:

Schleidt, W. M. 1964. Uber die Spontaneitat von Erbkoordinationen. Z.f. Tierpsychol. **21**:235–256.

Schleidt, W. M. 1974. How "fixed" is the fixed action pattern? Z.f. Tierpsychol. **34**:189–211.

26

Reprinted from *J. Anim. Behav.* 6:319–324 (1916)

NOTE ON THE BEHAVIOR OF CAPONS WHEN BROODING CHICKS

H. D. GOODALE

Massachusetts Agricultural Experiment Station

References to the brooding of chicks by capons are often found in the literature but descriptions of their behavior do not seem to be on record. During the past season five tests with capons have been made. Two of the capons used were Rhode Island Reds, a race in which the females are very much inclined to broodiness and make excellent mothers, while three were Brown Leghorns, a race the females of which rarely become broody. The chicks used have all been Rhode Island Reds. They readily accept the capon as a foster-parent. Much of the success of the experiments is due to the excellent care given the birds by the foreman of the poultry yards, Mr. John Sayer.

First instance.—This capon was a Rhode Island Red hatched in April, 1913, and caponized in late July in the usual commercial manner by an expert caponizer. He had all the usual characteristics of his kind. The chicks were four in number, hen-hatched and also brooded by the hen for the first two or three days of their lives. The capon was placed in a coop on the night of May 20, 1914, and two of the chicks placed beneath him. He accepted them without trouble and hovered them from the start. The remainder of the chicks were given to him the next morning. During the day he was taken a little distance from the coop, leaving the chicks behind. On being released, he returned at once to his charges. The capon and his flock were then removed from the coop and given the run of a large yard. The chicks followed the capon about, who clucked to them occasionally in an *imperfect* way as though he did not quite know how to do it, and found bits of food for them. Although somewhat disturbed when approached, he did not ruffle up his feathers and spread his wings as a hen would do, but simply retreated. Neither did he cluck as often nor as regularly as a hen.

By this time it was evident that the chicks would be cared for by the capon and the brood accordingly was removed to an out-door yard. The capon showed no disposition to leave his charges, though adult birds, including his recent companions, were in neighboring yards. In a few days the cluck of the capon became more like that of a hen, the bird giving voice to it frequently as he walked about the yard. He continuously hunted food for the chicks and on finding a morsel called the chicks to it as a hen would do, but less vociferously. He also attacked anyone who attempted to interfere with the chicks but did not give a call like the warning call of the hen as far as observed. When these chicks were about three weeks old, another lot of chicks a week younger was added to the flock. The capon accepted these without trouble, though confined in a relatively large yard, 18 x 50 feet. Hens do not as a rule accept chicks much younger than their own. The new chicks, however, were somewhat slow to follow their foster-parent, partly because they were weaker. It was necessary on this account to round up the flock once or twice the first day. The next few days were uneventful, but before the new chicks were a week old a hen in an adjoining yard came off with two ducklings. About one-half the younger chicks promptly deserted the capon and adopted the hen.

Thus far the behavior of the capon, as far as caring for the chicks is concerned, has been essentially like a hen. There are, however, two points in which he differed decidedly from hens. First, the capon began caring for the chicks without becoming broody or without previous training. The chicks were simply placed beneath him and he began to care for them. The hen becomes first broody, i.e., stops laying, remains continuously on the nest, clucks and ruffles her feathers when disturbed, and sometimes, at least, will not take chicks until she has been broody for some time. The second point was wholly unexpected and adds to the general anomaly of the capon's behavior.

This point of behavior may be described as follows: At one moment the capon is moving sedately about. Suddenly he seizes one of the chicks by the nape of the neck and dangling it in the air squats and goes through sidewise shaking movements of the posterior part of the body, as a cock does when treading a hen.   The orgasm is soon over; the struggling chick is released and the capon resumes his rôle of caretaker.   Sometimes, though not often, the act is repeated two or three times with the same or different chicks within a short time—ten or fifteen minutes.   On one or two such occasions, I have seen him employ a trick used by the cock in getting a hen indifferent to his attentions within reach.   This trick consists in uttering the food call, at the same time standing with head down, repeatedly pretending to pick up and drop an imaginary bit of food, but at the same time with an eye on the hen.   When she comes within reach, he attempts to seize her.   A hen with chicks calls them to the food but in a different way.   The capon's behavior toward the chicks leaves little room for doubt that he is attempting to tread them.   Though hens have since been kept in the same pen, this capon has not been observed treading them.   This capon reared all the chicks given him.

In the second instance the capon was a Brown Leghorn, hatched June 29, 1913, and caponized by the writer August 8, 1913.   The same procedure was adopted with the Brown Leghorn as with the Rhode Island Red.   The hen-hatched chicks were accepted just as readily as before.   However, he was extremely wild, so that whenever anyone approached he promptly attempted to escape from the coop or yard in which he was confined.   He made no attempt to fight off an intruder but promptly deserted his charges for the far corner of the pen.   As soon, however, as the intruder withdrew he returned to the chicks.   This bird clucked very little, even when he had a tidbit to offer.   His movements were very quick.   After a little he learned to fly out of the pen, and as the meshes of the wire were coarse, the chicks were able to follow him.   Only two of the four originally given him survived beyond the first few days.   The trio kept together all summer until the capon moulted.   Soon after his new feathers were about two-thirds grown he was observed at several different times on top of a post, crowing.   In appearance he is as typical a capon as any other.

The third instance was a Rhode Island Red capon of similar history to the first.   The chicks used were incubator-hatched.   The bird was shut up in the evening of July 12 and two chicks placed beneath him.   He took them readily.   The following night the rest of the hatch, some twenty-five in number, was placed beneath him.   The entire flock was accepted and cared for. This bird from the start clucked more like a hen than the first capon used.   The chicks and capon were kept in a loft for a few days, and then placed in a large "A" coop out-of-doors, the chicks being allowed to run out between the slats.   This capon took most excellent care of the chicks.   He was never observed making any attempt to tread them.

The fourth capon was a Brown Leghorn of similar history to the others.   He was given incubator-hatched chicks in early September.   They were poor chicks and died after a few days. This capon was fairly quiet and as far as could be seen gave the chicks good care while they lived, which was less than a week. Before the bird's band number was taken, a heavy wind blew open the door of the coop, permitting his escape.   He returned of his own accord to the pen which contained the flock of capons, so that it is uncertain whether this or another Leghorn was used in the fifth trial.

Much the same procedure was employed in the fifth trial as with number four.   The capon, a Brown Leghorn, was placed in a coop in a loft and given five chicks, three of which soon died.   A few days later he was removed out-of-doors into a large "A" coop.   This bird, while somewhat wild at first, soon quieted down and paid little attention to people who came near the coop.   He wandered up and down the coop very much like a hen, with an eye on the chicks which ran at large.   His cluck, made at intervals, sounded much like that of a hen.   The chicks spent most of the time racing about outside the coop. As far as observed he did not attempt to tread the chicks.   If released from his coop he led the chicks about like a hen, scratched for them and called them to him.   If one picked up a chick, the bird spread himself like a hen and advanced to the attack.

The general behavior of capons is anomalous.   In many respects quieter than normal males, they do not as a rule give evidence of sexuality.   Usually, they impress one as being essentially neutral in sexual behavior, very much like any immature bird.   Darwin states that they are said to incubate eggs as well as care for the chicks.   We have not yet tested this report.   Their behavior with chicks, in the main, is very much like that of the hen, such differences, with the exception of the actions of the first capon in treading the chicks, being of a minor character and difficult to distinguish clearly from those of a broody hen. On the other hand they sometimes crow and may give attention to the hens, their behavior at such times being as complex as that of the cock.

It is remarkable that the Leghorn capons should receive the chicks quite as readily as the Rhode Island Reds. S nce the females rarely become broody (only two or three per cent) one would not expect to find such strong evidences of the broody instinct in capons of this race. The differences in the races with regard to broodiness is well marked. Of the Reds, only about two per cent *fail* to become broody at least once during the first year, while many individuals become broody eight or ten times within a year from the time they begin to lay. We have no statistics of our own on the Leghorns, but a breeder of white Leghorns reports that his records show that of nearly three hundred pullets only twelve became broody (most of them only once) but one became broody five times and one three times. My own experience with Brown Leghorns is of the same general character. As a rule, Leghorns that show signs of broodiness rarely make good mothers, for the exhibitions of broodiness are very transitory.

Although the tests are not extensive enough to warrant definite conclusions, it would appear possible that the brooding instincts of the capon are after all not necessarily a female character. Many male birds, e.g., pigeons, assist in brooding and rearing the young. The domestic cock sometimes assists the hen in finding a nest. An instance of a broody cock which hatched a clutch of eggs was recently published in " Fur and Feathers." Darwin states in "Animals and Plants," as follows: " The capon takes to sitting on eggs, and will bring up chickens. . . . Reaumur asserts that a cock, by being long confined in solitude and darkness, can be taught to take charge of young chickens; he then utters a peculiar cry, and retains during his whole life this newly acquired maternal instinct."

Tests of non-broody hens, castrated hens, normal males and chicks of various ages from various races will have to be made in order to throw more light on the question of the capon's behavior. Older chicks are sometimes used as leaders in teaching very young chicks the way in and out of a brooder. A single test that was made of a normal male gave negative results.

Obviously, until more studies have been made of the behavior of fowls, the brooding instinct of the capon cannot be cited as proof of the assumption of a female secondary sexual character by a castrated male.

NOTE.—In a recent paper in this journal Pearl makes the following statement: " It appears to be the case that in the domestic fowl the brooding instinct has to a large degree disappeared along with the fact of domestication." He points out that at the Maine Station they have found great difficulty in finding hens that would incubate eggs through the full period of twenty-one days. My own experience with several varieties of poultry, extending over several years, does not coincide at all with Pearl's. Instances in which the birds fail to complete the full period have not been frequent. At various times, by providing a fresh lot of eggs, hens have been made to incubate eggs for six weeks or longer. In the Little Compton district, a noted poultry center, hens are depended upon almost entirely for hatching and rearing chicks. The natural method is also used extensively by farmers and poultrymen in general. The interesting results reported by Pearl would seem to apply only to *some* flocks of birds belonging to the so-called broody races (Rocks, Wyandottes, Orpingtons, etc.), but can hardly be considered of general applicability to such races. Quite the reverse, however, would be true of the Leghorns and similar non-broody races where, as stated above, only a small per cent show any traces of the broody instinct.

27

Reprinted from *Behaviour* 19:76–116 (1962)

EGG SHELL REMOVAL BY THE BLACK-HEADED GULL, *LARUS RIDIBUNDUS* L.; A BEHAVIOUR COMPONENT OF CAMOUFLAGE

by

N. TINBERGEN, G. J. BROEKHUYSEN [1]). F. FEEKES [2]),
J. C. W. HOUGHTON [3]), H. KRUUK [2]) and E. SZULC [4])

(Department of Zoology, University of Oxford) [5])

(With 11 Figs)

(Rec. 28-IV-1961)

I. INTRODUCTION

Many birds dispose in one way or another of the empty egg shell after the chick has hatched. A shell may be built in or trampled down; it may be broken up and eaten; or, more usually, it is picked up, carried away and dropped at some distance from the nest. C. and D. NETHERSOLE THOMPSON (1942), who have given a detailed summary of our knowledge of egg shell disposal in birds, emphasise the inter- and even intra-specific variability of the responses involved. Since, in addition, the actual response is often over in a few seconds, and happens only once or twice for each egg, it is not surprising that our knowledge is still fragmentary. On the whole, the presence or absence of the response and its particular form seems to be typical of species or groups of species; for instance, it seems to be absent or nearly so in Anseres and in Gallinaceous birds; Accipitres often break up and eat the shell; Snipe are said to be "particularly lax" (NETHERSOLE THOMPSON); Avocets, *Recurvirostra avosetta* L., remove discarded egg shells anywhere in the colony (MAKKINK, 1936). In the many species which carry the egg shell away, the response, occurring as it does just after hatching, when the young birds need warmth and protection from predators, must be supposed to have considerable survival value.

The Black-headed Gull invariably removes the egg shell in a matter of

1) Dept. of Zoology, University of Cape Town.
2) Dept. of Zoology, University of Utrecht.
3) City of Leeds Training College.
4) Nencki Institute of Biology, Warsaw.
5) We are indebted to Sir WILLIAM PENNINGTON RAMSDEN, Bart., and the Cumberland County Council for permission to camp and to carry out our work on the Drigg sand dunes; to the Nuffield Foundation and the Nature Conservancy for generous support; and Dr J. M. CULLEN for constructive criticism.

hours after hatching (Pl. XI, fig. 1); it is extremely rare to find an egg shell in the nest once the chicks have dried. We have only a few direct observations on the time lapse between hatching and carrying in undisturbed gulls, but the 10 records we have (1′, 1′, 15′, 55′, 60′, 105′, 109′, 192′, 206′, 225′) suggest that the response is usually not very prompt. The carrying is done by the parent actually engaged at the nest, never (as far as we know) by the non-brooding partner which may be standing on the territory, even when it stands next to the sitting bird. At nest relief either the leaving partner, or, more often, the reliever carries the shell. Often however it is the sitting bird who starts looking at the shell, stretches its neck towards it, takes it in its bill and nibbles it (sometimes breaking off fragments while doing so, which then are swallowed), and finally rises and then either walks or flies away with the shell in its bill. The shell is dropped anywhere between a few inches and a hundred yards from the nest. We have also observed birds which flew off with the shell, made a wide loop in the air, and descended again at the nest with the shell still in their bill which they then either dropped on the nest or carried effectively straight away. There is no special place to which the shell is carried, though there may be a slight tendency to fly against the wind, or over an updraught, or where the carrier is less likely to be harrassed by other gulls; almost always the shell lands well beyond the territory's boundary. On rare occasions a shell may land in a neighbour's nest—where the latter then treats it as one of its own shells, *i.e.* removes it.

II. STATEMENT OF THE PROBLEMS

During our studies of the biology and the behaviour of gulls this response gradually began to intrigue us for a variety of reasons. (1) The shell does not differ strikingly from an egg, since it is only the small "lid" at the obtuse end which comes off during hatching; yet it is treated very differently from an egg, and eggs are never carried away. This raised the qu ~tion of the stimuli by which the gulls recognise the shell. Systematic test. with egg shell dummies could provide the answer. (2) What could be the survival value of the response? (a) Would the sharp edges of the shell be likely to injure the chicks? NETHERSOLE THOMPSON raises this possibility, adding that poultry breeders know this danger well. (b) Would the shell tend to slip over an unhatched egg, thus trapping the chick in a double shell? (c) Would the shells interfere in some way with brooding? (d) Would the moist organic material left behind in the shell provide a breeding ground for bacteria or moulds? (e) Would egg shells, if left near the nest, perhaps attract the attention of predators and so endanger the brood?

The following facts, obtained earlier by our co-workers, seemed to give some clues.

(A) C. BEER (1960) found that Black-headed Gulls do not merely carry shells but a great variety of other objects as well if they happen to be found in the nest. Some of these objects are shown in Pl. XI, fig. 2. It seemed that the best characterisation of this class of objects would be: "Any object—perhaps below a certain size—which does not resemble an egg, or a chick, or nest material, or food"; in short: "any strange object". The very wide range of objects responded to suggests that the birds respond to very few sign stimuli; it might be that the response was adapted to deal with a much wider range of objects than just the egg shell.

(B) C. BEER (1960), testing the gulls' readiness to show this response at different times of the season, offered standard egg shell dummies (halved ping-pong balls painted egg shell colour outside, Pl. XI, fig. 2) to a large number of gulls once every day from the moment nest scrapes were formed (which is up to about three weeks before the laying of the first egg) till well beyond the hatching of the chicks. He found that under these conditions of standard (and near-optimal) stimulation the response could be elicited from at least 20 days before laying till 3 weeks after hatching. In this respect the response behaves rather like typical incubation responses such as sitting and egg retrieving which also develop gradually in the pre-egg period (BEER, 1960). In view of the heavy predation to which eggs are subjected (see below), and of the fact that the eggs are otherwise carefully guarded, this fact suggests that the response is important throughout the incubation period, and not merely during the few days when the chicks hatch.

(c) Finally, E. CULLEN (1957) found that the Kittiwake, *Rissa tridactyla* L. never carries the egg shell. The shells are just left in the nest until they are accidentally kicked off. It is true that this often happens in the first few days after hatching, but shells occasionally stay in the nest or on the rim for weeks, and at any rate they remain in or on the nest much longer than is the case with the Black-headed Gull. The Sandwich Tern, *Sterna sandvicensis* Lath. does not remove the egg shells either (J. M. CULLEN, 1960).

These observations combined suggest that neither the avoidance of injury, nor of parasitic infection, nor of interference with brooding are the main functions of egg shell removal — if this were so, then the Kittiwake as the most nidicolous species of gull would not lack the response. The most likely function seemed to be the maintenance of the camouflage of the brood — neither Kittiwake nor Sandwich Tern can be said to go in for camouflage to the extent of the other gulls and terns.

Fig. 1. Black-headed Gull about to remove the empty egg-shell. (G. J. BROEKHUYSEN phot.)

Fig. 2. A sample collection of various objects which Black-headed Gulls remove from the nest. Top centre: BEER's standard model (a halved ping pong ball); bottom right: real egg-shell. After C. BEER, 1960. (N. TINBERGEN phot.)

Fig. 4. A small angle presented on a nest's rim. (N. TINBERGEN phot.)

Fig. 5. Some of the dummies used. Left, top to bottom: large angle, small angle, halved ping pong ball; centre top to bottom: khaki hens' shell, gulls' egg shell, black egg; right top to bottom: large ring, small ring. (N. TINBERGEN phot.)

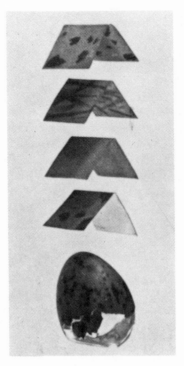

Fig. 10. Some of the small angles of Experiment 8, together with a gulls' egg shell.
Top to bottom: "all egg', "cam", "sage green", and "shell". (J. HAYWOOD phot.).

Fig. 11. Black-headed Gull trying to swallow rectangle measuring ½ × 18 cm.
(G. J. BROEKHUYSEN phot.).

Thus these observations naturally led to an investigation into the function of the response and to a study of the stimuli eliciting it. In the following we shall deal with the problem of survival value first.

III. THE SURVIVAL VALUE OF EGG SHELL REMOVAL

The assumption that egg shell removal would serve to maintain the camouflage of the brood presupposes that the brood is protected by camouflage. This basic assumption, usually taken for granted but — as far as we know — never really tested, was investigated in the following way.

First, we collected whatever observations we could about predation in the colony. While these observations are largely qualitative, they show convincingly that predation is severe throughout the season.

Very many eggs disappear in the course of spring. We did not make systematic counts but can give the following qualitative data. In the Ravenglass colony Carrion Crows, *Corvus corone* L., did not account for many egg losses, because, as we could observe time and again, they are easily chased away by the mass attack of the Black-headed Gulls. In fact we never saw a Crow alight in the colony. However, attacks by one, two or three gulls (which often occurred in our tests) did not deter Crows, and we must assume that nests on the fringe of the colony, and the dozens of nests we regularly find outside the colony and which do not survive, often fall victims to the Crows.

Egg predation within the colony was due to the following predators. Three pairs of Herring Gulls, *Larus argentatus* Pont., which bred in the gullery levied a constant toll of eggs, and later of chicks. Although the Black-headed Gulls attacked them, they could not altogether stop them from snatching eggs and chicks. We observed hundreds of occasions on which non-breeding Herring Gulls and Lesser Black-backed Gulls, *Larus fuscus* L., (many of them immature) passed near or over the colony or over our tests. On only one occasion did we see any of these taking an egg; they usually were totally uninterested in the colony.

Black-headed Gulls prey on each other's eggs to a certain extent. Most of those who visited our experiments did not attack an undamaged egg (the few exceptions are mentioned in our tables) but finished an egg once it had been broken by other predators. Later in the season individual Black-headed Gulls specialised on a diet of newly hatched chicks (see section VI).

Foxes, *Vulpes vulpes* L., which regularly visited the colony, and killed large numbers of adults early in the season and many half or fully grown chicks towards the end, did visit the gullery in the egg season, but we have no direct evidence of the amount of damage done by them in this part of the year. The gulls were greatly disturbed whenever a Fox entered the colony, but they did not attack him as fiercely as they attacked Herring Gulls and Crows. They flew over the Fox in a dense flock, calling the alarm, and made occasional swoops at it.

Stoats, *Mustela erminea* L., and Hedgehogs, *Erinaceus europaeus* L., visited the gullery and probably accounted for some losses. The gulls hovered over them in a low, dense flock but did not quite succeed in deterring them. In spite of several thousands of man-hours spent in hides in the gullery by us and by our colleagues Dr C. BEER and Dr G. MANLEY we never saw either a Fox, a Stoat or a Hedgehog actually taking an egg; they stayed in the dense cover, and though we have been able to read a great deal from their tracks whenever they had moved over bare sand, this method is of no avail in the egg season since the gull's nests are situated in vegetation.

Surprisingly, the Peregrine Falcons, *Falco peregrinus* L., which were often seen near or over the gulleries left the Black-headed Gulls in peace, though they regularly took waders and carrier pigeons. The gulls panicked however when a Peregrine flew past.

In order to eliminate the effect of the gulls' social nest defence we put out eggs, singly and widely scattered (20 yards apart) in two wide valleys outside the gullery proper. These valleys had a close vegetation of grasses, sedges and other plants not exceeding 10 cm in height. Each egg was laid out in a small depression roughly the size of a Black-headed Gull's nest. Two categories of eggs were arranged in alternate pattern, and in successive tests exchanged position. While predators, particularly Carrion Crows, showed remarkable and quick conditioning to the general areas where we presented eggs, there was no indication that the exact spots were eggs had been found were remembered, and in any case such retention, if it would occur, would tend to reduce rather than enhance differential predation. We assumed (erroneously) that it might well be days before the first predator would discover the eggs, and in our first tests we therefore did not keep a continuous watch. We soon discovered however that the first eggs were taken within one or a few hours after laying out the test, and from then on we usually watched the test area from a hide put up in a commanding position, allowing a view of the entire valley. The tests were usually broken off as soon as approximately half the eggs had been taken.

Experiment 1.

Not wishing to take eggs of the Black-headed Gulls themselves (the colony is protected) we did our first test with hens' eggs, half of which were painted a matt white, the other half painted roughly like gulls' eggs. To the human eye the latter, though not quite similar to gulls' eggs, were relatively well camouflaged. It soon became clear that we had underrated the eye-sight of the predators, for the "artificially camouflaged" eggs were readily found.

TABLE I

Numbers of artificially camouflaged hens' eggs ("Artif. cam.") and white hens' eggs ("White") taken and not taken by predators.

| | Artif.cam. | | White | |
	taken	not taken	taken	not taken
Carrion Crows	16		18	
Herring Gulls	0		1 (+3)	
Total	16	36	19 (+3)	33 (−3)

N presentations: 2 × 52. Difference between "artif.cam." and "white" not significant. [1]

---

1) The P-Values given in the tables were calculated by the χ^2-method, except where stated otherwise.

In 9 sessions, lasting from 20 minutes to 7½ hours, we saw Carrion Crows and Herring Gulls take the numbers of eggs (out of a total presented of 104) given in Table I.

We were astonished to see how easily particularly the Carrion Crows found even our "camouflaged" eggs. Each test area was usually visited every few hours by a pair of Crows [1]. They would fly in at a height of about 20 feet, looking down. Their sudden stalling and subsequent alighting near an egg were unmistakable signs that they had seen it; they usually discovered even camouflaged eggs from well over 10 metres distance. Often the Crows would discover every single egg, whether white or camouflaged, in the area over which they happened to fly. They either carried an egg away in their bills without damaging it to eat it elsewhere, or opened it on the spot and ate part of the contents, or (usually after they had first eaten four or five eggs) they carried it away and buried it. In some cases we saw Crows uncover these buried eggs one or more days after they had been cached.

Of the numerous Herring Gulls and Lesser Black-backed Gulls living in the general area, many of which flew over our test area every day, interest in eggs was shown only by three resident pairs of Herring Gulls. Their eyesight was undoubtedly less keen than that of the Crows (as expressed in the scores in table I, and particularly table II); further they were remarkably timid. For instance in experiment 1 we were certain on three separate occasions that a Herring Gull had discovered a white egg but did not dare approach it (the "3" in brackets in table I refers to these occasions). Yet as soon as Crows began to search the area, Herring Gulls would appear and attack them. Often Crows and Gulls attacked each other mutually, swooping down on their opponents from the air, for minutes before either of them alighted near an egg. Usually the Gulls succeeded in claiming an egg first. There were also occasions on which the Crows had the area to themselves.

It seemed obvious that our "camouflage" was not effective at all. Now the artificially camouflaged eggs differed from real gulls' eggs in four respects: (1) the ground colour, though to the human eye matching the overall colour of the rather brownish background very well, was slightly different from the ground colour of most of the gulls' eggs; (2) the dark grey dots which we painted on the eggs were more uniform in size and distribution than those on the real gulls' eggs; (3) unlike the natural dots they were all of one hue; and (4) hens' eggs are considerably larger than Black-headed Gulls' eggs, and hence probably more conspicuous.

---

1) We have good reasons to believe that each of our two test areas were visited by one pair.

Experiment 2.

Therefore, we next tested real, unchanged Black-headed Gulls' eggs against Black-headed Gulls' eggs painted a matt white. The tests were conducted in the same way as the previous ones. The results, obtained in 12 sessions lasting from 20 minutes to 4 hours, in which 137 eggs were presented, are summarised in table II.

TABLE II

Numbers of normal Black-headed Gulls' eggs ("Natural") and Black-headed Gulls' eggs painted white ("White") taken and not taken by predators.

| | Natural | | White | |
	taken	not taken	taken	not taken
Carrion Crows	8		14	
Herring Gulls	1		19	
Black-headed Gulls	2		7	
Unknown	2		3	
Total	13	55	43	26

N presentations: 68 + 69: Difference between "Natural" and "White" significant at .1% level.

Experiment 3.

In order to get an impression of the parts played by size and by the nature of our "artificial camouflage", we painted Black-headed Gulls' eggs in the same way as the "camouflaged" hens' eggs, and compared their vulnerability with that of Black-headed Gulls' eggs painted white. The results of this test, to which we devoted only 5 sessions of from 1 to 3 hours, and in which 48 eggs were presented, are given in table III.

TABLE III

Numbers of artificially camouflaged Black-headed Gulls' eggs ("Artif. cam.") and Black-headed Gulls' eggs painted white ("White") taken and not taken by predators.

| | Artif.cam. | | White | |
	taken	not taken	taken	not taken
Carrion Crows	4		5	
Herring Gulls	4		8	
Black-headed Gulls	1		1	
Total	9	15	14	10

N presentations: 24 + 24. Difference between "artif. cam." and "white" not significant (20 % < p < 30 %).

From these experiments, and particularly from experiment 2, we conclude that the natural egg colour of the Black-headed Gulls' eggs makes them less vulnerable to attack by predators hunting by sight than they would be if they were white; in other words that their colour acts as camouflage. The difference between the results of experiments 1 and 3 on the one hand and experiment 2 on the other indicates that we had underrated the eyesight of the predators, and also that their reactions to the different aspects of camouflage deserve a closer study: the parts played by over-all colour, by pattern and hue of the dotting, perhaps even by the texture of the eggs' surface we hope to investigate later — it was a pleasant surprise to discover that large scale experiments are possible. For our present purpose we consider it sufficient to know that painting the eggs white makes them more vulnerable.

Experiment 4.

We can now turn to the question whether or not the presence of an egg shell endangers an egg or a chick it is near. For obvious reasons we chose to investigate this for eggs rather than for chicks. The principle of this experiment was the same as that of the previous ones. We laid out, again avoiding site-conditioning in the predators, equal numbers of single Black-headed Gulls' eggs with and without an empty egg shell beside them. The shells used were such from which a chick had actually hatched the year before and which we had dried in the shade and kept in closed tins. The shells were put at about 5 cm. from the eggs which were again put in nest-shaped pits. Predators were watched during 5 sessions lasting from 45 minutes to 4½ hours. Sixty eggs were used. The results are given in table IV.

TABLE IV

Numbers of Black-headed Gulls' "eggs-with-shell" and "eggs-without-shell"
taken and not taken by predators. Eggs not concealed.

	eggs-with-shell		eggs-without-shell	
	taken	not taken	taken	not taken
Carrion Crows	6		7	
Herring Gulls	9		7	
Total	15	15	14	16

N presentations: 30 + 30. Difference between egg with shell and egg without shell not significant.

We did not consider this result conclusive because the circumstances of the experiment differed from the natural situation in two respects. (1) Although a nest in which a chick has recently hatched may contain unhatched eggs, there are chicks equally or rather more often; and chicks, apart from having a less conspicuous shape than eggs, do at a quite early stage show a tendency to crouch at least half concealed in the vegetation when the parent gulls call the alarm. The Crows and Gulls might have less difficulty finding eggs than seeing chicks. (2) Both predators were always vigorously attacked by many Black-headed Gulls whenever they came in or near the gullery. In avoiding these attacks their attention is taken up for the greater part of the time (as judged by their head movements and evading action) by keeping an eye on the attackers; in the natural situation they never have the opportunity to look and search at their leisure. And to predators searching for camouflaged prey leisure means time for random scanning and opportunity for undivided attention — both probably factors enhancing discovery i.e. fixation of non-conspicuous objects. In other words our experiment had probably made things too easy for the predators, even though in the colony the nests themselves are often visible from a distance.

Experiment 5.

We therefore decided to repeat this experiment with slightly concealed eggs. This was done by covering each egg (whether or not accompanied by an egg shell) with two or three straws of dead Marram Grass, a very slight change which nevertheless made the situation far more similar to that offered by crouching chicks. Most tests of this experiment were done without watching from a hide; we knew by now who the main predators were, and for our main problem it was not really relevant to know the agent. In 8 tests, lasting from 2 hours 40 minutes to 4 hours 40 minutes, 120 eggs were offered, of which 60 with a shell at 5 cm. distance. The results are given in table V.

TABLE V.

Number of Black-headed Gulls' eggs-with-shell and eggs-without-shell taken and not taken by predators. Eggs slightly concealed.

eggs-with-shell		eggs-without-shell	
taken	not taken	taken	not taken
39	21	13	47

N presentations: 60 + 60. Difference between egg with shell and eggs without shell: $p < .1\%$.

The conclusion must be that the near presence of an egg shell helps Carrion Crows and Herring Gulls in finding a more or less concealed, camouflaged prey, and that therefore egg shells would endanger the brood if they were not carried away.

Experiment 6.

This was a series of pilot tests designed to examine whether the Carrion

Crows would be readily conditioned to shells once they had found eggs near shells, and whether, if they would find shells without eggs, they would lose interest in the shells. These tests began on April 21st, 1960, after the Crows operating in the valley had already gained experience with white and camouflaged eggs, and, since they had broken many of those while eating them, could have learned that egg shells meant food. In the very first test, egg shells were laid out in the Western area of the valley, where in previous tests with whole eggs the Crows had never been seen to alight. At their first visit the Crows alighted near these shells, pecked at them, and searched in the neighbourhood. They left after a few minutes whereupon we took the shells away. A few hours later we again laid out egg shells without eggs, this time scattered over the whole valley. When the Crows next returned they flew round over the valley, looking down as usual, but they did not alight. Next morning semi-concealed eggs were laid out over the whole valley, each with an egg shell at the usual distance of 5 cm. This time the Crows did alight and took a number of the eggs. A few hours later we once more laid out shells only, scattered over the valley, and at the next visit the Crows came down near several of them and searched. The morning after this we laid out just shells, this time in the Eastern part of the valley. This time the Crows did not alight. When next, between this and their next visit we gave each shell an egg at 5 cm., the Crows alighted again when they returned and took several eggs. Later that same day just shells were given in the Western part, which could not induce the Crows to alight. Such tests were continued until April 30, and while they did not give sufficient information, they strongly suggest: 1) that one experience with shells-without-eggs was sufficient to keep the Crows from alighting near shells the next time; and b) that renewed presentation of eggs with the shells attracted them again. Further, we had the impression (c) that the Crows later learned that egg shells in a part of the valley where shells only had been presented several times meant no food — in other words that they associated "shells only" with the locality. The full record of these experiments is given in table VI.

TABLE VI

Record of "conditioning test" explained in text.

Es + = eggs + shells offered, eggs taken; Es — = eggs + shells offered, eggs not taken; S + = shells alone offered, crows alighted and searched; S — = shells alone offered, crows did not alight.

Eastern area					S —	Es +		Es +		
Entire valley		S —	Es +	S +						S —
Western area	S +						S —		S —	S —
Date (April)	21	21	22	22	23	23	23	24	24	25	25

Eastern area	Es +	Es +			Es —	Es —	Es +		Es +		
Entire valley										S +	
Western area			S —	S —				S —			S +
Date (April)	25	26	26	27	27	28	28	28	29	29	30

Experiment 7.

We next investigated the effect of distance between egg and shell. This was done in a mass test without direct observation from the hide, and in the same valley where we had regularly seen Crows and Herring Gulls take eggs. Half concealed gulls' eggs were laid out; one third of them had a shell at 15 cm. distance, one third at 100 cm., and one third at 200 cm. A total of 450 eggs were presented in 15 tests, with the following results (table VII).

Table VII

Number of eggs taken out of the 450 offered with a shell 15, 100, and 200 cm. away.

15 cm.		100 cm.		200 cm.	
taken	not taken	taken	not taken	taken	not taken
63	87	48	102	32	118

N presentations: 3 × 150. Difference between 15 cm. and 200 cm. significant at .1% level. Significance of the total result: $p < 1\%$.

Part of each group of eggs may of course have been found without the aid of the shell (and this may explain why so many eggs were found even of the 200 cm.-group), but the figures show that the "betrayal effect" is reduced with increased distance.

A second, similar test was taken with the shells at 15 cm. and 200 cm.; altogether 60 eggs were presented in 3 tests; the results are given in table VIII and show a similar result.

TABLE VIII

Numbers of eggs-plus-shell taken | not taken for different distances (in cm.) between egg and shell. 60 eggs were offered.

15 cm.		200 cm.	
taken	not taken	taken	not taken
13	17	4	26

N presentations: 2 × 30. Difference between 15 cm. and 200 cm. significant at 2.5% level.

The following test, to be mentioned only briefly because it did not contribute to our results, was in fact done before any of the tests mentioned so far in an attempt to short-circuit our procedure. We put out 50 hens' eggs painted in what we then hoped would be a good camouflaged pattern (see above), and gave each egg an egg shell dummy at 15 cm. distance. These dummies were metal cylinders made by bending a strip of metal sheet measuring 2 × 10 cm. as used in our later tests with Black-headed Gulls, see experiments 11, 17 and 18 (Fig. 5). Half of these were painted like the eggs, inside and outside (they were in fact satisfactorily cryptic to the human eye); the other half were painted white. We assumed that predators would be slow to come, and that in the beginning one check per day would be sufficient. Upon our first check, about 24 hours after we laid out the eggs, we found that all 50 eggs had disappeared. Our first conclusion was that a whole horde of predators, such as a flock of Herring Gulls, had raided the valley, but we later saw that this valley was searched mainly by one pair of Carrion Crows and one, two, or sometimes three pairs of Herring Gulls. During the following weeks we discovered that the Crows kept digging up these eggs which they must have buried on the first day.

This failure forced us to check the effectiveness of our "camouflage" paint first (see experiments 1-3) and to test the effect of the natural egg shell (experiments 4-7); by the time these questions had been settled we had to start our tests on the stimuli eliciting egg shell removal in the Black-headed Gulls themselves, and so had to abandon for the moment any tests on the effect of the colour of the egg shell on the predators. However, the method having now been worked out, it is hoped to investigate this more fully.

The conclusion of this part of our study must therefore be that the eggs of the Black-headed Gulls are subject to predation; that in tests outside the colony the number of eggs found by Carrion Crows and Herring Gulls is lower than it would be if the eggs were white; that the proximity of an egg shell endangers the brood; and that this effect decreases with increasing distance. While it will now be worth investigating the predators' responsiveness to eggs and shells in more detail, the facts reported leave little room for doubt about the survival value of egg shell removal as an antipredator device. Whether or not the response has other functions is of course left undecided.

IV. THE STIMULI ELICITING EGG SHELL REMOVAL

BEER's demonstration that a great variety of objects can elicit the response naturally led to the question whether all these objects did so to exactly the same extent. This was investigated by presenting dummies of different types, one at a time at each nest, to hundreds of nests, and checking after a certain period (constant for each experiment but varying for different experiments according to the overall stimulating value of the dummies compared) whether or not the dummy had been removed. Nests were marked with numbered wooden pegs, and, unless otherwise mentioned each nest was used for only one experiment. Each experiment was arranged according to the latin square method, as set out in fig. 3. The nests were divided in as many equal groups as there were dummies, and each nest was offered each dummy once; the sequence of presentation to each group was arranged in such a way that each dummy was presented equally often in the first, second, third etc. position in the sequence.

Since each experiment involved repeated presentation of dummies, even though no nest had the same dummy twice in one latin square experiment, the problem of waning of the response was important. As will be discussed in detail in section V the waning within the time limits of one latin square was so slight as to be negligible, but with repetition of entire experiments with the same group of birds there was definite waning from one experiment to the next.

The dummies were put down in a uniform way on the nest's rim, namely approximately 12 cm. from the centre of the nest (in 80 nests measured the

	a	b	c	d	e	f	g	h
A	1	8	2	7	3	6	4	5
B	2	1	3	8	4	7	5	6
C	3	2	4	1	5	8	6	7
D	4	3	5	2	6	1	7	8
E	5	4	6	3	7	2	8	1
F	6	5	7	4	8	3	1	2
G	7	6	8	5	1	4	2	3
H	8	7	1	6	2	5	3	4

Fig. 3. A latin square experiment sequence, in which 8 models (1-8) were offered in 8 successive tests (a-h) to 8 groups of nests (A-H).

average outside diameter was 27 cm., the average inside diameter 13 cm.) (Pl. XII, fig. 4). Their orientation was likewise standardised as much as possible. Watch was kept from a hide nearby in a number of cases. From these direct observations we know that both male and female remove egg shells; we also know that egg shell dummies were not moved by the wind (quite strong gales had no effect even on some of our lightest models if they were lightly anchored on the nest's rim; yet for other reasons we avoided testing in rough weather). Small movements of the dummies might be the result of the bird accidentally kicking it; this meant that very slight changes in the position of the dummies could not be interpreted with certainty; this category however was small and we tried to score them in a uniform way, displacement over more than one inch away from the nest being taken as evidence of carrying.

Because both sexes carried the egg shell, and because intervals between nest reliefs varied widely, it was impossible to keep records of individual birds, at least in the mass tests. This is an inaccuracy which, while probably not very important, has to be borne in mind in judging the results.

Since we had reasons to concentrate on the antipredator aspects of the response and particularly on camouflage, the question of visual conspicuousness, and therefore of the effect of different colours, was taken up first.

The effect of colour was tested by offering rectangles of flat sheet metal bent at right angles in the middle ("small angles", Pl. XII, fig. 5). In 1959 these rectangles measured 1 × 2 inches; in 1960, when we compared some of the angles with cylindrical rings of the same diameter as Black-headed Gulls' eggs, our angles were either made of strips of the same surface area as the rings, or half that size; the "small angles" measured 2 × 5 cm.; the "large angles" 2 × 10 cm. Within each colour experiment the size of the angles was of course uniform. As colours we used Bergermaster Magicote (matt) in 1959, and Rowny Fixed Powder Colours mixed with water or, for other dummies, Du-Lite emulsion paint. All paints gave a uniform matt

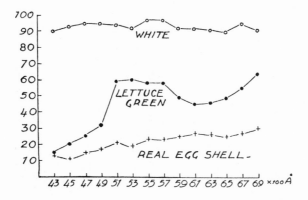

Fig. 6. Reflection graphs of white, lettuce green and real egg shell.

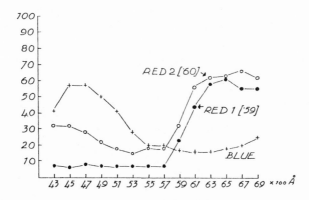

Fig. 7. Reflection graphs of blue, red 1959 and red 1960.

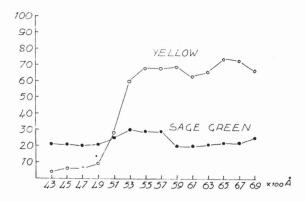

Fig. 8. Reflection graphs of yellow and sage green.

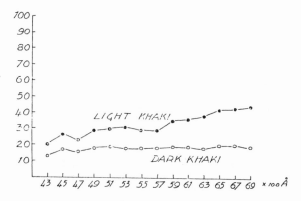

Fig. 9. Reflection graphs of light khaki and dark khaki.

surface. The reflected colours within the visible spectrum were measured
with a Keuffel & Esser "Colorimeter", Model E; Figs 6 to 9 incl. show the
reflection graphs. Any interpretation of the results obtained with such models
must depend on knowledge of the sensitivity of the Black-headed Gulls' eye
to light of various wave lengths. Although no data are available of this
species, all species of birds examined so far are sensitive to the same spectral
range as human beings; there are indications of a slightly reduced sensitivity
to blue in day birds, and owls are less sensitive to red than we are, but no
bird so far investigated sees infrared or ultraviolet (for a recent summary,
see SCHWARTZKOPFF, 1960). Several species have been shown to distinguish
well between the main colours. WEIDMANN (in litt.) analysing the begging
response of the Black-headed Gull chick, obtained scores for a variety of
colours which are not compatible with colour blindness. We will assume that

reflected colours which are similar to us are so to Black-headed Gulls as well, a conclusion substantiated by our own results.

Experiment 8.

In our first experiment, done in 1959, the following colours were used: "shine" (blank, shiny metal), white, red, black, "shell" (white inside, outside sage green with dark grey dotting), sage green, "all egg" (sage green with dark grey dotting both inside and outside), and "cam" (sage green with dark grey irregular striping inside and outside) (Pl. XIII, fig. 10). Shine was taken because, while its overall appearance was a rather dull grey (much darker than white) when seen from certain angles it reflected the sun and thus appeared much brighter than white. Since a bird, when moving about and when sitting on the nest, would be exposed to the bright flash far less often than to the overall greyish colour of the model, its score might help us find out when the stimulus eliciting shell removal had to be received: acting on the hypothesis (which proved correct) that white would have a high score, shine should have a lower score than white if the stimulus had to be received by the sitting bird (since the chances were then that it would usually see it as a grey angle), but if shine would have a high score, this would mean that it stimulated the bird particularly strongly whenever the bright reflection was briefly visible. White was chosen because to the human eye it was very conspicuous, and because real egg shells, particularly those which have dried for a little while, often show white patches on the rim where the pigmented shell has broken off and the egg membranes were showing (the inside of the shell is rarely white; also, most of it is in the shade). Red was chosen because there were indications in other gulls (see TINBERGEN, 1953) that red objects in the nest are pecked at more often than objects of other colours. Black was originally chosen in order to check whether, if white would have a high score, this was due to whiteness or to contrast with the environment. However, it was remarkable how cryptic black really was on the nest rim; even the deep black of our dummies blended easily with the shadows of the nest rim, and as soon as one or two straws blew over a dummy or were kicked over it by the bird, a black model blended with the shades. Sage green was chosen because it was, of the available standard colours, the one most similar to the ground colour of at least some of the gulls' eggs. As is well known, the eggs of all gulls and terns show considerable variation in colour; this, in combination with the fact that the Herring Gull (and according to some of our own pilot tests, the Black-headed Gull) respond by egg rolling and by incubation to egg dummies of an extremely wide range of colours (TINBERGEN,

1953, but see also BAERENDS, 1959) made us confident that sage green was
near enough to the natural egg colour to justify its use as such — later experiments showed us how wrong we were.

This experiment was done with 96 nests, and was repeated three times
with the same group. The results are given in table IX.

TABLE IX

*Number of small angles of different colours (experiment 8) carried and
not carried in three successive latin square tests. (In all tables, successive
tests with the same group of birds will be indicated by roman numerals
I, II, III; tests taken with different groups of birds will be indicated by
capitals A, B, etc.)*

	I carried	II carried	III carried	Total carried	not carried
shine	38	34	21	93	195
white	44	25	20	89	199
shell	38	25	18	81	207
red	32	23	16	71	217
black	25	24	14	63	225
all egg	27	19	12	58	230
cam	28	13	13	54	234
green	23	14	8	45	243

N presentations: $3 \times 8 \times 96$. Significance of differences discussed in text.

Applying the Mann-Whitney test to the whole range of the individual trial
scores (SIEGEL, 1956) we find the following p-values for what are clearly
borderline cases: Shell-red: 15%; Shine-red: 2%; White-red: 11%; Shell-
black: 8%; White-black: 4%; White-all egg: 1%; Cam-green: 23%;
Black-green: 3%.

We conclude that "shine", "white" and "shell" elicit more responses than
red and black, and that these have a higher valence than "all egg", "cam",
and "green". This suggested that the contrast in brightness between the
dummy and its background entirely determined its releasing value. Because,
as mentioned above, there is a considerable colour variation in the eggs of
the gulls, and because previous experiments on egg-recognition in Herring
Gulls had shown colour to play a relatively minor part, we considered that
sage green was probably similar enough to the colour of real eggs to justify
the conclusion that brightness was the only colour character which controlled
egg shell removal, and that the colour of the outside was not specifically
responded to. The intermediate position of red and black seemed to support
this, for red was physically less bright than white (figs 6 and 7) and black

was, to the human eye, remarkably cryptic. Experiments 11, 12, and 13 showed that this conclusion was incorrect. However, before discussing them, two experiments must be mentioned which were intended to examine the possible effect of contrast between the white and darker parts of the egg shell.

Experiment 9.

The fact that "shell" was not treated differently from either white or shine suggested that contrast between parts of one dummy did not play a part. In order to test this, we made a series of angles of which one "wing" was white inside and outside, and the other wing was coloured on both sides. The additional advantage of this type of dummy compared with those of the previous experiment was that the bird would always see both colours, whereever it was; the old shell models were all white when seen from the inside, all egg when seen from the outside. We used white/white, shine/white, shell/white, sage green/white and red/white. In this experiment, 80 nests were used, and the experiment was repeated three times with these birds. Table X gives the results.

TABLE X

Numbers of small angles of different colours (experiment 9) carried and not carried in three successive latin square tests.

	I carried	II carried	III carried	Total carried	Total not carried
shine/white	41	32	30	103	137
red/white	34	30	22	86	154
white/white	33	30	19	82	158
shell/white	34	26	19	79	161
green/white	32	29	13	74	166

N presentations: $3 \times 5 \times 80$. Differences discussed in the text.

The Mann-Whitney test gave the following results: Shine/white — shell/white: $2.5\% < p < 5\%$; shine/white — green/white: $1\% < p < 2.5\%$; shine/white—white/white $p > 5\%$; red/white—green/white $p > 5\%$.

Since in table IX "shine" received a higher score than "shell", and red had a higher score than green, the results of table X are consistent with the conclusion reached in the previous experiment, and there is no reason to assume that contrast within the shell-dummy plays a part. The high score of "shell" in experiment 8 must therefore have been due to its white surface alone.

329

Experiment 10.

Before we did the previous experiment, we had subjected 80 birds to one experiment with slightly different dummies. Two colours were used on each dummy as in experiment 9, but while colour A was on the inside of one wing, it was on the outside of the other: this ensured that both colours were always visible to the bird sitting on the nest. The combinations used were red/white, green/white, shine/white and shell/white. The scores, which we will not give in detail, are again similar except for shine/white which is a little higher than the others. Since shine had the highest score in all three experiments, it seems likely that this is a real effect; it seems most probable that this is due to its greater brightness when seen in such a way that the sun is directly reflected by it. Since this must be supposed to happen only during a very brief fraction of the time in which the bird sees the model, and in very different positions for the different birds, this again suggests that a very brief stimulus is sufficient to make the bird carry the shell away.

Experiment 11.

The first indication that our interpretation had been in part incorrect came from the results of this experiment, which was taken mainly for the purpose of investigating characters of shape. We had already found (see below, experiment 16) that white angles were far less stimulating than real egg shells, and that the halved ping-pong balls which BEER had used received a score halfway between real shell and white angle. We suspected that the rounded shape of the shells had to do with this, and we therefore made cylindrical rings by bending sheet metal strips measuring 2 × 10 cm. The rings were not quite closed but had a gap of about 1 cm. between the two ends (Pl. XII, fig. 5). Because we had, after the 1959 season, compared the colour "sage green" more carefully with the colours of real eggs and had seen that even the greener eggs were rather different from sage green, we had decided to use a new colour in 1960. This colour was mixed so that to the human eye it exactly matched the ground colour of the majority of eggs, which corresponds to "OOY, 11, 4°" in VILLABOLOS & VILLABOLOS (1947). This colour was called "khaki". As substitutes for real egg shells we used hens' egg shells with the "lid" broken off. The models used in this experiment were: hens' egg shells painted khaki outside and left white inside; small rings painted khaki inside and outside, white small rings, and white angles. The results of this test, taken with two groups of nests (group A, consisting of 80 nests, was tested twice; group B, also of 80 nests, was subjected to one experiment) are given in table XI.

TABLE XI

Numbers of four models (experiment 11) carried and not carried in three tests with two groups of birds.

	A$_I$	A$_{II}$	B	Total carried	not carried
hen's shell khaki outside	59	45	53	157	83
all white ring	36	30	44	110	130
all khaki ring	49	22	37	108	132
white angle	31	19	28	78	162

N presentations: $3 \times 4 \times 80$. Difference between khaki shell and all rings: $p < .1\%$. Difference between khaki ring and angle: $p < 1\%$.

The relevance of this test for the question of response to shape will be discussed below, but here the scores for khaki rings and white rings must be considered. They are not sigificantly different and in fact almost equal. If khaki represented no better "egg colour" than sage green, there was a contradiction between this and the previous experiments. It became likely that the birds responded not only to white but also specifically to the exact colour of the eggs.

Experiment 12.

We therefore decided to compare the effects of khaki and sage green in one experiment. Khaki angles were compared with green angles in 120 nests. The results are given in tableXII.

TABLE XII

Numbers of small angles of different colours (experiment 12) carried and not carried in one test.

	carried	not carried
khaki angles	52	68
sage green angles	29	91

N presentations: 2×120.

Our suspicion was strikingly confirmed: sage green received a much lower score ($p < .5\%$). This result raised two points: (1) how would khaki compare with the other colours? Perhaps it would turn out to be equally effective as white because it was similar to the real ground colour of the eggs; and (2) why was green so low — in experiment 8 it had received a decidedly lower score that either red or black. We thought that this was due to sage green being inconspicuous, but khaki was, if anything, less conspicuous on the nest's rim. We decided to run an entirely new and more comprehensive colour test.

Experiment 13.

Because their dotting makes the overall colour of gulls' eggs darker than their ground colour, we prepared two khaki colours, one resembling the ground colour, the other, made by mixing this with some black, a darker shade of the same colour, roughly matching the overall colour of eggs seen from a distance. Unfortunately, although we had prepared a large quantity of the standard khaki colour, much of this was accidentally spilled, and we had to mix a new khaki. This was again done in such a way as to match the natural eggs' ground colour, but although we used the same ingredients our new khaki was not exactly the same as the original khaki. We do not think it likely, however, that this very slight difference has affected our results significantly but experiments using the "new khaki" are mentioned separately. Apart from dark and light khaki our other colours were: yellow, white, a very bright yellowish green ("lettuce green"), sage green, blue, and red 2. We used small angles of 2 × 5 cm. The reflection graphs of these colours are given in Figs 6 to 9 incl. Table XIII gives the results obtained in one experiment with 240 nests; table XIV gives the figures of an experiment with four of these colours, run with 80 new nests.

TABLE XIII

Numbers of small angles of different colours (experiment 13) carried and not carried in one test.

	carried	not carried
dark khaki	92	148
white	85	155
light khaki	77	163
yellow	73	167
blue	62	178
red	57	183
sage green	45	195
lettuce green	19	221

N presentations: 8 × 240.

TABLE XIV

Numbers of small angles of different colours (experiment 13) carried and not carried in one test.

	carried	not carried
dark khaki	35	45
white	35	45
yellow	28	52
blue	22	58

N presentations: 4 × 80.

The Mann-Whitney test gave the following p's: dark khaki-light khaki: 5 %; white-yellow: 35 %; dark khaki-yellow: 1 % $<$ p $<$ 5 %; yellow-blue 5 %.

Together with the results obtained in 1959 these facts point to the following conclusion: angles of any colour, however bright and unnatural, are carried to a certain extent. There are peaks for white and for khaki. Since khaki is, on the nest rim which is made of dead grass mainly, by far the most cryptic of all, and since dark khaki is carried more often than light khaki; since further the very light and conspicuous lettuce green received a very low score, the light khaki's lightness cannot be responsible for its high score, and this must be due to its special colour. In other words the birds responded to wavelength rather than brightness. Yellow has, among the bright colours with intermediate scores, the highest position, undoubtedly due to the fact that in mixing the khaki a very large amount of just this yellow had to be used; red and blue have intermediate scores, and sage green and particularly lettuce green are exceptionally low.

It thus seems that the gulls' responsiveness to colours of egg shells fits the demands very well: the two colours of the natural egg shell, khaki and white, are most stimulating, while the gulls respond least to green, the colour of the vegetation in the immediate surroundings of the nests — it would be not only wasteful, but decidedly harmful to keep carrying away the leaves of the cover round the nests.

Experiment 14.

This experiment was done with small rings instead of angles. We used four colour combinations: (new) khaki inside and outside, white inside and outside, (new) khaki outside, white inside, and blank, shiny metal. The first three models were taken in order to investigate once more the possible effect of contrast within the dummy; the fourth model was added in the hope that we might find a "supernormal" stimulus (TINBERGEN, 1951). The results, with 80 nests, are given in table XV.

TABLE XV

Numbers of four models (experiment 14) carried and not carried in one test.

	carried	not carried
khaki ring	37	43
white ring	39	41
khaki/white ring	42	38
shiny ring	40	40

N presentation: 4 $\times$ 80. No significant differences.

This again confirms that contrasts within the egg shell dummy have no discernable effect (experiments 8, 9 and 10). The shiny rings, while in parts brighter than white, did not give a higher score than white. New khaki and white have again approximately the same score.

Experiment 15.

In this experiment, done before we began to use new khaki, old khaki, which was a little darker than new khaki, was compared, on angles, with white. Table XVI, of two successive tests with 80 nests, gives a higher score **for white.**

TABLE XIV

Numbers of small angles of different colour (experiment 15) carried and not carried in two successive tests.

| | I | II | Total | |
	carried	carried	carried	not carried
white angle	36	36	72	88
khaki angle	25	18	43	117

N presentations: $2 \times 2 \times 80$. Difference is significant at the .5% level.

In this admittedly small scale test the superiority of white over khaki, slight in experiments 13 and 14, is more pronounced. We do not know why this is so.

Experiment 16.

Both in 1959 and 1960 we compared some of our models with real gulls' shells, in order to get an impression of the overall valence of the dummies used. In the experiment mentioned here done in 1959, real gulls' shells were compared with halved ping-pong balls: still ignorant of the intricacies of responsiveness to colour discussed above, we painted them sage green with dark grey dots — roughly the same pattern as that used for "all egg" and "shell" angles of experiment 8. We further used white small angles of 1 × 2 inches, and flat paper rectangles as used in experiment 19 reported on below. These paper strips measured $2 \times 4\frac{1}{2}$ cm. and were a pale buff. Table XVII, obtained in one experiment with 160 nests, summarises the results.

TABLE XVII

Numbers of four models (experiment 16) carried and not carried in one test.

	carried	not carried
real shell	155	5
halved ping-pong balls	120	40
white angle	91	69
paper rectangle	53	107

N presentations: 4×160. All differences are significant at the .1% level.

White angles therefore, while better than the buff flat strips, were clearly inferior to the halved ping-pong balls (in spite of the latters' inferior colour); and these received a lower score than real shells. This however could at least in part be due to the sage green colour of the halved ping-pong balls.

Experiment 17.

Hens' shells of which the blunt end was broken off so that they resembled gulls' eggs in shape while being a little larger, were painted all white, and these were compared with hens' shells painted (old) khaki outside and white inside, with real gulls' shells, and with all white small rings. This experiment differed from all others, for although each model was presented to 40 nests, each group of 40 nests received the same model on all four occasions (this was done to gain an impression of the degree of waning which will be discussed later). The results are in table XVIII.

TABLE XVIII

Numbers of four models (experiment 17) carried and not carried in four successive tests. No latin square arrangement; each nest received the same model throughout.

	I	II	III	IV	Total carried	not carried
white hens' shell	34	29	28	32	123	37
khaki hens' shell	35	34	34	33	136	24
real gulls' shell	36	35	35	33	139	21
white ring	29	28	22	28	107	53

N presentations: 4 × 4 × 40.

There is no significant difference between real shell and khaki hens' shell; white hens' shell received a score significantly lower than real gulls' shells at the 5 % level; the difference between the white ring and the white hens' shell was almost significant (p just above 5 %), and that between white ring and the other shells was thoroughly significant (p < .1 %).

The conclusions to be drawn from the two last experiments are that khaki hens' shells are not inferior to real gulls' shells: that white shells are superior to white rings; and that shells are considerably superior to angles. We have already seen (experiment 11) that rings are superior to angles, and (experiment 16) that halved ping-pong balls are superior to angles. Although the colour of the paper strips was slightly different from khaki, there seems little doubt that such flat rectangles are inferior to angles.

Our next task was now to examine whether the size difference between angles and rings (which were made of metal strips twice the size of the angles) could be responsible for their different effects.

Experiment 18.

We presented the following four models, all painted all-khaki (new): small ring (bent strips measuring 2 × 10 cm.), large ring (made of a strip alluminium sheet measuring 4 × 20 cm.; its weight was 16 grams, that of the small ring was 10 grams); the small angle made of a strip of 2 × 5 cm., and "large angles" made of the same strips as small rings but folded at right angles in the middle (Fig. 5). Three groups of birds were used: group A of 80 nests, group B of 80 nests, and group C of 160 nests; each group was subjected to one complete latin square. The results are in table XIX.

TABLE XIX

Numbers of four models (Experiment 18) carried and not carried by three groups of birds.

	A	B	C	Total carried	not carried
large ring	34	30	56	120	200
small ring	38	40	74	152	167
large angle	34	20	47	101	219
small angle	31	24	52	107	212

N presentations: 4 × 3 × 320 (− 2).

As expected, the small rings were carried more than small angles ($p < .1\%$). Small rings were carried more than large rings ($p < 2.5\%$); small rings were significantly superior to large angle ($p < .1\%$), but there was no significant difference between the two angles.

The relatively low score of the large rings was rather unexpected because we knew from past experience (TINBERGEN, 1951, 1953, BAERENDS, 1959) that the egg retrieving response of gulls responds better to outsize eggs than to normal eggs. However, since we also knew that the responses to such large eggs were a combination of retrieving and avoidance (which often kept the birds away from the nest in the beginning of a test), we suspected that we might have to do with a similarly ambivalent response towards the large rings, which might, either in part or even entirely, be responsible for the difference in removal scores between large ring and small ring.

We therefore observed from a hide the responses of ten individual birds to small rings and to large rings. Six of these birds received the small ring first, the large second; one bird was presented the large ring first, the small second; and three birds had the small ring first, then the large ring, then the small ring once more. Their fear responses, expressed by delay in returning to the nest, alarm calls, stretching of the neck, repeated withdrawal after initial approach, were roughly classified as: no signs of fear (o), and a scale of four intensities (1, 2, 3, 4) of avoidance behaviour, 4 being the highest score. Each test lasted until the birds settled on the eggs. Table XX summarises the results of this admittedly crude assessment.

TABLE XX

Fear responses shown by ten birds presented with either a large ring or a small ring on the nest's rim. Explanation in the text.

Bird No.	Small Ring	Large Ring	Small Ring
1	0	0	
2	0	0	
3	1	1	
4	1	4	
5		3	1
6	2	3	
7	0	3	0
8	0	0	
9	0	4	0
10	0	4	0

On 9 occasions the fear score for large ring was higher than for small ring; it was the same (either 0 or 1) on 4 occasions, while small ring never elicited a higher score.

The lower score of these large rings is therefore in part due to the interference by another response (fear, or avoidance) with the removal response. This was the first indication of a fact which became increasingly clear as the work proceeded, *viz.* that we were not measuring the direct expression of the "releasing mechanism" of just the removal response, but rather the effect of the interaction of this response with, or its dominance over, a variety of other responses. Applied to the present experiment, it might be that the large ring does stimulate the removal response as strongly as or even more strongly than the small ring. Our scores, in other words, could be said to indicate the "resultant removal tendency".

Experiment 19.

This was done in 1959 with the special aim of investigating how the gulls distinguished between egg shells and nest material. The immediate impetus to this was given by BEER's observations, illustrated in Fig. 2, which had shown (a) that some objects were very similar to objects which the gulls were sometimes seen to carry to their nests while building; and (b) that a bird would sometimes actually alternate between carrying an object to the nest and carrying it away. Further, some of our "small angles" were found at the end of a test in the nest rim in such a position that it seemed as if the birds had made "sideways building" movements with them. From a comparison of the classes of objects carried away and those found built into the nests, the main criterion of objects eliciting nest building movements seemed to be the elongate shape. This led to the use of four flat models which differed only in their proportions. Flat rectangles of thin cardboard were used of the following dimensions: 3 × 3 cm. (model 1); 2 × 4½ cm. (2); 1 × 9 cm. (3); and ½ × 18 cm. (4). The surface area of all these models was the same. They were painted a buff colour rather similar to light khaki. Direct observations, and checks in which models found moved at the end of a test were classified according to the direction of the wind proved that even these light models were not blown away, not even on days with winds up to about Beaufort force 8. However, to prevent confusion through birds taking up models in their bills and dropping them, which would expose them to the wind, we decided to be on the safe side and did not run tests when the wind was stronger than about force 5.

In the first series of tests 40 nests (group A) were offered the models in latin square arrangement four times in succession; in a second series, another group (B), also consisting of 40 nests, was subjected to three consecutive latin square tests. These tests were repeated so often because the absolute level of the score was low — this was no doubt due to the fact (see table XVII) that important characters of shape were missing in all four models. Table XXI summarises the results.

TABLE XXI

Numbers of four models (experiment 19) carried and not carried in seven tests with 2 groups of birds.

	A_I	A_{II}	A_{III}	A_{IV}	B_I	B_{II}	B_{III}	Total carried	not carried
1 (3 × 3 cm.)	11	5	11	7	14	7	11	66	214
2 (2 × 4.5 cm.)	8	5	10	8	15	5	7	58	222
3 (1 × 9 cm.)	15	11	12	12	14	9	9	82	198
4 (½ × 18 cm.)	12	9	12	7	10	10	6	66	214

N presentations: 4 × 7 × 40.

While there is no significant difference between the scores for models 1, 2 and 4, model 3 was carried significantly more often than model 2 ($p < 2.5$ %). This was a rather unexpected result, because model 3 could not be said to approach the dimensions of the egg shell most closely; both 2 and 1 would seem to conform better. We believe that the clue to this problem was provided by the fact that on many occasions where one of these flat models had not been removed it was found in the nest, partly covered by nest material. Table XXII shows how often a model of each category which was not carried was found in this position.

TABLE XXII

Numbers of models found in nest after experiment 19 compared with numbers not carried ("minus responses").

1 (3 × 3 cm.)	52 out of 214	"minus responses"	24 %
2 (2 × 4½ cm.)	62	228	27 %
3 (1 × 9 cm.)	59	198	30 %
4 (½ × 18 cm.)	94	206	46 %

The differences between 1 and 4 and between 2 and 4 are significant at the .1 % level; that between 3 and 4 at the .5 % level.

At first we thought that the birds had in such cases invariably treated the model as nest material and had performed sideways building movements with it. However, when we decided to check this by direct observations from a hide, the situation appeared to be even more complicated. We found that a model could land among the material of the nest either because it was taken up and dropped (which often made the impression of an incomplete carrying response), when it could subsequently be covered by the scraping and egg shifting movements of the bird; or because the bird performed actual nest building movements with it. Further, as table XXIII shows, a third response was involved as well: some of the models elicited feeding (Pl. XIII, fig. 11). It is true that in all but one of the observations (when a model 4 was fully swallowed) the bird did not succeed in actually eating the model, but attempts at swallowing were unmistakable.

TABLE XXIII

Direct observations of responses to the four models of experiment 19.

	no notice	carried	attempts at swallowing	built in	Total obs.
1 (3 × 3 cm.)	22	5	0	1	28
2 (2 × 4½ cm.)	16	4	1	2	23
3 (1 × 9 cm.)	10	7	2	4	23
4 (½ × 18 cm.)	8	7	4	5	24

It is no doubt significant that all three responses (removing, swallowing, and building) started off with the same movement: taking the model in the bill. It was further revealing that on more than one occasion we saw a bird alternate between attempts at swallowing, incomplete building movements, and incomplete carrying movements (standing up with the model in its bill, stretching the neck, but then settling down again). What happened was therefore something like this: a model might be picked up because it proveded stimuli for either of the three responses. Once the bird "found itself" with the model in its bill, this stimulus might set off either the same response as was started by the visual stimulus which elicited picking-up, or one of the two other responses. Since feeding and nest building are obviously more readily elicited by the longer models, particularly model 4, their "carrying score" might have become higher than expected if merely egg shell removal were aroused. Model 3 might have given the highest score because it elicited more initial feeding and building responses than 2 and 3, and because it was still less exclusively eliciting feeding and nest buiding than the extremely long model 4. A further reason for its high score could have been that it proved more difficult to use as building material than 4, and that the response was therefore more often continued into carrying.

Whatever the exact explanation however, the observations once more reveal the fact that the carrying score as obtained in the mass experiments reflects, not the functioning of the carrying response alone, but its preponderance over other responses, with which it interacts.

Experiment 20.

We had found (experiment 7) that the "betrayal effect" of shells decreased rapidly with increased distance between shell and egg. It was therefore of interest to test the effect of distance on the gulls' response as well. We offered hens' shells painted khaki outside at four distances from the centre of the nest: on the rim (appr. 10 cm. from the centre), and at 15, 25 and 35 cm. distance from the centre. This was done with 80 nests: the results are given in table XXIV.

TABLE XXIV

Numbers of hens' eggs painted khaki outside, carried and not carried when offered at different distances from nest (experiment 20)

	carried	not carried
10 cm.	64	16
15 cm.	54	26
25 cm.	27	53
35 cm.	11	69

N presentations: 4 × 80. Difference between 10 and 15 cm. not significant; that between 15 and 25 cm. significance at .1% level; that between 25 and 35 cm. significant at 1% level. Significance of the total result: p < 1%.

It is obvious that the response decreases very rapidly with increased distance. The sharpest drop is just outside the nest's rim, between 15 and 25 cm. It is regrettable that we did not use the same scale of distances in the predation tests. It looks from the predation tests as if it might "pay" the gulls to show a better response to shells further than 15 cm. from the nest, but as we shall see later there are good reasons to believe that here again another response interferes with shell removal: the need of covering the brood.

Experiment 21.

When we compared the results obtained so far in analyses of stimulus situations which specifically release a single response (TINBERGEN, 1951; MAGNUS, 1958; WEIDMANN in litt.) we were puzzled by the fact that, while on the one hand shell removal could be elicited by a very wide range of objects (neither shape, nor size, nor colour being strikingly specific) yet there were these two rather sharply defined peaks for khaki and white, and the very specific trough for green. We wondered to what extent this response was normally conditioned, and to what extent the selective responsiveness was independent of previous experience with eggs or egg shells. One effect of experience will be discussed below: having carried a shell slightly reduces the likelyhood of carrying one again.

The present experiment was a gamble, and was done on a small scale; but the rather striking results justified it.

We knew from U. WEIDMANN's work (1956) that Black-headed Gulls could be prevented from laying by offering them eggs on the empty scrape well before the first egg was due. We therefore laid out a black wooden egg in each of a number of scrapes in early April, some weeks before the majority of birds laid. As it turned out, some of these nests were abandoned; others were occupied by birds which did not sit on the black egg dummies but built nests on top of them; in others again eggs were laid soon after we gave the black eggs. However, fourteen pairs accepted our black eggs, began to incubate on them and were thereby stopped from laying eggs of their own. When these birds began to incubate, we added two more black eggs to each clutch so as to ensure the best possible situation, and these fourteen pairs were allowed to incubate these black clutches for approximately 5 weeks (well beyond the normal incubation period, which according to BEER, 1960, and YTREBERG, 1956, is on the average about 24 days). We then presented to these birds, in the normal latin square arrangement, black angles and khaki angles. Though we had never compared the effects of these two colours directly, experiment 8 had shown that black has a low valence compared with white, shine, and shell, and was about equal to red. Nevertheless we selected fourteen normal pairs as controls, which had been sitting on their own eggs, and ran this experiment with them simultaneously. The results of seven latin squares in succession, run over 5 consecutive days, are given in table XXV.

TABLE XXV

*Numbers of small angles of different colours carried and not carried by
birds which had been sitting on black eggs ("experimentals") and birds
which had been sitting on normal eggs ("controls") (experiment 21).*

	black angle		khaki angle	
	carried	not carried	carried	not carried
Experimentals	12	37	13	36
Controls	3	46	12	37

N presentations: $2 \times 7 \times 14$.

The difference between Experimentals and Controls in their scores for black angles is
significant at the 2.5 % level.

The conclusion must be that having incubated black eggs increases the
response to black egg shell dummies. We do not know the age of these birds,
and hence cannot say whether they may have had experience with normal
eggs in a previous season; in inexperienced birds the effect may well be
still more pronounced. It is worth pointing out that these birds had had no experience with carrying black egg shells; the experience they had was with
another response: incubating black eggs.

There is of course the possibility that this is not the result of a learning
process at all, but that the birds at the moment of carrying matched the
colour of egg shell dummy and egg: we hope to follow this question up
next year.

Experiment 22.

In experiment 17 we had compared real gulls' shells with hens' shells
painted either khaki or white, and had been surprised to see that the scores
for these three models were about equal. We though it possible that they
were not really equally effective but that even the least effective of them
was still near-optimal and thus sufficient to give, under the conditions of
our tests, a near 100% score. The response to shells was usually so prompt
that the birds who carried at all did so immediately upon alighting on the
nest. Shortening the duration of the tests was therefore hardly feasible,
and we decided, after having found that the effect of a shell decreased with
increasing distance, to run an experiment with several highly effective
models at a greater distance from the nest. The following dummies were
therefore presented at 15 cm. from the centre of the nest: real gull shell;
all white hens' shell; all (old) khaki hens' shells; and all lettuce green
hens' shells. The results, obtained in one latin square with 160 nests, are
set out in table XXVI.

TABLE XXVI

Numbers of four different types of egg shell offered at 15 cm. from nest (experiment 22) carried and not carried in one test.

	carried	not carried
real gulls' shell	114	46
white hens' shell	80	80
khaki hens' shell	36	124
lettuce hens' shell	31	129

N presentations: 4×160.

The difference between real shell and white is significant ($p < .1\%$); as is the difference between white and khaki ($p < .1\%$); khaki and lettuce are not significantly different.

While, according to expectation, real gulls' shells were removed from more nests than any of the other models, the khaki shells were not carried more than the lettuce shells, and considerably less than white shells. [1] White was superior to khaki in our previous tests too, though usually not to this extent, but in the angle tests khaki had been far superior to lettuce green. The only suggestion we can offer is that with increasing distance from the nest the khaki shells, because of their strikingly cryptic colour, failed to attract the birds' attention; while a khaki model on the nest's rim cannot be missed, its inconspicuousness might have effect at greater distances, but it is certainly puzzling that this should already be noticeable at 15 cm.

V. CHANGES IN RESPONSIVENESS IN TIME

As mentioned in Section II C. BEER (1960) had found that Black-headed Gulls are ready to remove an egg shell throughout the incubation period and even well before the first egg is laid. Our experiments were done at different times in the breeding season, and although our conclusions are not based on inter-test comparisons, it is of interest to know as precisely as possible how the overall readiness behaves in the course of time. Dr BEER has kindly allowed us to publish the details of the tests he did to this purpose in 1958 and on which his general cosclusion was based. A large number of nests were offered a halved ping-pong ball painted egg shell colour outside once every day, from 20 days before egg laying occurred till 13 days after hatching. The models were left at the nest for 6 hours, at the end of which a check was made, as in our experiments, and the models still present removed. Table XXVII summarises the results.

---

1) It should be noted however that in experiment 17 we offered hens' shells which were khaki outside but white inside whereas the present "khaki hens' shells" were khaki on both surfaces.

TABLE XXVII

Summary of C. Beer's results with a standard dummy (halved ping-pong ball) offered once daily throughout the breeding season. Eo: day on which first egg was laid; Ho: day on which first chick hatched. Data lumped for periods of different lengths. Further explanation in the text.

	E—20/—10	E—9/—1	Eo	E1	E2	E3
shell carried	174	584	97	88	91	90
not carried	92	210	12	13	12	9
total tests	266	794	109	101	103	99
% carried	65.4	73.7	89.0	87.1	88.3	91.0

	E4/10	E11/17	E18/25	Ho/6	H7/13	H14/25
shell carried	437	222	230	187	111	105
not carried	47	22	29	51	80	76
total tests	484	244	259	238	191	181
% carried	90.3	91.0	88.8	78.5	58.1	58.0

The number of pairs tested during the first few days is naturally low since at this stage birds may stil abandon their scrapes. The score for days E—20/—10 is significantly lower than that for days E—9/—1 ($p < 2.5\%$). During laying the score rises still more (p diff. E—9/—1 and Eo $< .5\%$). From then on it stays practically constant until the time when the chicks begin to hatch (p diff. between E18/25 and Ho/6 $< .5\%$), when it begins to drop.

Before we can accept this as a fully reliable index of the readiness to carry at different times, we must consider the possibility that the responsiveness changes as a result of repeated presentation; it could wane, increase, stay constant, or fluctuate in a complicated way. While we have not run systematic tests with the same models throughout the season and using new birds for each test (which would have monopolised the entire colony just for this experiment alone), we have two sets of data which throw some light on this question.

First, our experiments give some information of waning under the conditions of our experiments. Table XXVIII gives the scores for consecutive tests within a latin square sequence for the first four presentations of each of 14 latin squares, with specification of the experiments of which the total scores have been given already.

In judging these figurese it should be remembered that in each of these experiments every nest received a new model on every occasion *i.e.* a model that had some but not all of the characteristics of the preceding model. Whatever stimulus-specific revival of the response (HINDE, 1954) there might be therefore would tend to counteract the waning of the scores. The

TABLE XXVIII

Scores of four successive tests in one latin square each, for 14 latin squares

Table	Test 1	Test 2	Test 3	Test 4	N nests
9-1	43	35	36	28	96
10-1	46	39	33	27	80
X 1)	53	35	38	39	80
11-A	46	45	43	41	80
11-B	39	43	39	41	80
13	93	62	69	68	240
14	35	28	27	30	80
15	44	36	37	41	80
17	99	114	105	101	160
26	71	66	60	64	160
18	134	126	119	126	160
19C	59	61	60	49	320
19A	39	36	29	33	320
19B	34	21	27	32	320
Totals	835	747	722	720	

differences between the models compared in the experiments however varied greatly: in some experiments some models scored much better than others; in others the various models used had approximately the same valence. It is therefore impossible to say more than that a certain degree of stimulus-specific change might, but must not, be reflected in these figures, and that this differs from one experiment to the other. It might be rewarding to plot waning against the degree of newness of each model as expressed through differences in absolute scores, but since we intend to continue this work we think it advisable to await more results. For our present purpose it is sufficient to point out that there is some evidence of waning, but this intra-experiment waning is at best slight.

Second, we can present some figures about waning with repeated presentation of exactly the same model, since on various occasions we repeated full latin squares one or more times with the same birds. This was done with seven groups of birds, and the total carrying scores ·of each latin square are given in table XXIX.

The table shows convincingly that repeated presentation of the same models does reduce the score. In some experiments this reduction is marked even when pauses of 8-10 days were given between two experiments, whereas in one of the series with (low-valence) cardboard rectangles the pause was followed by a revival. It is further perhaps significant that the figures of experiment 17 obtained with high-valence objects (mainly egg

1) This concerns an experiment not mentioned in the text.

TABLE XXIX

Positive scores in seven series of successive latin squares with the same sets of models. A double vertical line indicates an interval of 8-10 days; all other repeats were run without interruption.

Experiment	I	II	III	IV	N presentations	Models used
8	255	177 ‖	122		2304	angles diff. colours
9	174	147	103		1200	angles diff. colours
11	175 ‖	116			640	shell, rings, angle
15	61	54			320	angles, diff. colours
17	134	126	119	126	640	shells and ring
19A	46	25 ‖	45	34	640	cardboard rectangles
19B	53	31 ‖	33		480	cardboard rectangles

shells) hardly drop at all after the first test, and show no significant fluctuation through the next three repeats.

Judging from these admittedly inadequate figures, repeated presentation of the same model must be supposed to cause some waning. This effect may have been relatively slight in BEER's work because he left the models at the nests for 6 hours—much longer than in our tests, which varied from 30 minutes to 4 hours—and because his standard model had a relatively high valence.

A second set of relevant data is given in table XXX.

TABLE XXX

Four series of scores obtained with the same models with different groups of birds at different times. Further explanation in the text.

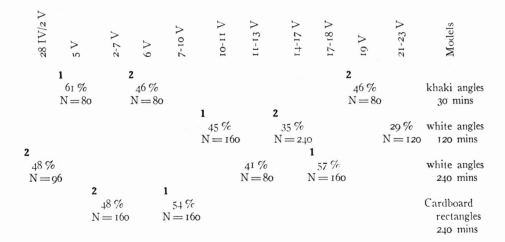

28 IV/2 V	5 V	2-7 V	6 V	7-10 V	10-11 V	11-13 V	14-17 V	17-18 V	19 V	21-23 V	Models
	1 61% N=80		2 46% N=80							2 46% N=80	khaki angles 30 mins
					1 45% N=160		2 35% N=240			29% N=120	white angles 120 mins
2 48% N=96							1 41% N=80	57% N=160			white angles 240 mins
			2 48% N=160	1 54% N=160							Cardboard rectangles 240 mins

346

On various occasions the same model was given at different times of the season to different groups of new birds, and the scores therefore reflect their valence at different times without interference by waning. Because the duration of exposure to the models varied from one experiment to another comparison is possible only between tests of the same exposure time, that is, between figures in the same horizontal row. Further, the tests are arranged according to calendar date. While the colony as a whole is relatively well synchronised (most birds laying their first egg between April 12 and early May, with a peak in the last week of April — G. BEER, 1960), late birds do come in, and birds which have lost their first clutch do relay; this blurs the colony's calendar; yet on the average an arrangement of experiments according to calendar dates roughly reflects a classification according to age of eggs. In each horizontal row of table 30 the highest score is marked 1 and the next lower score is marked 2; it will be seen that while in two rows 1 precedes 2 in time, in the two others 2 precedes 1. Whatever differences in responsiveness there are between the different groups of birds within one horizontal series, they do not seem to be correlated with age of eggs; this, as far as it goes, is in accordance with BEER's findings.

It seems most likely therefore that the readiness to remove egg shells remains roughly constant through the incubation period, with perhaps a slight tendency to increase with time, which in BEER's figures might just have been offset by whatever slight waning may have occurred. Another possibility is that waning occurs only in the first few repeats, after which the response stays at a slightly lower but further constant level. If this were so, BEER's figures would mean that the build-up before egg laying is slightly steeper than apparent, and that the lack of increase in his figures after egg laying is real.

VI. THE LACK OF PROMPTNESS OF THE RESPONSE

The predator tests reported in section III demonstrated the intense pressure exerted at least by Carrion Crows and Herring Gulls against leaving the egg shell near the nest. Admittedly our observations refer to only one colony (which however contains a sizeable part of the British breeding population), but Carrion Crows are practically omnipresent and are notorious egg robbers. One should therefore expect that the Black-headed Gull would have developed a very prompt response and would remove the shell immediately after the chick has hatched. Yet, as we have seen, this is not usual. We cannot believe that the species has not been able to achieve promptness — egg shell removal is so widespread taxonomically that it must be an old response. The most likely reason is that there is a counteracting selection pressure — that too prompt removal is in some way penalised. At first we thought that the risk of carrying the chick with the shell before it had hatched completely might be responsible, but this risk is the same for gulls and other species. Yet we have observations (admittedly few in number) which show that Oyster-catchers and Ringed Plovers carry the shell with far less delay — in spite of the fact that their chicks stay in the nest for a mere couple of hours, and shell removal therefore must be less urgent. In the course of 1959 and 1960 the reason for the delay became gradually clear: in both years there were a number of Black-headed gulls in the colony which preyed selectively on nearly hatched eggs and on wet chicks. Although we are certain that not all gulls engage in this "cannibalism", this type of predation is very common, particularly towards the end of the season. In fact many of our efforts to observe the development of the behaviour during the first few hours of a chick's life (which were usually done late in the season) were time and again frustrated by the wet chicks being snatched away by such robber gulls immediately after hatching. We have not made systematics notes on this, but twenty is a conservative estimate of the number of occasions on which we actually observed such chicks being taken in front of the hide, while dry chicks only hours older (equally or even more available) were left alone. The number of occasions on which we lost wet chicks without actually seeing it happen is much higher still. On only three occasions did we observe a Black-headed Gull trying to swallow a dry chick. While a wet chick is usually swallowed in a few seconds (often even in flight) dry chicks of less than a day old took approximately ten minutes to swallow. There can be no doubt that chicks are practically safe from predation by neighbours (though not from Herring Gulls) as soon as their plumage becomes dry and fluffy.

It is interesting to see the behaviour of parent gulls sitting on hatching chicks while there are robber gulls about. The parents are aware of the latters' intentions; they show signs of increased hostility whenever the robber comes near, and they are extremely loath to leave the nest. As soon as the gulls are disturbed by other predators or fly up for human visitors robbers snatch up the wet chicks in a fraction of a second. We had the impression that the robber gulls kept an eye on many nests and knew where chicks were hatching. On one occasion we saw a still wet chick being taken during the few seconds when the parent carried away the shell.

We feel justified therefore to ascribe the lack of promptness of the response to this tendency of some members of the colony to prey on wet chicks.

DISCUSSION

Removal of an egg shell lasts a few seconds. It is normally done three times in a year. Nothing would seem to be more trivial than this response, which at first glance might seem to be no more than fussy "tidying-up" by a "house proud" bird. Yet we have seen that it has considerable survival value, and that the behavioural organisation is complicated, and well adapted to the needs. In addition, our study has given us some insight in some more general problems of an ecological and evolutionary nature. The following discussion owes much to the stimulating studies of E. CULLEN (1956) and M. CULLEN (1960).

Territory. In 1956, N. TINBERGEN listed the alleged functions of territory in gulls as follows. One component, site attachment, assists in pair formation, in homing to the nest site, and in providing known shelter for the chicks. Inter-pair hostility, the other aspect of territoriality, prevents interference with pair formation, and, by forcing breeding pairs and thus nests apart, renders mass slaughter by predators less likely (see also TIN-BERGEN, 1952b). We believe that the facts mentioned in this paper show a hitherto unrecognised function of territorial hostility in the Black-headed Gull: reducing the likelihood of predation by neighbouring gulls. We distinguish this effect from the effect of spacing-out on inter-specific predation for the following reason. L. TINBERGEN (1960) has shown that some predatory birds can be guided for shorter or longer periods, by a "searching image": they can, through an unknown process, concentrate their attention on one particular type of prey while being less responsive to other types. The stability of this state of narrowed responsiveness seems to be controlled in part by the density of the prey species: a series of successes in a short time seems to strengthen or at least to maintain specialisation, but lack of success tends to widen their responsiveness again. This property of predators, which may well be wide-spread, puts a premium on spacing-out of potential prey animals, because it increases searching time between successes. In this connexion it is distance which counts.

We suggest that predation by neighbouring Black-headed Gulls is reduced by another aspect of territoriality. Each gull learns, during the prolonged period of territorial fighting in the early part of the season, not to intrude into the territories of its neighbours. The factor which reduces predation in this case is the existence of barriers, irrespective of distances between the broods (density).

Of course the inhibition of trespassing as a consequence of acquired knowledge of the boundaries does not totally prevent predation, but the fact that

349

gulls do not trespass except very briefly and on rare moments (such as during a general alarm caused by another predator, *e.g.*, Man or Fox which require adult gulls to leave the ground) obviously reduces the amount of intra-species predation considerably.

The compromise character of colony density. As we have seen, the mass attack by Black-headed Gulls discourages at least one predator, the Carrion Crow, from penetrating into the colony. This demonstrates the advantage of colonial nesting — Crows were not deterred by attacks of one, two or even three Black-headed Gulls. On the other hand, spacing-out of breeding pairs within the colony and the establishment of knowledge of boundaries, both achieved by territorial hostility, have also distinct advantages. Thus the density of a Black-headed Gull colony has the character of a compromise between at least these two opposing demands. As E. CULLEN (1957) and J. M. GULLEN (1960) have shown, a study of the anti-predator devices of other species might help in elucidating the adaptedness of interspecific differences in colony density.

Synchronisation of the breeding calendar. Some of our data strengthen the conclusion that predator pressure may be an ultimate factor in the synchronisation of breeding. BEER's data (1960) show that the scatter in time of the appearance of fledged young on the beach does not differ strikingly from the scatter of egg laying. Yet there are every season a large number of late broods, partly those of birds arriving later than the main body, partly repeat clutches of birds who have lost their first clutch. Our observations show that at least intra-specific predation of wet chicks is particularly severe towards the end of the season, and it is striking how many of the late broods disappear. It is clear that most of the successful broods come from pairs which arrive early and which have not been forced to relay; failure to synchronise is heavily penalised.

The anti-predator system as a whole. At this stage of our studies it seems worthwhile to review what is now known of the ways in which the Black-headed Gull protects itself against predators. Some of these devices protect the individual— for brevity's sake we will call these "egoistic" devices, even though they may at the same time protect others. Others protect the brood, often at the cost of danger to the individual, and such devices we will call "altruistic". Naturally these terms refer to the function, not to motivation.

The most obvious, and seemingly trivial response is escape. Even this however takes different forms, dependent on the nature of the predator and on the age of the bird. A gull can simply fly away, as it does at the approach of a human being. The response to a Peregrine Falcon, which we have ob-

served in detail several times, is different: the gulls fly up, form dense
flocks and fly at great speed low over the ground or the water, executing
quick zigzag manoeuvres. We believe that "erratic flights", in which individual gulls separate themselves from the flock and fly away, often downwards, at great speed and with very quick and sharp turns, are likely to be
elicited by a Peregrine approaching gulls that are flying high. Chicks, in
response to the alarmed behaviour of the adults in a colony, crouch, at first
on the spot, but already after one day and occasionally even earlier they
walk a little distance away from the nest and towards the surrounding vegetation. Each chick becomes soon conditioned to one or more individual hiding
places; this conditioning allows them to reach safety more quickly than they
would if they had to search for suitable cover.

Outside the breeding season Black-headed Gulls select wide open spaces:
marshy meadows, open seashores, or water. This no doubt allows them to
see an approaching predator in time.

In flocks of adult birds there operates at least one signaling system: the
alarm call alerts other individuals.

The "altruistic" habitat selection in the breeding season shows signs of anti-
predator adaptedness: in the open sand dunes gulls avoid nesting on the bare
sand, even though males may start by taking up a pairing territory there.
Once paired however they select a nest site in moderately dense vegetation.
Black-headed Gulls breeding on inland lakes usually select islands; either
large ones which accomodate large numbers, or smaller islands such as individual *Molinia* bushes which offer space for one nest only. Where Black-
headed Gulls nest on tidal saltings, this inclination to select islands can be
the undoing of their broods in high spring tides (TINBERGEN, 1952a).

Several other behaviour patterns appear in the breeding season which,
while endangering or at least not protecting the lives of the parents, do
contribute to the safety of the brood. First, the scattering of the broods,
which provides a certain degree of protection both from inter- and from
intraspecific predation, is effected by the balanced attack-escape system,
with its components of actual attack, withdrawal, and agonistic displays
(TINBERGEN, 1959; MANLEY, 1960). Further, unlike camouflaged species
such as ducks, curlews and several other waders, and pheasants, the incubating gull leaves the camouflaged brood at the first sign of danger. The
camouflage of the eggs depends on a specialised pigmentation system in
the upper reaches of the oviduct.

Parent gulls attack predators; the fierceness of the attack, and the degree
to which it is counteracted by escape tendencies depends on the type of
predator and the resultant seems highly adaptive. From the moment the

first egg is laid, at least one parent stays on the territory and guards the brood. As we have seen, egg shell removal is also effective as an anti-predator device. There can finally be little doubt that the chick's plumage protects its bearer by being camouflaged.

Thus the picture that emerges is one of great complexity and beautiful adaptedness. It has further become clear that at least some of the different means of defence are not fully compatible with each other, and that the total system has the character of a compromise between various, in part directly conflicting demands. These conflicts are of different types. First, the safety requirements of the parents may differ from those of the brood. Thus the parent endangers itself by attacking predators. This is suggested by the fact that Foxes succeed in killing large numbers of adults in the colony. Though we have never seen a Fox killing an adult, their tracks in the sand cannot be misinterpreted. Often they kill many more birds than they eat. Some of these birds were "egg-bound" females (MANLEY, 1958), but in 1960, when we sexed 32 gulls killed by Foxes we found that 21 of these were males. Many of these gulls have their tails torn off and/or their legs broken. We believe that a Fox sometimes kills such birds by jumping at them when they "swoop". All this suggests that a certain balance between the tendency to attack a Fox and the tendency to flee from it is selected for.

The conflict between "egoistic" and "altruistic" behaviour is also very obvious in the time when the winter preference for wide, open spaces changes into the preference for the breeding habitat which, as we have just seen, is dangerous to the adults. The switch towards the breeding habitat selection is not sudden; there is a long period in which the birds show that they are afraid of it; even when, after long hesitation, they settle in the colony, there are frequent "dreads" when the birds suddenly fly off in panic; these dreads gradually subside (see also TINBERGEN, 1953 and CULLEN, 1956). Towards the end of the breeding season the adults begin to desert the colony in the evening to roost on the beach, leaving the chicks at the mercy of nocturnal predators.

Second, there are conflicts between two "altruistic" modes of defence, each of which has its advantages. Crowding, advantageous because it allows social attacks which are effective against Crows, has to compromise with spacing-out which also benefits the broods.

Finally there may be conflicts between the optimal ways of dealing with different predators. Herring Gulls and Crows might be prevented entirely from taking eggs and chicks if the gulls stayed on the nests, but this would expose them to the Foxes. While Herring Gulls and Crows exert pressure towards quick egg shell removal, neigbouring gulls exert an opposite pressure; the timing of the response is a compromise.

We cannot claim to have done more than demonstrate that egg shell disposal is a component of a larger system, nor are we forgetting that much in our functional interpretation requires further confirmation. It seems likely however that a more detailed study of all the elements of anti-predator systems of this and other species, and of the ways they are functionally interrelated, would throw light on the manifold ways in which natural selection has contributed towards interspecific diversity.

VIII. SUMMARY

The Black-headed Gull removes the empty egg shell shortly after the chick has hatched. The present paper describes some experiments on the function of this response, and on the stimuli eliciting it. Carrion Crows and Herring Gulls find white eggs more readily than normal gulls' eggs; it is concluded that the natural colours of the eggs afford a certain degree of cryptic protection. When normal eggs are given an egg shell at 15 cm. distance their vulnerability is greatly increased; this "betrayal effect" decreases rapidly with increased distance between egg and shell. We therefore conclude that egg shell removal helps to protect the brood from predators.

As reported by C. Beer (1960) the Black-headed Gull removes a surprisingly wide range of objects from the nest. Large scale tests with egg shell dummies in which colour, shape, size and distance from the nest were varied showed that objects of all colours are carried but that "khaki" (the normal ground colour of the egg) and white are particularly stimulating, while green elicits very few responses. Egg shells elicit more responses than cylindrical rings of the same colour, and these are responded to better than "angles". Size can be varied within wide limits; very large rings elicit fear which interferes with removal. Various other indications are mentioned which show that the score as obtained in the mass tests does not accurately reflect the responsiveness of the reaction itself but rather the result of its interaction with other behaviour tendencies. The eliciting effect decreases rapidly with increasing distance.

On the whole, the gulls' response is very well adapted to its main function of selectively removing the empty shell, but the relatively high scores for objects which have very little resemblance to egg shells suggest that it is adapted to the removal of any object which might make the brood more conspicuous.

A pilot test showed that gulls which have incubated black eggs respond better to black egg shell dummies than normal gulls.

The lack of promptness of the response as compared with non-colonial waders (Ringed Plover and Oystercatcher) is adaptive, since it tends to reduce predation by other Black-headed Gulls, which are shown to prey selectively on wet chicks. A hitherto unrecognised function of territory is suggested.

In a discussion of the entire anti-predator system of the Black-headed Gull its complexity and its compromise character are stressed: the safety demands of the individual clash with those of the brood; there are conflicts between the several safety devices which each benefit the brood; and there are clashes between the ideal safety measures required by each type of predator.

353

REFERENCES

BAERENDS, G. P. (1957). The ethological concept "Releasing Mechanism" illustrated by a study of the stimuli eliciting egg-retrieving in the Herring Gull. — Anatom. Rec. 128, p. 518-19.

—— (1959). The ethological analysis of incubation behaviour. — Ibis 101, p. 357-68.

BEER, C. (1960). Incubation and nest-building by the Black-headed Gull. — D. phil. thesis, Oxford.

CULLEN, E. (1957). Adaptations in the Kittiwake to cliff-nesting. — Ibis 99, p. 275-302.

CULLEN, J. M. (1960). Some adaptations in the nesting behaviour of terns. — Proc. 12th Internat. Ornithol. Congr. Helsinki 1958.

—— (1956). A study of the behaviour of the Arctic Tern, *Sterna paradisea*. — D. phil. thesis, Oxford.

HINDE, R. A. (1954). Factors governing the changes in strength of a partially inborn response, as shown by the mobbing behaviour of the Chaffinch (*Fringilla coelebs*). II. The Waning of the response. — Proc. Royal Soc. B 142, p. 331-58.

MAGNUS, D. (1958). Experimentelle Untersuchungen zur Bionomie und Ethologie des Kaisermantels *Argynnis paphia* L. (Lep. Nymph.). I. — Zs. Tierpsychol. 15, p. 397-426.

MAKKINK, G. F. (1936). An attempt at an ethogram of the European Avocet (*Recurvirostra avosetta* L.). — Ardea 25, p. 1-62.

MANLEY, G. H. (1957). Unconscious Black-headed Gulls. — Bird Study 4, p. 171-2.

—— (1960). The agonistic behaviour of the Black-headed Gull. — D.phil. thesis. Oxford.

NETHERSOLE THOMPSON, C. and D (1942). Egg-shell disposal by birds. — British Birds 35, p. 162-69, 190-200, 214-24, 241-50.

SCHWARTZKOPF, J. (1960). Physiologie der höheren Sinne bei Säugern und Vögeln. — Journ. f. Ornithol. 101. p. 61-92.

SIEGEL, S. (1956). Nonparametric Statistics for the Behavioral Sciences. — New York.

TINBERGEN, L. (1960). The natural control of insects in pine woods. I. Factors influencing the intensity of predation by songbirds. — Arch. néerl. Zool. 13, p. 265-336.

TINBERGEN, N. (1951). The Study of Instinct. — Oxford.

—— (1952a). When instinct fails. — Country Life, Feb. 15, p. 412-414.

—— (1952b).On the significance of territory in the Herring Gull. — Ibis 94, p. 158-59.

—— (1953). The Herring Gull's World. — London.

—— (1956). On the functions of territory in gulls. — Ibis 98, p. 408-11.

—— (1959). Comparative studies of the behaviour of gulls (Laridae); a progress Report. — Behaviour 15, p. 1-70.

VILLABOLOS, C. and J. VILLABOLOS (1947). *Colour Atlas.* — Buenos Aires.

WEIDMANN, U. (1956). Observations and experiments on egg-laying in the Black-headed Gull (*Larus ridibundus* L.) — Brit. Journ. Anim. Behav. 4, p. 150-62.

YTREBERG, N. J. (1956). Contribution to the breeding biology of the Black-headed Gull (*Larus ridibundus* L.) in Norway. — Nytt Magas. Zool. 4, p. 5-106.

28

Reprinted from *British Birds* 36:186–188 (1942/43)

THE INSTINCTIVE NATURE OF NEST SANITATION
PART II
BY
STUART SMITH, B.SC., PH.D.
(PLATE 5.)

IN a previous article (*antea*, Vol. xxxv, pp. 120-4), attention was drawn to the fact that nest-sanitation manifests itself as an instinctive action on the part of the parent birds. The instinct to remove fæces from the area of the nest was shown to be capable of stimulation by artificial means, and in addition, opening up of the nest-site enlarged its sphere of influence so that the birds under observation removed any white objects resembling the genuine fæces from a comparatively wide area around the exposed nest. These observations did not however, enable any decision to be made about the nature of the controlling stimulus, which might be the nest alone, or the young, or these two acting together as a unit. During the past breeding season, elucidation of the problem has been carried a stage further by experimental observations made during an extensive study of the Meadow-Pipit (*Anthus pratensis*) and the Yellow Wagtail (*Motacilla f. flavissima*).

A Meadow-Pipits' nest was found in a small grass bank, the clutch of four eggs being completed on May 20th. Incubation started with the last egg, and the clutch hatched on June 3rd (period—14 days). During the first 24 hours, no defecation by the young was observed and the hen bird brooded them most of the time. When the young were two days' old, defecation commenced and the fæces were removed by both adult birds direct from the cloaca and carried away to be dropped at a distance. On the third day, one of the young which appeared to be the strongest in the brood, began to make attempts to deposit the fæcal sac on a small flat area at the back of the nest. Such attempts at first proved abortive, due no doubt to lack of the necessary physical strength, but finally towards the end of the day this young bird succeeded in depositing on the " latrine." During the fourth day after the hatch, all four young were using the latrine. Stimulation of the young by the adult birds was, as is usual, by prodding and by tugging of the down. Fig. 1 shows one of the adult pair stimulating the young, five days' old ; Fig. 2 shows a fæcal sac deposited on the latrine, and Fig. 3 shows the removal of the sac by the adult bird.

When the young were six days' old, experiments were commenced using artificial fæces made from white plasticene moulded to the shape and size of a fæcal sac. The adult birds readily removed any of these that were placed on the latrine

MEADOW-PIPIT AND NEST SANITATION.

Upper.—Parent stimulating nestling five days old.
Middle.—A fæcal sac deposited by nestling on latrine.
Lower.—Parent removing fæcal sac.

(*Photographed by* Stuart Smith.)

but—and this is the first important point—*they never did so if one of the young commenced to make any of the movements associated with defecation ;* movements which involve a shuffling backwards towards the rim of the nest, with raising of the cloaca. Under these circumstances, the fæces in the latrine were ignored by the adults, who " froze," tense and rigid, as soon as the young commenced the typical movements. If the young did not attempt these movements however, the adult bird removed the fæces from the latrine. Prodding and tugging stimulation only took place when the latrine was empty. Hence it appears that the main factor controlling the instinct of nest sanitation is the series of stereotyped pre-defecation movements of the young, but a secondary factor might be the young themselves with or without the nest-site incorporated as a visual unit. Associated with the nest-site as an integral part might be added the latrine with its deposited fæcal sac. That the nest-site *per se*, plays no part in the stimulation of the instinct was shown by the following experiment. When the young were 10 days' old, and the adult birds well trained in removing artificial fæces from the latrine, the young were taken from the nest while the adults were away, and several artificial fæces were placed prominently on the latrine. An adult pipit returned with food and alighted at the edge of the nest. Finding it empty, the bird swallowed the food in its beak and stood in a puzzled sort of attitude for a while before flying to the top of the bank in which the nest was situated, where it called for a few moments. The fæces in the latrine were completely ignored. The second adult now arrived and behaved in a somewhat similar manner, although it returned to the nest-site several times and walked around in an aimless way before swallowing the food in its beak. Once again the fæces were ignored. When both birds had left, a single young one was replaced in the nest, and immediately the adult pair recommenced removal of the artificial fæces from the latrine. It was found possible to invoke this sequence of behaviour at will, simply by removal and subsequent replacement of a single young bird in the nest. It appears therefore, that the nest-site, together with the latrine and fæces, have no power as a visual unit to invoke the response which we call nest sanitation. The young bird and its movements make up the unit.

On the twelfth day, the young left the nest and were found huddled together at a distance of about a yard from the nest. The adult Pipits were feeding them regularly *and still removing fæces from the area near the young.* Here again we find striking confirmation of the fact that the instinct is strictly governed by the presence and actions of the young, and not in any way by the nest-site. Adopting the " lock and key " simile of

Lorenz*, we may say that for nest-sanitation, the dormant inlet or *innate perceptory pattern*, called by him the "lock" has as its *releaser*, or "key" a highly specific visual stimulus associated primarily with the very definite pre-defecation movements of the young and to a secondary extent with the presence of the young themselves.

In the case of the Yellow Wagtail, similar, though not identical behaviour was observed. No definite "latrine" could be identified, deposition being on the rim of the nest at one side. Such deposition did not commence until the sixth day after hatching, attempts by the young after four and five days failing to achieve their purpose. Once again it was possible to train the adults to remove artificial fæces of plasticene, although the hen Wagtail showed no great inclination to accept them, and only removed them on rare occasions. The cock however accepted them most readily, and this difference between the sexes agreed with that noted in a previous year for the Yellow Wagtail. Once again it was possible to demonstrate, by removal and replacement of the young that they, and they alone, are the releasers. One incident however, was instructive in showing how closely the instinctive actions involved in nest sanitation approximate to those of conditioned reflex action. It so happened that on one occasion, both the cock and hen Wagtails arrived simultaneously with food, and perched on opposite sides of the nest. Both fed the young and both did a little stimulating. One of the young then commenced the pre-defecation movements and both adults became tense at once and waited. The young one raised its cloaca towards the cock, and he proceeded to take the fæcal sac as it emerged. The effect on the hen bird was immediate; it was as though a coiled spring had been suddenly released. She flew violently at the cock, showing all the actions associated with threat display, and attempted to seize the fæcal sac from his beak. Few actions could have been more strikingly illustrative of the wholly instinctive nature of nest sanitation.

---

* "Der Kumpan in der Umwelt des Vögels." *J. f. Orn.*, vol. 83. pts. 2-3, 1935.

29

Reprinted from *Ibis* 109:412–424 (1967)

SOME ECOLOGICAL ASPECTS OF THE FEEDING BEHAVIOUR OF THE OYSTERCATCHER *HAEMATOPUS OSTRALEGUS* ON THE EDIBLE MUSSEL *MYTILUS EDULIS*

M. NORTON-GRIFFITHS

Received on 6 May 1966

INTRODUCTION

Descriptions of the behaviour of the Oystercatcher *Haematopus ostralegus* when feeding on the Edible Mussel *Mytilus edulis* have been published by Dewar (1908), Drinnan (1954, 1958 *a*, *b*), Tinbergen & Norton-Griffiths (1964) and Hulscher (1964). I studied the feeding behaviour of Oystercatchers on the mussel beds in the estuary of the River Esk at Ravenglass, Cumberland and on the Barnscar mussel bed at Drigg, Cumberland, and it appeared that the birds were feeding differently both on these two mussel beds and on different parts of the same bed. In addition, their feeding behaviour in these places differed strikingly from the published observations made elsewhere. The present study was aimed at finding out how these various mussel beds differed from one another, how the differences affected the bird's feeding behaviour and the extent to which individual birds could modify their behaviour to deal with different situations.

METHODS OF OPENING MUSSELS

The Oystercatcher has two methods of gaining entry into a mussel shell. The first is used against mussels which are exposed to the air and which are therefore tightly closed, and consists of hammering the shell until it breaks with the bill. The second method is used against mussels which are under water and slightly open. Here the bill is stabbed into the mussel between the gaping valves.

THE HAMMERING METHOD

When Oystercatchers hammer mussels, they select patches of hard sand in a way similar to that described by Drinnan (1957) when they hammer Cockles *Cardium edulis*. Should the mussel sink into the sand while being hammered the bird will carry it to another patch of sand, and repeat the process until sand of sufficient firmness is found. It will then tend to return to that area repeatedly with other mussels to hammer open. Before hammering, the mussel is orientated with its flat ventral surface upwards; although this is rather an unstable position, the bird maintains it by re-positioning the mussel if it falls over. The blows are delivered on the mid-ventral region of the shell (rarely anywhere else) with the bill held vertically downwards, the neck held stiff, and the body pivoting from the pelvis. The toes tightly grip the sand to maintain purchase. Anything between one and twenty-six blows may be delivered before the shell is fractured, the average for 55 observations being five blows. The usual effect of the blows is to break off a semi-circular chip from one or both of the ventral valve margins which is then driven inside the shell. Sometimes, however, the dorsal part of the shell shatters into small fragments leaving the ventral part intact, giving the misleading appearance that the shell has been struck from the dorsal side and not from the ventral. In some cases a fracture line runs from the point of impact on the ventral margins to the dorsal margins so that the whole of the posterior part of one, or both, valves breaks off. The blows often seem to have the effect of " knocking out " the mussel, for occasionally a mussel will spring open after only a few blows and before the valves are damaged.

359

Once a hole has been made in the shell, the Oystercatcher thrusts its bill up into the posterior region of the shell and cuts through the adductor muscle. It then prizes the valves apart and removes the flesh.

The great majority of those shells collected by Dewar, Drinnan and by myself, which had been hammered open by Oystercatchers, had been attacked through the ventral margins (Dewar 100%, Drinnan 95%, present study 92%), and both Dewar and Drinnan commented on the orientation of the mussel prior to hammering and on the restriction of the attack to the valve margins. A similar orientation of the shell, or restriction of attack to a particular part of the shell, of other molluscs has been described by Dewar (1908), Drinnan (1957) and Tomkins (1948). Drinnan (1958a) suggested that the ventral valve margins of the mussel might be the weakest part of the shell because of the persistent gape and because of the obtuse angle between the valve edges which therefore lend each other little support. I decided to test this idea by cracking the shells of live mussels under controlled conditions to see whether different regions of the shell varied in strength.

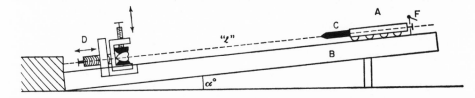

Figure 1. Mussel cracking machine. (A) Wheeled trolley; (B) smooth plane; (C) brass Oystercatcher bill; (D) clamp for holding mussel; (F) release mechanism.

The apparatus used was an adaptation of a Fletcher's Trolley. The basis of the machine (Fig. 1) is a wheeled trolley (A) running in grooves down a smooth plane (B). The mussel is held in a clamp (D) and is struck with a brass rod (C) which is attached to the trolley. The last 10 mm. of the brass rod was shaped to correspond with the bill tip of a stuffed adult Oystercatcher in the University Museum, Oxford. The trolley was released by a simple mechanism (F). The mussel was held in the clamp between plastic moulds so that, while being firmly supported, the shell could not derive additional strength from the valves being pressed together. The clamp could be moved both up and down, and forwards and backwards, so that it was possible to orientate the mussel with its dorsal-ventral axis parallel to, and continuous with, the longitudinal axis of the " bill ". The weight (w) of the trolley, the length (l) of the run and the angle (α) of the slope were kept constant throughout the tests so that the blows delivered to the mussels were always the same. Only two regions of the shell were tested, the mid-ventral region, and the highest point on the dorsal margins. These were chosen because they could be exactly located on every shell. Blows were delivered to the shell until the shell fractured and the tip of the bill penetrated inside the shell, and thus the number of blows necessary to penetrate the different regions gives a measure of their relative strengths. Mussels of the following lengths were tested: 20 mm., 25 mm., 30 mm., 35 mm., 40 mm., 45 mm., 50 mm., 55 mm. and 60 mm. The mussels were picked at random off the mussel bed at Ravenglass and were kept alive in sea-water prior to testing. Mussels from the different size classes were tested in a random order and only one test was done on each mussel.

The average number of blows taken to penetrate the different regions of the different sizes of mussel is shown in Fig. 2. The difference between the strength of the dorsal and ventral margins is at once apparent, and it is interesting that, while the scores for the ventral margin remain low and relatively constant, those for the dorsal margin increase sharply with size until, with shells larger than 50 mm., 50 blows to the dorsal margin produced no effect at all (the tests were stopped at this stage). The functional significance of the orientation of the mussel shell by the Oystercatcher prior to hammering can now be fully seen, for not only is the ventral margin of a mussel absolutely weaker than the dorsal margin, but also the ventral margins of even the largest shells are relatively weak. In addition, the ventral margin offers an almost flat surface at which to hammer while the dorsal surface is usually sharply pointed and so a much more difficult target.

The damage done by the machine to the ventral margins of the shells was almost indistinguishable from that done by Oystercatchers (see Plate 12 (a)). Usually a semi-circular chip was knocked out from one of the valve margins, but in some cases either the dorsal region collapsed in fragments or a fracture line ran from the point of impact to the dorsal region so that the whole of the posterior part of one of the valves would be broken off. The " knocking out " effect was observed, usually after three or four blows, and only when the blows were to the ventral region. The valves would suddenly gape 1 or 2 mm. and the following blow would always break the shell, thus demonstrating the additional strength gained by the valve margins from being tightly held together by the force of the adductor muscle.

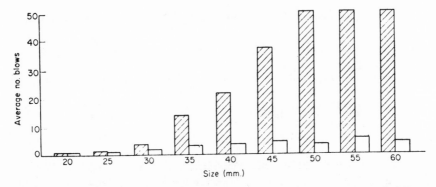

FIGURE 2. Ventral and dorsal scores tested against one another using a one-tailed Wilcoxon matched-pairs signed-ranks test where

$$Z = \frac{T - \frac{N(N+1)}{4}}{\sqrt{\left[\frac{N(N+1)(2N+1)}{24}\right]}} .$$

$N = 36$, $T = 23$ so $Z = -3 \cdot 6$, $P = < 0 \cdot 001$.

▨ , Dorsal margin; ▭ , ventral margin.

THE STABBING METHOD

The stabbing method is used against mussels which are covered with sea-water and therefore slightly open. A hard downward stab with the bill, accompanied by rapid levering and twisting of the neck and head, serves to gain entry into the shell. The rapid levering and twisting may serve either in forcing the bill between the valves or in tearing the mussel off the substrate once entry has been gained. The fact that (1) the mussels are open under water and that (2) they are gaping when brought to the surface by the Oystercatcher suggests that the stab enters between the open valves and severs the posterior adductor muscle, thus preventing the mussel from closing. On three occasions Oystercatchers were chased away from mussels which they had just stabbed but which they had not yet fully opened, and each time it was found that the posterior adductor muscle was cut through. Plate 12 (b) and (c) show details of the damage to the largest of these mussels. The torn mantle attachment in Plate 12 (b) shows that the bill must have entered from just below the siphon and between the posterior valve margins, where the gape of an open mussel is widest. Plate 12(c) shows that the bill cut through the adductor and ended up against the left-hand valve where it tore the mantle.

The posterior margins of mussels which have been opened by stabbing are sometimes very slightly chipped, but there are never any other signs of damage to the shell. Once the gaping mussel is brought to the surface the valves are prized apart and the flesh removed.

The different conditions under which these two methods of opening mussels are used are clearly shown from a collection of freshly opened mussel shells from two nearby areas of the Barnscar mussel beds at Drigg. Only those shells were collected which still had traces of fresh meat inside them and which were surrounded with Oystercatcher footprints. The shells were classified as " fractured " (i.e. they had been hammered) or as " undamaged " (i.e. they had been stabbed). In Area A (see Table 1) the birds were hammering open mussels which had been left exposed by the tide, while in area B, some 40 yards away, the birds were stabbing mussels which were still covered with water.

TABLE 1. *Damage to mussel shells opened by Oystercatchers* Haematopus ostralegus *on the Barnscar Beds.*

	% FRACTURED	% UNDAMAGED
AREA A (totally exposed at low tide)	85·5% (231)	14·5% (39)
AREA B (covered with water at low tide)	10·1% (46)	89·9% (411)

THE FEEDING BEHAVIOUR ON THE RAVENGLASS MUSSEL BED

Oystercatchers were observed feeding on the largest of the mussel beds on the Esk, which is situated well out towards the mouth of the estuary. It is about half a mile long and some 200 yards wide, with a substrate of rocks and hard sand. The bed can be divided up into two regions: the Upper bed is flat with numerous small pools of water and occasional rocky lumps, and is covered with a dense unbroken carpet of small mussels of average size 20·1 mm. ($s=4·2$ mm.); the Lower bed consists of a strip twenty yards wide along the low water mark and is broken up by ridges of large rocks festooned with mussels, as well as by expanses of bare sand. There are many mussels on the Lower bed of a much larger size, up to 65 mm., the average size being 22·5 mm. ($s=12·2$ mm.). At low tide, the Upper bed is always completely exposed, while the Lower bed, except at low water spring tides, is covered with water to a greater or lesser extent.

The empty shells of mussels which had been opened by Oystercatchers were collected off the Upper and Lower beds, classified as " fractured " or " undamaged ", and also measured. From Fig. 3, it can be seen that the birds had fed differently on the Upper and Lower beds; on the former most mussels had been hammered open (Table 2), while on the Lower bed, most had been stabbed. This can perhaps be explained by the fact that on the Upper bed the birds have to hammer open the exposed mussels (except for a few caught open as the tide recedes, or in the small pools), whereas on the Lower bed they can feed on either exposed or covered mussels. If, however, the average size of all the mussels taken by Oystercatchers on both the Upper and Lower beds is used to distinguish between " large " and " small " mussels, then it appears that the Oystercatchers are only hammering open the " small " ones (Table 3). In addition, as Fig. 3 shows, mussels larger than 40 mm. do not seem to be hammered at all. This is difficult to understand, in view of the cracking tests (described above), on mussels off this same bed, which showed that the ventral margins of mussels of all sizes are almost equally easy to break.

TABLE 2. *Methods used by Oystercatchers* Haematopus ostralegus *to open mussel shells on the Upper and Lower beds at Ravenglass.*

	FRACTURED	UNDAMAGED	
UPPER BED	97	35	$\chi^2 = 13 \cdot 6$
LOWER BED	43	165	$P = <0 \cdot 001$

TABLE 3. *Methods used by Oystercatchers* Haematopus ostralegus *at Ravenglass to open " large " and " small " mussels (mean size of all mussels opened (35 mm.) is used to distinguish " large " from " small ").*

	FRACTURED	UNDAMAGED	
SMALL MUSSELS	129	72	$\chi^2 = 117 \cdot 6$
LARGE MUSSELS	9	165	$P = <0 \cdot 001$

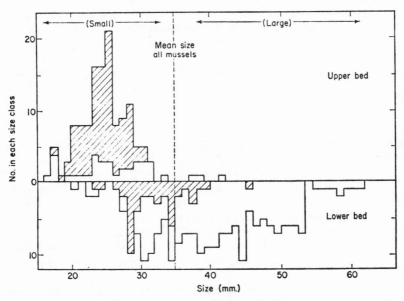

FIGURE 3. Distribution by size of mussels opened at Ravenglass. ▨, Fractured shells; ▭, undamaged shells.

COMPARISONS BETWEEN THE FEEDING BEHAVIOUR AT RAVENGLASS, IN THE FIRTH OF FORTH AND THE CONWAY

Dewar, working in the Firth of Forth, and Drinnan, working in the Conway estuary, also collected shells which had been opened by Oystercatchers and their results indicate that the birds are feeding differently in each of these places (see Table 4).

RAVENGLASS UPPER BED; MORPHA BED AT CONWAY

Here the birds are feeding in essentially the same way except that the proportion of shells hammered at Morpha is significantly larger than the proportion hammered on the Upper Bed.

RAVENGLASS LOWER BED; PENSARN BED AT CONWAY; FIRTH OF FORTH

The birds seem to be feeding in the same way on the Lower bed as they are in the Firth of Forth, except that the proportion of stabbed shells is higher at the Firth. The birds are, however, feeding very differently at the Pensarn bed in the Conway. Here, not only are the majority of shells hammered open, but also shells of all sizes are treated in this way. This is in complete contrast to the situation on the Lower bed at Ravenglass where, on a bed similarly exposed at low tide, the majority of shells are stabbed, and shells longer than 40 mm. are not hammered at all.

THE SUBSTRATE OF THE MUSSEL BEDS

One important factor which might cause the difference in feeding behaviour on the different mussel beds is the nature of the substrate. Drinnan (1957) stressed the dependence of Oystercatchers on areas of hard sand on which they can hammer open cockles, and he showed also that birds select patches of hard sand to which they carry cockles. Oystercatchers similarly select hard sand to which they carry mussels too, so they must be dependent on a firm substrate for success with this prey.

Dewar did not give an adequate description of the mussel beds in the Firth of Forth, but his frequent references to " sand and mud " indicate that the substrate may be rather soft. If this is so, it could explain why so few shells were found hammered open there, since the only way possible for the birds to open the mussels under these circumstances would be to stab them while they were still covered with water.

Drinnan described the substrate of the Pensarn bed in the Conway as ranging from soft mud with occasional stones, to sand, sand and rocks, and rocks; in general, therefore, it is very firm. The mussels are exposed (and closed) at low tide, as in the Firth of Forth, but at Pensarn the birds are able to hammer them open as well, because of the firm substrate. There is, however, one area on the Pensarn bed which is so muddy that the birds do not feed there at all.

At Ravenglass, the substrate of both the Upper and Lower beds is very hard indeed, for the whole mussel bed is formed on packed rocks. All the mussels are firmly attached to this substrate, for they are continually swept by strong waves and currents. There is, therefore, a very firm substrate, and mussels are therefore found hammered open. Yet on the Lower bed, in a similar situation to the Pensarn bed, relatively few mussels are found hammered open, and certainly no large ones. There is, however, one important difference between the Lower and Pensarn beds. Thus while the former lies at the mouth of the estuary and is continually swept by strong currents, the Pensarn bed lies almost as far upstream as it is possible for marine animals to live, and where the mussels are protected from strong tidal currents.

It has already been shown that the ventral margins of a mussel shell are weaker than the dorsal ones and that for this reason Oystercatchers have to tear a mussel off the substrate before hammering it. Thus while the very hard substrate at Ravenglass provides ideal condition for hammering, these favourable feeding conditions could be to some extent countered if the mussels were difficult to detach. I therefore carried out some tests to see whether the larger mussels on the Lower bed were more firmly attached to the substrate than were the small mussels, and whether the mussels on the Lower bed were in general more firmly attached than were the mussels on the Pensarn bed. The apparatus used was a simple spring balance with a sliding scale attached to a clamp made from a small test tube holder. Mussels were chosen at random on the two beds and the clamp was attached around the shell, great care being taken not to damage any of the byssal threads. The mussel was then pulled off the bed with a smooth vertical pull and the force needed to detach it read off the sliding scale. Figure 4 shows the force needed to detach mussels of different sizes from the two beds.

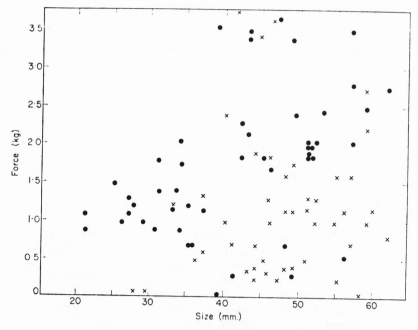

FIGURE 4.   Force (kg.) needed to detach mussels of different sizes from the Lower and Pensarn beds.   (**X**) Pensarn shells; (●) Lower bed shells.

The data for the Lower bed shells, taken separately and tested with a median test, show a good correlation between an increase in size of the shell and an increase in the force needed to detach it ($\chi^2 = 22\cdot3$, $P = <0\cdot001$).   A median test on the combined data from the Lower and Pensarn beds shows that the Lower bed shells are more firmly attached than are Pensarn shells of equivalent size ($\chi^2 = 85\cdot0$, $P = <0\cdot001$).   These results could well account for the fact that while mussels of all sizes are found hammered open on the Pensarn bed, only small ones are found hammered on the Lower bed, the large ones being too firmly attached.   From Fig. 4, it can be calculated that the average force needed to detach mussels of 35–45 mm. from the Lower bed is 1·7 kg., and further that 80% of the Pensarn shells were detached with less effort.   One might therefore predict that 80% of the mussels eaten on the Pensarn bed would be hammered and that 20% would be stabbed.   Table 4 shows that the observed figures are 79% hammered and

TABLE 4.   *Methods used by Oystercatchers* Haematopus ostralegus *to open mussels at Ravenglass, Conway and Firth of Forth.*

AUTHORITY	Drinnan	Norton-Griffiths	Drinnan	Norton-Griffiths	Dewar
LOCALITY	Morpha (Conway)	Ravenglass Upper Bed	Pensarn (Conway)	Ravenglass Lower Bed	Firth of Forth
SIZE RANGE	10–35 mm.	10–35 mm.	20–60 mm.	20–60 mm.	"All sizes"
NO. IN SAMPLE	500	132	350	208	Not given
UNDAMAGED	12%	26·5%	21%	79·3%	91%
FRACTURED	88%	73·5%	79%	20·7%	9%
	$\chi^2 = 16\cdot1$		$\chi^2 = 179\cdot2$		
	$P = <0\cdot001$		$P = <0\cdot001$		

21% stabbed. This therefore suggests that both the observed difference between the feeding behaviour of the Oystercatchers on the two mussel beds and the difference in the size range of mussels found hammered can be accounted for by the varying strength with which the mussels are attached to the two beds.

ECOLOGICAL FACTORS AFFECTING THE MUSSEL SHELLS

There is considerable evidence that the laying down of calcium in the shell of marine shellfish is susceptible to ecological conditions. Orton (1925), Fox & Coe (1943), Robertson (1941), Bevelander (1948), Wilbur & Jodrey (1952) and Rao & Goldberg (1954) all reported that the calcium incorporated into marine molluscan shells was taken directly from the sea-water. Collip (1920), Dugal (1939), Dugal & Forteir (1941) and Baird & Drinnan (1957) reported that marine bivalves, exposed to the air (or to fresh water) and tightly closed, respire anaerobically; the resulting acidity in the body tissue fluids is buffered by Ca^+ ions which are released from the shell. Finally, Fox & Coe (1943) reported that exposure of mussels to wave action results in a stunting of shell growth and in thick shells.

The conditions of the Lower Ravenglass bed and the Pensarn bed differ in many ways. The mussel bed in the Ravenglass estuary is exposed to high wave action and the sea-water covering the beds is little diluted by fresh water from the river. Indeed, the mussels close only when they are exposed by the tide, so here the loss of calcium from the shell due to anaerobic respiration is at a minimum. The protected position of the Pensarn bed, on the other hand, means that wave action is at a minimum; but since it lies far upstream the sea-water covering the bed is more diluted by river flow. The salinity at low water drops to a very low level (especially after long rainfall), so tidal exposure to both air and fresh water is much longer. It therefore seemed possible that the shells of the two populations of mussels would differ in strength.

The first factor to be investigated was the effect of wave action on mussel shells. A sample of mussels from the Lower and Pensarn beds was taken and the height, breadth and length of each shell was measured. The ratio of height to breadth was plotted against length (Fig. 5), a ratio of more than 1 indicating that the shell was higher than

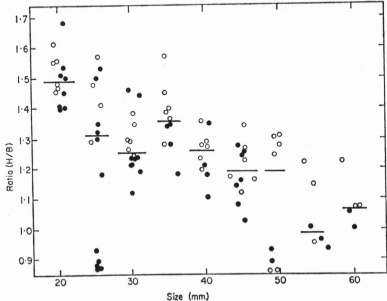

FIGURE 5. Ratio of height/breadth of Ravenglass and Pensarn shells. (◯) Ravenglass shells; (●) Pensarn shells. (Median lines for each size class marked.)

it was broad.   For any given length of shell, a Pensarn shell has a higher ratio of height to breadth than does a Lower bed shell ($\chi^2 = 12.58$ $P = <0.001$).   Unfortunately, the opportunity to weigh the shells and measure their thickness was overlooked, but it seemed by inspection that the Ravenglass shells were thicker.   The stunting of growth and thickening of the Lower bed shells is most likely an effect of strong wave action, which might result in the Lower bed shells being stronger than the Pensarn ones.

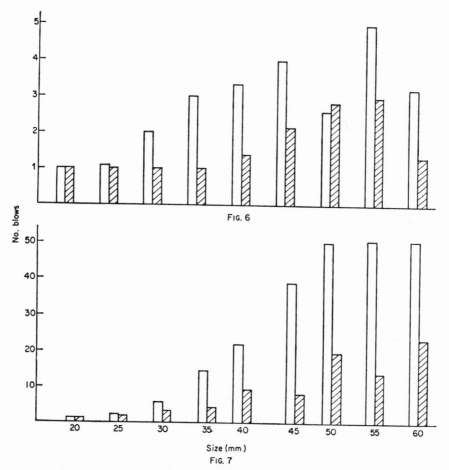

FIGURE 6.   Average number of blows to crack ventral margin of Ravenglass and Pensarn shells. [IIIIII], Pensarn shells; [ ], Ravenglass shells. $Z = -3.6$; $P = <0.001$.

FIGURE 7.   Average number of blows to crack dorsal margin of Ravenglass and Pensarn shells. [IIIIII], Pensarn shells; [ ], Ravenglass shells. $Z = -4.1$; $P = <0.001$.

A sample of shells from the Pensarn bed was therefore tested with the same cracking apparatus to see if they were weaker than the Lower bed shells.   The same nine size classes were used and tested in the same way.   In Fig. 6 the average number of blows taken to crack the ventral margins of the Ravenglass and Pensarn shells is compared, and it shows that the Ravenglass shells are in fact significantly stronger than the Pensarn ones.   In Fig. 7, the difference is even more striking, for even the largest of the Pensarn shells were fractured by blows to the dorsal margin.   The difference in strength could, of course, greatly influence the feeding behaviour of the birds.   There is unfortunately no way of relating either the force needed to detach the mussels from the substrate or the force needed to crack the shells to the physical strength of the Oystercatchers, and this makes it difficult to assess which of these two factors is the more important in affecting the feeding behaviour.

THE PROBLEM OF SIZE SELECTION

There is presumably an "optimum" size of mussel which gives Oystercatchers maximum return for minimum expenditure of energy, and if Oystercatchers had evolved some mechanism for size selection it would show up by their selecting mussels of a particular size range, irrespective of the mussel population on which they were feeding. Such a mechanism would, however, be highly disadvantageous, for not only does the frequency of occurrence of mussels of a particular size vary in different populations, but also, as we have seen, mussels of the same size in different populations vary in the strength of their shells and, in addition, in the amount of meat they contain.

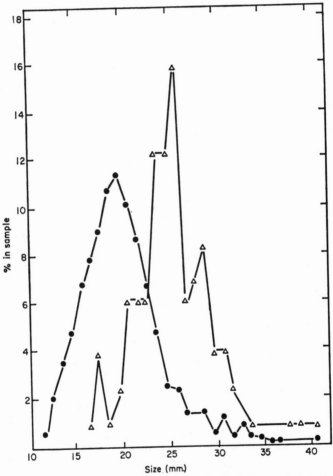

FIGURE 8.   Apparent size selection on the Ravenglass Upper bed.

Mussel population (—●—●—):   mean size 20·1 mm.
                                        s.   4·2 mm.
Taken by Oystercatchers (—Δ—Δ—): mean size 25·39 mm.
                                        s.   3·86 mm.

$$d = 14·52; \ P < 0·001.$$

The references to size selection by Oystercatchers in the literature present a rather confused picture. Dewar (1908) reported that on the Morpha bed Oystercatchers "prefer" mussels of 35 mm. Drinnan (1958a) found that on the Morpha bed, too, 35 mm. mussels were taken more than any other (35 mm. being the largest available there), while on the Pensarn bed the birds took the various sizes in proportion to their occurrence in the population. Hulscher (1964) studied breeding birds which were taking mussels from the beds to their young. He found apparent size selection when comparing those mussels taken with the mussel population of the whole bed, but no selection when comparing those taken with the population on the restricted part of the bed on which the birds were feeding. These observations suggest that the Oystercatchers are taking whatever is available, with perhaps a bias against small mussels. At Ravenglass, on both parts of the bed the birds seem to be selecting the larger mussels (Figs. 8 and 9), this preference being more marked on the Lower bed. The general morphology of the two beds must, however, be taken into consideration. It has already been described how, on the Lower bed, the mussels are clumped together in large masses on the rocks, with the larger mussels on the outside of the clumps. In order for an Oystercatcher to take the small mussels, it would first have to pull the clump to pieces. It is much more reasonable to suppose that it would instead feed on the outer mussels because of their greater accessibility, in which case the population at risk would be very different to the population of the bed as a whole. On the Upper bed, the mussels are evenly spread so all sizes are equally available, yet here again, the larger mussels are taken more frequently. Of all the shells which have been opened by Oystercatchers on the Ravenglass beds, none have been smaller than 16 mm. in length. There thus seem to be two factors operating to produce the size selection on the Ravenglass beds: first the way in which the mussels are arranged on the bed, and second their size, in that the very small mussels may not present the correct stimuli to elicit feeding behaviour on the part of the Oystercatcher.

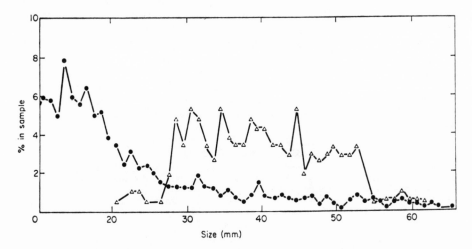

FIGURE 9. Apparent size selection on the Ravenglass Lower bed.

Mussel population (—●—●—):   mean size 22·49 mm.
                                s.   12·16 mm.
Taken by Oystercatchers (—Δ—Δ—): mean size 39·08 mm.
                                s.   8·49 mm.

$d = 23·95$; $P < 0·001$.

DISCUSSION

To fully understand the feeding behaviour of the Oystercatcher on mussels it is as important to take the mussels into consideration as it is to study the behaviour of the birds. This can be shown in two ways. The feeding methods are adapted to the structure of the mussels. For instance, the hammering is not directed at random against the shell but only at the weakest part, and the bird first positions the shell accordingly. The stab, on the other hand, is directed towards the posterior region where the gape is widest and is so aimed that it passes straight through the main posterior adductor muscle. Secondly, the way in which the mussels are grouped on the mussel bed, the strength of their shells and the firmness with which they are attached to the subtrate can influence the way they are taken. On the Ravenglass Lower bed, for example, the mussels have strong shells, are firmly attached to a hard substrate and are grouped in large clumps; here few are hammered and the larger mussels are apparently preferred. In contrast, the mussels on the Pensarn bed have weaker shells, are more loosely attached and more evenly spread; here the greater part are hammered open and there is no apparent size selection.

This, however, presupposes that any individual Oystercatcher is able to both hammer and stab mussels, and that the conditions prevailing on any mussel bed determine which method it will use. This indeed was my belief, until I started to work with individually marked adults, when it then appeared that this was not the case at all: an individual bird will either hammer mussels or stab them, but will not do both. This throws an entirely new light on the problem for it means that individual birds select mussel beds, or parts of mussel beds, which have conditions suitable for their particular feeding method. However, two new problems are now raised: how do individuals become hammerers or stabbers, and how do they find suitable feeding grounds?

By individually colour ringing breeding adults and their young, it has been shown that young Oystercatchers develop the same feeding technique as their parents. In other words, young whose parents stab mussels develop the stabbing technique while those whose parents hammer mussels develop the hammering technique. My detailed work on this is still in progress, but it seems most unlikely that this is due to genetic differences between the different parents. While the young are developing their feeding technique they are fed by their parents and accompany them onto the mussel beds. In this way they can learn at least one locality which is suitable for their particular feeding technique, but what happens to them when they migrate at the end of their first summer is still not clear. The large scale ringing studies sponsored by the Ministry of Agriculture, Fisheries and Food at Conway will soon produce information on this point. It is worth noting, however, that Oystercatchers have many other foods, besides mussels.

ACKNOWLEDGMENTS

My thanks are due to my supervisor, Professor N. Tinbergen, F.R.S. I also want to express my gratitude to Sir William Pennington Ramsden, Bart., and the Cumberland County Council for permission to work in the Ravenglass sanctuary. Hides and other equipment used in this study were provided by funds put at the disposal of Professor Tinbergen by the Nature Conservancy.

SUMMARY

Oystercatchers have two methods of opening mussels, both neatly adapted to the structure of the prey. To open mussels exposed by the tide, the bird hammers a hole along the ventral margin of the shell, whereas to open mussels under water, the bird drives its bill into the gape of the valves to cut through the posterior adductor muscle. Large and strong-shelled mussels can be opened by stabbing but not by hammering.

Feeding methods of Oystercatchers vary from mussel-bed to mussel-bed. These variations in behaviour are attributable to differences in the strength of the shells of the prey and to the firmness of their attachment to the substrate.

Diversity of ecological conditions on the mussel-beds causes an apparent size selection of prey by Oystercatchers.

REFERENCES

BAIRD, R. H. & DRINNAN, R. E. 1957.   Ratio of shell to meat in *Mytilus*.   J. Cons. perm. int. Explor. Mer: 22, No. 5.

BEVELANDER, G. 1948.   Calcification in marine molluscs.   Biol. Bull. 94: 176–183.

COLLIP, J. B. 1920.   Studies on Molluscan coelomic fluid.   J. biol. Chem. 45: 23–49.

DEWAR, J. M. 1908.   Notes on the Oystercatcher (*Haematopus ostralegus*) with reference to its habit of feeding on the mussel (*Mytilus edulis*).   Zoologist 12: 201–212.

DRINNAN, R. E. 1954.   Investigation of Oystercatchers feeding in Morecambe Bay.   Merseyside Naturalists Bird Report 1954/55: 30.

DRINNAN, R. E. 1957.   The winter feeding of the Oystercatcher (*Haematopus ostralegus*) on the cockle (*Cardium edule*).   J. Anim. Ecol. 26: 441–469.

DRINNAN, R. E. 1958 a.   Winter feeding of the Oystercatcher on the edible mussel in the Conway estuary.   Fish. Invest. Ser. 2, 22: 1–15.

DRINNAN, R. E. 1958 b.   Observations on the feeding of Oystercatchers in captivity.   Br. Birds 51: 139–149.

DUGAL, L. P. 1939.   The use of calcareous shell to buffer the products of anaerobic glycolysis in *Venus mercenaria*.   J. cell. comp. physiol. 13: 235–251.

DUGAL, L. P. & FORTEIR, G. 1941.   Le métabolisme anaerobique chez les Mollusques, 2.   Variations du calcium et de l'acide lactique chez les huîtres.   Ann. A.C.F.A.S., 7: 112.

FOX, L. & COE, W. R. 1943.   Biology of the Californian Sea Mussel, 2.   Nutrition, metabolism and growth.   J. exp. Zool. 93: 205–249.

HULSCHER, J. B. 1964.   Scholeksters en Lamellibranchiaten in de Waddenzee.   Levende Nat. 67: 80–85.

ORTON, J. A. 1925.   The conditions for calcareous metabolism in oysters and other marine animals. Nature, Lond. 116: 13.

RAO, K. P. & GOLDBERG, E. D. 1954.   Utilization of dissolved calcium by a Pelecypod.   J. cell. comp. Physiol. 43: 283–292.

ROBERTSON, J. D. 1941.   The function and metabolism of calcium in the Invertebrata.   Biol. Rev. 16: 106–133.

TINBERGEN, N. & NORTON-GRIFFITHS, M. 1964.   Oystercatchers and mussels.   Br. Birds 57: 64–70.

TOMKINS, I. R. 1947.   The Oystercatcher of the Atlantic coast of North America and its relation to Oysters.   Wilson Bull. 59: 204–208.

WILBUR, K. & JODREY, H. 1952.   Studies on shell formation 2.   Measurements of rate of shell formation using Ca 45.   Biol. Bull. 103: 269–276.

M. Norton-Griffiths, Department of Zoology, University of Oxford.

30

Reprinted from *Condor* 69(4):323-343 (1967)

TRANSPORT OF WATER BY ADULT SANDGROUSE
TO THEIR YOUNG

Tom J. Cade and Gordon L. Maclean

In 1896 the English aviculturist Meade-Waldo published an astonishing and seemingly incredible account of how the males of sandgrouse that he successfully bred in captivity carried water to their young in their breast feathers. To quote from his original report:

As soon as the young were out of the nest (when twelve hours old) a very curious habit developed itself in the male. He would rub his breast violently up and down on the ground, a motion quite distinct from dusting, and when all awry he would get into his drinking water and saturate the feathers of the under parts. When soaked he would go through the motions of flying away, nodding his head, etc. Then, remembering his family were close by, would run up to the hen, make a demonstration, when the young would run out, get under him, and suck the water from his breast. This is no doubt the way that water is conveyed to the young when far out on waterless plains. The young . . . are very independent, eating hard seed and weeds from the first, and roosting independently of their parents at ten days old (Meade-Waldo, 1896). See also Meade-Waldo (1921).

Despite the fact that Meade-Waldo (1897; 1921) observed 61 broods from three different species of sandgrouse hatched in his aviaries between 1895 and 1915, and soon received confirmation from another breeder for two species (St. Quintin, 1905), and despite the fact that field naturalists and native hunters have frequently observed wild male sandgrouse wetting their breast feathers at water holes in the way described (Meade-Waldo, 1906; Buxton, 1923; Heim de Balsac, 1936; Hoesch, 1955), the idea that the young do receive water in this exceptional way has met with a great deal of scepticism (Archer and Godman, 1937; Meinertzhagen, 1954, 1964; Hüe and Etchécopar, 1957; Schmidt-Nielsen, 1964). Schmidt-Nielsen, however, had his doubts shaken when Dr. Mendelssohn of Tel Aviv University told him "that chicks of sand grouse that he reared would die from thirst even if drinking water was available to them, but that they would take water from wet cotton (p. 217)."

First of all, the doubters argued that such a method of transport is too inefficient to survive the rigors of natural selection. How could an adult sandgrouse traveling from the water hole at 30 to 40 miles per hour in hot, dry air arrive back on the breeding grounds with sufficient moisture in his feathers to nourish the young? More importantly, some recent authors have assumed—incorrectly as it turns out—that sandgrouse feed and water their young by regurgitation like doves and pigeons. This notion probably first arose as an extension of the equally erroneous belief that sandgrouse drink by "sucking," with their beaks continually immersed like the columbids, and that they are specialized desert representatives of the Columbiformes (but see Goodwin, 1965; Cade, Willoughby, and Maclean, 1966). Only Meinertzhagen (1954) claims actually to have seen watering by regurgitation—in captive *Pterocles exustus* kept in Hampshire between 1895 and 1897.

We believe, for reasons stated elsewhere (Maclean, 1967), that sandgrouse are phylogenetically near the Charadriiformes and that they are not at all closely related to doves and pigeons. In any case, their young are highly precocial and are not fed by regurgitation or by any other method.

Until recently, field observations on the parental care of young sandgrouse in the wild have been notably lacking. Then Marchant (1961, 1962) reported on some observations he made in Iraq on broods of *Pterocles alchata* and *P. senegallus*. Although he was uncertain about some of the details, the activities that he describes and calls "litter of puppies" behavior clearly support the correctness of Meade-Waldo's original observations. During the 1965 breeding season in the Kalahari Gemsbok National Park, we were also able to confirm, in detail, the fact that the male sandgrouse (*Pterocles namaqua*, in this case) transports water to his young in his belly feathers and that the young extract this water in some way by stripping the wet feathers with their beaks, exactly as Meade-Waldo (1921) said.

OBSERVATIONS AND DISCUSSION
ADULTS DRINKING AND WETTING THEIR FEATHERS AT WATER HOLES

Since October 1964, once or twice weekly, observations have been made on sandgrouse watering at the Houmoed game well located in the Kalahari Gemsbok Park about 10 miles up the dry bed of the Auob River from Twee Rivieren, the park headquarters. Less-regular visits have also been made to other water holes and wells both in the Kalahari and in the Namib Desert of South West Africa. These observations have provided a great deal of accurate information on the times of watering, seasonal changes in numbers watering, effect of rainfall on watering, method of drinking, and seasonal changes in the percentage of watering birds that also wet their ventral feathers, for two locally abundant species, *Pterocles namaqua* and *P. burchelli*. Only information pertinent to the question of water transport to the young is presented here. For details about drinking behavior, see Cade *et al.* (1966).

Table 1 shows seasonal data on the numbers of sandgrouse watering and wetting their feathers at Houmoed. Unfortunately, actual counts of birds wetting their feathers were not made during the period from December through April, but our general impression was that few males soaked their feathers at this time. Although some males of both species are to be seen soaking their feathers right through the year (a few pairs may also be found breeding at any time, too), there was a greatly increased incidence of such behavior beginning in late May and continuing through the winter into early summer. These months correspond with the time when *P. namaqua* was found breeding in greatest numbers in the park. The first nest was found on 1 June 1965, and five nests or broods were found in June, seven in July, eight in August, only three in September, a month in which continuous observations were not made, five in October, seven in November, eight in December, two in January 1966, one in February, none in March, and one again in April, plus one of *P. burchelli*. Presumably the latter species was breeding all through this period, too (collected specimens showed active or recently active gonads), but its nests, which are located in the sand dunes rather than on the outcroppings of calcrete adjacent to the river valleys, the typical nesting habitat of *P. namaqua*, are difficult to find.

Thus, there was a close association between the incidence of males wetting their feathers at water holes and the peak breeding period. It is also worth noting that these peaks occurred in winter, exactly the opposite season to be expected if the wetting behavior were mainly adaptive for evaporative cooling. Further, it is nearly always the males that wet their feathers. Only six times out of hundreds of cases have we seen females wetting their feathers.

TABLE 1

SEASONAL DIFFERENCES IN PER CENT OF WATERING SANDGROUSE THAT SOAK
THEIR VENTRAL FEATHERS

Date	Pterocles namaqua				Pterocles burchelli			
	Number drinking	Soaking males	Soaking females	%	Number drinking	Soaking males	Soaking females	%
11/5/65	300	1	0	0.3	40	0	0	0
14/5/65	500	1	0	0.2	200	0	0	0
21/5/65	550	3	0	0.5	250	6	0	2.4
28/5/65	800	4	0	0.5	600	20	0	3.0
3/6/65	650	0	0	0.0	160	9	0	5.6
18/6/65	300	7	0	2.3	360	4	0	1.1
25/6/65	770	13	0	1.7	650	6	0	0.9
29/6/65	600	15	0	2.5	950	30	0	3.2
6/7/65	340	5	0	1.5	400	8	0	2.0
9/7/65	480	16	0	3.3	900	21	0	2.3
13/7/65	380	30	0	7.9	750	40	0	5.3
16/7/65	270	7	0	2.6	350	8	0	2.2
3/8/65	200	3	0	1.5	340	1	0	0.3
17/8/65	140	6	0	4.3	220	4	0	1.8
1/10/65	100	3	1	4.0	55	1	0	1.8
2/11/65	650	43	4	6.6	670	35	1	5.2

When about to soak, the male sandgrouse walks into the water until it touches his belly. He then squats down on his tarsometatarsals with his feet crossed over each other (a detail which has been observed several times in *P. burchelli* from a distance of nine feet with 8-power binoculars through clear water); he then rocks his body on an axis about the pelvis in short, rapid bursts of five to six rocks at intervals varying from a few seconds to a minute or more. During rocking, the head is held up high, the tail is elevated, the abdominal feathers are raised away from the body, and the body is lifted so that only the tips of the feathers touch the water (see fig. 1). The feathers are shaken about, and the water is thereby thoroughly worked into them. The soaking process lasts from a few seconds (abortive attempts) to more than 15 minutes in some cases. Some individuals remain soaking in the water long after the rest of the flock has departed, and such birds appear to be in a kind of dazed or trancelike state, similar to that often seen in sun-bathing or anting birds (Burton, 1959); however, they immediately fly up if disturbed.

TRANSPORT OF WATER TO YOUNG

In order to observe how the adults bring water to the young, it is necessary to establish contact with a family group before 0630 hours in summer and before 0900 hours in winter, as these sandgrouse visit water in the morning as a rule. Routine surveys of suitable nesting habitat were made in a Land Rover between 0600 and 0900 hours in the hope of seeing sandgrouse delivering water to their young. Four unequivocal instances of such behavior have now been witnessed by Maclean for *P. namaqua*, and the details are presented below.

On 10 August 1965 a male with two chicks was found on the calcrete one and one-half miles south of the Houmoed water hole at 0855 hours, before any of the adults had begun to fly to water.

Figure 1. Adult male sandgrouse (*Pterocles namaqua*) soaking his belly at a water hole in the Kalahari Desert. The bird is resting after a period of rocking.

The chicks were only a foot from the nest where they had hatched and probably were no more than one day old. Even at this age they were feeding themselves and preening in typical adult fashion. The female was not seen at first. The following observations were made from the Land Rover with 8-power binoculars at a distance of 75 feet from the chicks.

At 0920 the male walked to the nest and brooded the empty nest for about 20 seconds; then he went to brood the chicks. At 1010 the male left the chicks and went back to brood at the nest for about five minutes. Then at 1015 the male walked away from the nest for about 12 feet and flew up, but landed about 50 yards away. Soon the male and female flew up together (she had evidently been foraging there), the female landing after about 20 yards, the male continuing alone in the direction of Houmoed. The female began to walk toward the chicks about 80 yards away from her, but then she flew up and landed about 15 yards from the chicks at 1020 hours. At 1027 the male flew in from the water hole, landed at the nest, and then walked to the chicks. He stood upright, showing clearly his wet, abdominal feathers. The chicks ran to him and at once raised their heads to his wet feathers and took the tips of the feathers in their beaks.

On 27 November 1965 at 0745 hours a pair of *P. namaqua* with chicks about three days old was found on the calcrete. The male had just returned from water, and his ventral feathers were soaked to a point well above his chest band. As the parents called to the chicks and led them away, the young birds had probably already taken water from the male's feathers. Another male sandgrouse landed nearby about the same time. His feathers were dry, and he went to a nest with eggs.

On 30 November 1965 at 0630 hours a female with three five-day-old chicks was encountered on the calcrete. The female flew up and joined the male 200 yards away. The chicks crouched at the base of some shrubs, and the following observations were made in the Rover at a distance of 20 feet from the chicks. At 0725 both parents returned from the water hole. Only the male

Figure 2. Young sandgrouse (*Pterocles namaqua*) clustered around the wet abdomen of adult male in the Kalahari Desert.

had wet ventral feathers. At 0750 the parents walked up to the chicks. All three chicks rushed to the male, which uttered the "kelkiewyn" call. To get to the male parent, the chicks actually had to run past the female. The chicks clamored around the male's breast and abdomen, but his feathers were almost dry. As the chicks tried to take water, the male called a low, staccato "kirri, kirri, kirri," and then led them away.

On 11 December 1965 at 0840 hours a male *P. namaqua* landed on the calcrete, and Maclean drove toward it, stopping the Land Rover 30 feet away. Both parents were present with three large chicks about two to three weeks old but as yet unable to fly (almost certainly the same family seen on 30 November above). The male had wet ventral feathers. The chicks came out of cover and rushed to the male, ignoring the nearby female. The chicks at once began to take water from the male's feathers by sharp, downward jerks of their heads, probably a "stripping" action with their beaks to get the water out. The male stood bolt upright, exposing the wet feathers. The chicks worked at the feathers for five minutes and then left to lie in the shade. The male immediately began to dry his feathers by rubbing his body on sandy ground for several minutes.

Although Marchant (1961, 1962) evidently watched his birds from considerably greater distances (ca. 150 meters) than we did and therefore missed some details, our field observations parallel his on *P. alchata* and *P. senegallus* in Iraq very closely. The departing of the male with dry feathers and returning to the brood with wet ones, the early-morning time of water transport, the behavior of the chicks in running from their hiding places to the male and ignoring the presence of the female, the male's upright posture with fluffed out feathers, the clustering of the chicks about his abdomen, and their obvious head movements around his wet feathers are all the same and lead only to the conclusion that the young obtain water in the way first described by Meade-Waldo. Figure 2 shows young *P. namaqua* in the Kalahari clustered about the wet abdomen of the male.

Mechanisms of Water Transport in Feathers

Physical problems of water transport in feathers. While the method of water transport in wet feathers is now established beyond doubt, all of the questions raised by this unique behavior are by no means answered. The major adaptive question, of course, is how far can a male sandgrouse transport usable amounts of water in his breast feathers? The answer to that question is dependent upon the answers to a number of subsidiary problems. What is the rate of evaporation from the wetted feathers of a sandgrouse flying through the early-morning desert air? How much water can be taken up and held in the ventral feathers of a sandgrouse? Are there special structural features of sandgrouse feathers that aid moisture retention? Are there behavioral specializations for trapping or holding water in the plumage? Let us begin with the more elementary questions and see how far we can develop a reasonably approximate answer to the first, which has important implications for breeding distribution, reproductive yield, and limitations on population size.

Behavioral adaptations for water transport in feathers. The details of feather soaking have already been presented. We have been impressed by the degree to which the ventral feathers—especially those on the belly—are erected as the male sandgrouse enters water, by the vigorous downward rocking action of his belly while soaking, and by the large amounts of water that drip off of the bird in the first few feet of flight—more it appears than can easily be accounted for on the assumption that the shower of drops merely represents excess draining off the surfaces of the feathers. When a sandgrouse lifts out of the water his ventral feathers are already closely appressed to his body. These observations suggested that a sandgrouse may be able to trap and in some way hold excess water in his plumage—more, that is,

TABLE 2

COMPARISON OF WATER-HOLDING CAPACITY OF FEATHERS AND OTHER MATERIALS

Material sampled	g dry wt.	g wet wt.	g H₂O/g dry wt.
Synthetic sponge	8.1	51.3	5.3
Paper towel	4.0	24.4	5.1
Passer melanurus breast feathers	0.35	2.15	5.1
Pterocles namaqua male breast feathers	0.90	8.10	8.0

than the absorbed amount in the feather material itself. This inference led us to examine the water-holding capacity and the structure of sandgrouse feathers and to make comparisons with other birds.

The water-holding capacity of sandgrouse feathers. Feathers are not very wettable. The imperviousness of the horny, outer sheath impedes penetration by water molecules, and the surface structure of feathers is admirably constructed to shed water. In addition, the preen-gland oil, which many birds apply to their feathers, provides a further hindrance to the absorption of water, as does the powder down of some other birds. One adaptive advantage of the "dusting" behavior that male sandgrouse sometimes perform before entering the water may be to remove oil from the feathers and thereby render them more wettable.

Once they do become wet, however, feathers have a good water-holding capacity. Table 2 presents data from a preliminary series of experiments comparing samples of feathers with other wettable materials. It can be seen that the breast feathers of the Cape Sparrow (*Passer melanurus*) have as good a water-holding capacity as an ordinary synthetic kitchen sponge or a piece of paper towel. The interesting discovery was that the breast feathers of a male *P. namaqua* held almost twice as much water as any of these other materials. This significant difference encouraged us to examine in detail the water-holding capacity of individual feathers from males and females of four species of sandgrouse (*P. namaqua, P. burchelli, P. bicinctus,* and *P. gutturalis*) and to compare the results with values for six species from other families and orders of birds.

The results are presented in figure 3. Feathers from the sandgrouse held about two to three times as much water when thoroughly soaked and allowed to drip until all surface excess was lost as did the feathers of any other species tested. Male sandgrouse feathers held significantly more water than female feathers. Even after repeated soaking and stripping with the fingers to remove water, the belly feathers of sandgrouse still retained their structural integrity; this was not true of the upper breast feathers, neck feathers, or dorsal body feathers of sandgrouse, or of the body feathers in general of other species tested. After being thoroughly soaked and stripped several times, these latter feathers became frayed, matted, and twisted into a cordlike structure while wet. On drying, it was often impossible to restore their webbing. These differences between the ventral feathers of sandgrouse and other kinds of feathers demand explanation in structural terms.

Structural specializations of sandgrouse feathers. The belly feathers of sandgrouse are elongate, averaging about six times as long as they are wide, with a slight curvature away from the midline of the body. The elongate shape seems to be advantageous for grasping and stripping by the beak of a young sandgrouse—certainly more so than the broader, shieldlike shape of the ventral feathers in many

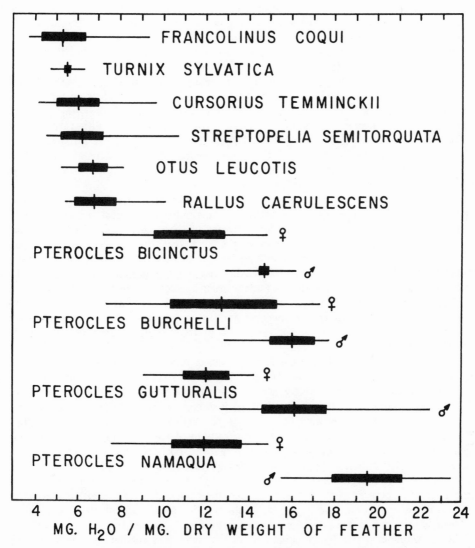

Figure 3. Water-holding capacity of sandgrouse belly feathers compared with that of the belly feathers of the Coqui Francolin (Galliformes), Kurrichane Button-Quail (Gruiformes), Temminck's Courser (Charadriiformes), Red-eyed Turtle Dove (Columbiformes), White-faced Owlet (Strigiformes), and Cape Rail (Gruiformes). Horizontal lines represent the ranges of each sample, vertical lines indicate the means, and the black rectangles delimit plus and minus 2 standard errors of the mean. N equals 10 in all samples. In all cases, individual feathers were allowed to become thoroughly wetted in water, then were removed with forceps, were suspended in air until all excess water dripped off, and then were immediately weighed on a Mettler balance accurate to 0.1 mg.

other kinds of birds. Also, it is interesting to note that female feathers average rather broader in proportion to length than those of males. The curvature in the shaft may help to retard downward movement of water along the length of the feather, or it may be related to the high density of feathers in the belly region and to their manner of overlapping each other so completely that only the distal fifth or less of each feather is exposed to the surface.

The belly feathers of all sandgrouse, as in many other birds, have downy bases. In addition, they bear a fringe of downy tufts on the ends of the barbs on both sides, extending about four-fifths of the way toward the tip. Only the distal fifth—the part exposed to the surface—has a typical body-feather construction with rather loose webbing in the vanes and slightly frayed tips. The webbing of the proximal four-fifths adjacent to the shaft is very strong and springy, and the barbs are not easily separated. The difference in structure between the proximal and distal areas is sharply demarcated and is quite clear to the unaided eye.

Most importantly, the ventral surface of this proximal zone is densely covered with fine, hairlike extensions of the barbules. This pubescence occurs on all the belly feathers, extends anteriorly into the lower breast feathers, to a reduced extent in the upper breast feathers, and disappears in the neck feathers. It does not occur on any of the dorsal body feathers, nor does its exact counterpart occur on any body feathers of other birds which we have examined, including species representing 10 orders, although the body feathers of some other species do bear a pubescence of a different structural arrangement on their ventral surfaces. Figures 4 and 5 compare sandgrouse feathers with the feathers of four other species, as they appear when thoroughly soaked and floating on water. All of the structures described above are clearly visible.

After a sandgrouse belly feather is fully wetted, a downward stroke of the feather in water, as when the male rocks his belly during the soaking process, fluffs out and expands the barbs, and the downy fringes are maximally spread. The ventral, fuzzy area appears turgid. Then, as the feather is removed from the water, the downy tufts are drawn inward toward the shaft on the ventral side, and in so doing they form a kind of groove or trough in which a column of water is trapped. The pubescence forms a matrix that aids in this retention, probably by supplying a large surface area for adhesion by interfacial tension between the droplets of water and the hairs. If the hairlike barbules are also hollow, as seems likely in conformity with the general structural plan of feathers, then they may also hold some of the water by capillary attraction. At any rate, while the dorsal surface of the feather only appears damp and bears no droplets after withdrawal, the ventral, proximal surface holds a sizable body of water enmeshed in the hairs and further held in along the sides by the downy tufts.

The exact mechanism of water retention is more clearly seen under a microscope with magnifications of 50 to 100 power. On a dry feather it can be seen that the proximal zone is made up of highly specialized barbules completely unlike those at the distal end of the feather, where the barbs have the typical overlapping arrangement of hooks on the distal barbules and grooves on the adjacent proximal ones. The proximal barbs bear a series of barbules that are flattened and riblike along their basal portions and that are coiled along the ventro-lateral sides of the barb to form a series of overlapping helices. Each helical portion consists of two or three open coils, after which the barbule becomes attenuated into a long, straight, hairlike structure. None of these specialized barbules bears hooks or grooves; instead, adjacent

barbs are held together by intertwining of coils between the distal barbules on one barb and the proximal barbules of the adjacent barb. The hairy ends extend distally along the axis of each barb for one millimeter and lie in a flattened position against the ventral surface of the feather. The general appearance under the microscope is of a series of double coils formed around each of the barbs. The basic structures of a single barb are shown in figure 6. It is this coiled construction that gives the inner proximal zone of the feather its springy character, and the interlacing of coils between adjacent barbs produces a strong web that is highly resistant to mechanical disruption.

When a drop of water is applied to the distal end of a sandgrouse belly feather on its under surface, the water is drawn up into the hairs, probably by capillarity or interfacial tension. As the water penetrates to the basal, tightly coiled portion of the barbules, the latter suddenly spring open and extend their hairy ends perpendicular to the plane of the main feather surface, thereby forming a dense stand of upright hairs about two millimeters deep in which the body of water becomes enmeshed and held by interfacial tension. Additional drops of water can be added until the entire specialized proximal zone is loaded, after which any excess drains off along the distal end. We are uncertain what causes the barbules to spring open and project their hairy ends at right angles to the plane of the feather, but it may result from the force of turgor built up inside the coiled, basal parts of the barbules when they imbibe water, or to some differences between the two flattened sides of the barbules in their expansional response to wetting. In the basal parts of the barbules there are some curious grooves that show as dark striations in figure 6 and which may be involved in the mechanics of opening. In any case, as the barbs dry out, a certain point is reached when the recoiling mechanism is set off, and the barbules rapidly reform tight coils around each other, reuniting adjacent barbs into a firm web as they do so.

We have found nothing like the coiled construction of these barbules in the feathers of any other African birds examined, but the ventral feathers of all four species of sandgrouse in southern Africa have barbules of this type. These peculiar structural modifications of sandgrouse belly feathers are undoubtedly what gives them their great water-holding capacity.

Female sandgrouse feathers have essentially the same structure as those of males, except that the area of specialized barbules is not as large on female feathers, and they tend to have a more highly developed, downy, basal area. Also, as one progresses anteriorly along the venter, there is a greater tendency for the female feathers to bear after-shafts than in males. Only a few of the female feathers hold as much water as the highly specialized male belly feather. The area of specialized feathers is smaller on the female than on the male and is mainly restricted to the upper belly, whereas it covers the entire belly area of the male and also extends well up into the breast. The feathers of the juvenal male are also much less specialized for holding water than in the adult.

Thus the highly specialized water-transporting behavior of the adult male sand-

←

Figure 4. Above. Wet sandgrouse belly feathers floating on the surface of water (ventral aspect). Left, female *Pterocles namaqua*; middle and right, male *P. gutturalis*. Below. Feathers from male *P. bicinctus*. The specialized proximal zone stands out as a light, fuzzy area fringed by darker, downy tufts.

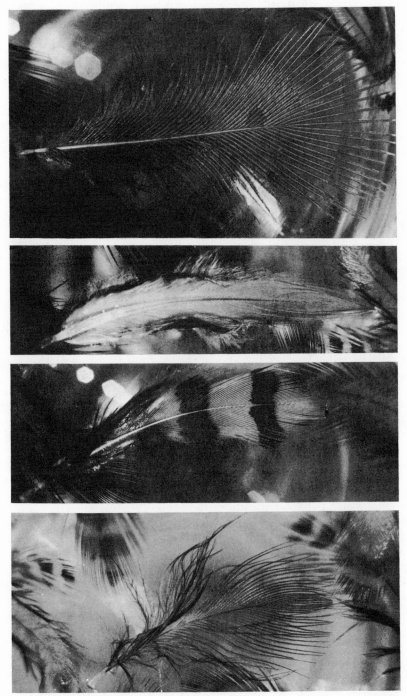

Figure 5. Wet belly feathers of the Cape Rail (uppermost), Red-eyed Turtle Dove (upper center), Coqui Francolin (lower center), and Temminck's Courser (lowermost), floating on the surface of water (ventral aspect). None of these feathers shows a specialized proximal zone like that of sandgrouse feathers.

grouse is associated with equally specialized structural modifications of his ventral body feathers. We conclude that these modifications are anatomical adaptations for holding large amounts of water in the plumage and for resisting structural breakdown from repeated wetting and "stripping" by the young sandgrouse.

A recent examination of all the skins of sandgrouse in the American Museum of Natural History revealed one curious exception. All 14 species of *Pterocles* have the highly specialized belly feather described above; so does *Syrrhaptes paradoxus*, but the closely related *S. tibetanus* does not. This high montane species is also exceptional in that the sexes are only weakly dimorphic in plumage. The ventral feathers are soft, extremely dense, with highly developed, downy bases, after-shafts, and other auxiliary downy tufts and edgings, and with no trace of the specialized proximal zone found in all other sandgrouse. Instead, the feathers of *tibetanus* give the appearance of being specially modified for maximum insulation. This species breeds at 12,000 to 16,000 feet and winters at only slightly lower elevations. It seems that the environment of this species has placed such a premium on the evolution of a ventral feather structured for maximum insulation that the specialization for water transport has been secondarily lost. (Alternatively, *tibetanus* may represent the survival of a primitive line of sandgrouse.) It would be most interesting to know whether the water-transporting behavior has also disappeared. Perhaps the young are led to water, or perhaps they eat food of high moisture content. All that seems to be known is that this species does not fly to water as regularly as other sandgrouse (Meklenburtsev, 1951:90).

Total amount of water held in the belly plumage. The amount of water that a male sandgrouse takes up in his ventral feathers depends not only on the maximum holding capacity of his feathers, but also on the length of time he soaks his plumage. Most individuals do not soak for more than five minutes. Several dead specimens of male *P. namaqua* whose bellies we soaked in water for a period of five minutes took up and held between 25 and 40 ml of water; however, it is difficult to arrange the feathers on a dead sandgrouse in the same way they are held and manipulated by a live bird. Experiments on pelts of belly skin are not much better but yielded the following results for water-holding capacity: average for four adult male skins, 22.1 ml, with a range from 18.4 to 27.5 ml; average for four adult female skins, 8.4 ml, with a range from 7.5 to 9.1 ml; average for three immature males molting into adult plumage, 12.1 ml, with a range from 6.7 to 14.9 ml. It appears from these data that an adult male sandgrouse usually carries about 25 ml of water in his plumage, depending on the condition of his feathers and on how long he soaks at the water hole.

Evaporation from feathers during flight. The rate of evaporation from any wet material is dependent upon the difference between the vapor pressure at the evaporating surface and that of the surrounding air. The vapor pressure in the air is determined by temperature and by the vapor density—the absolute amount of water vaporized in the air. The rate of evaporation is also greatly affected by wind velocity. The general relationships can be summarized in a simple equation (as presented by Rohwer, 1931, and adapted by Kleiber, 1961):

$$E = (0.44 + 0.118W)(e_s' - e_d'),$$

where $E =$ evaporation in inches of water level per 24 hours, $W =$ wind velocity in miles per hour, $e_s' =$ water vapor pressure in saturated air at the surface temperature in inches of Hg, and $e_d' =$ vapor pressure in surrounding ambient air.

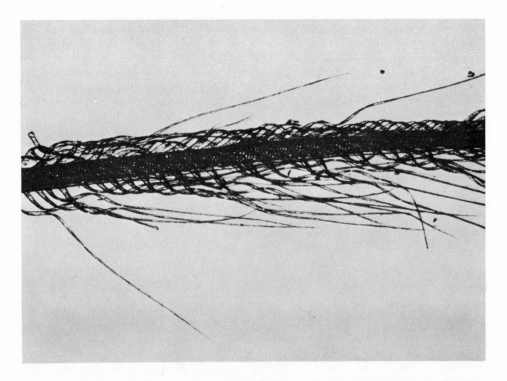

For our purposes it is convenient to convert inches of water level per 24 hours to grams of water per cm² per hour. This is done by multiplying the first part of the equation by 2.54 and dividing the answer from the whole equation by 24. Knowing or approximating the wet surface area of a sandgrouse's venter, the amount of water contained in the feathers, and the bird's flying speed, we can then make estimates of the rate of evaporation for a hypothetical bird flying from the water hole to his chicks under various environmental conditions.

Partly erroneous assumptions involved in these calculations are: (1) That water is continuously supplied to the surface of the belly from the wet feathers underneath to maintain a constant saturation vapor pressure over the area considered. Because of the way the feathers overlap, exposing only the distal fifth or less, and because of the special structures described above that hold most of the water in the proximal parts, water may not move freely to the tips of the feathers to provide a constantly wet surface. Attractive forces, such as capillarity and interfacial tension, associated with the fine hairlike barbules on the ventral surfaces of the feathers, may further retard movement to the exterior and allow for considerable drying of the surface while moisture is retained within. These are points that require checking. (2) That the temperature of the surface is the same as the ambient temperature. In fact, because of the cooling effect of evaporation, it will probably be less, although the body heat should warm the underlying moisture before it arrives at the surface. We know of no easy way to integrate these two opposing influences on the surface temperature; therefore, for simplicity we have assumed no temperature difference between the surface and the ambient air.

Table 3 presents some calculations of water loss for different flying conditions. Assuming a wet surface area of 50 cm²—about right for adult male *P. namaqua*— and a moisture content of 25 g of water in the feathers, if the sandgrouse flies for half an hour at 40 miles per hour under ambient conditions of 30° C and relative humidity of 15 per cent, it will have lost 14.6 gm of water; under ambient conditions of 30° C and 30 per cent relative humidity, 12 g of water; 20° C and 15 per cent relative humidity, 9.5 g of water; 20° C and 30 per cent humidity, 6.6 g. From these data, it appears that a male sandgrouse flying under the relatively moderate conditions of the early-morning desert atmosphere should be able to deliver from 10.4 to 18.4 g of water for a distance of 20 miles, starting with an initial amount of 25 g. Since it is quite easy to squeeze out 5 or 6 ml of water from a sandgrouse pelt that has soaked up 10 ml, it is reasonable to believe that young sandgrouse can obtain significant amounts of water from feathers carrying between 10 and 18 ml of water.

These data also demonstrate how necessary it is for sandgrouse to transport water under the most favorable physical conditions of the desert year and of the 24-hour daily cycle. Winter is the most favorable season, and early morning is an advantageous time; the air temperature has not yet reached a maximum, and humidity is likely still to be high from the night-time effect of cooling and condensation. For species such as *Pterocles bicinctus*, which water at night, when temperatures are still

---

←

Figure 6. Photomicrographs of the basal part of single barbs from the specialized proximal zone of the belly feather of a male sandgrouse (*P. namaqua*). The ventral surfaces are oriented toward the plane of view; some barbules have been disrupted by slide preparation. Upper figure shows the tightly coiled barbules of a dry barb; the lower shows barbules extended after contact with water. Note dark striations in basal parts of barbules. Total length of one barbule equals about one millimeter.

TABLE 3

RATES OF EVAPORATION FROM A WET SURFACE EXPOSED TO A WIND VELOCITY OF 40 MILES PER HOUR

Air temp. °C	Relative humidity %	Saturation vapor pressure, inches Hg	Vapor pressure at given R.H., inches Hg	Mg H_2O evap./ cm²—hour
30	15	31.82	4.77	582
30	30	31.82	9.55	480
20	15	17.54	2.63	377
20	30	17.54	5.26	264

lower and humidity higher, the problems involved in water transport in wet feathers are greatly reduced. It would be most interesting to know whether such species do, in fact, water their young, as well as drink, at night.

Although the rate of evaporation increases drastically with wind velocity (air speed of the bird), it is nevertheless advantageous for a sandgrouse to fly fast in terms of the amount of water lost per distance traveled, as shown in figure 7. The greatest percentage savings per unit increase in speed occur at speeds under 20 miles per hour, and above about 50 miles per hour there is little further decrease in the amount of water lost for each 10-miles-per-hour increase in speed. Sandgrouse are notably fast fliers, and we believe that 40 miles per hour is a conservative average speed for flocks seen flying to and from water. Perhaps this fast flight has evolved in association with the mechanism of water transport in feathers; or it may have been one of the pre-existing, permissive attributes that have made this rather inefficient mode of water transport possible.

Physiological aspects of the problem. How much free water does a young sandgrouse require each day? Unfortunately, we do not know the answer to this question, but we suspect it may not be very much. It has generally been assumed that sandgrouse require relatively large amounts of water—like doves and pigeons, which are also obligate drinkers—to compensate for the fact that their food consists almost exclusively of air-dried seeds with little moisture content. There are some records in the literature that indicate large consumptions of water by sandgrouse; see, for example, the translation from a Russian source in Salt and Zeuthen (1961:403), where it is stated that individuals of *Pterocles orientalis* shot at watering holes held up to "a cup" of water in their crops. We found, however, that adult *P. namaqua* drank no more than 15 ml of water after having been deprived of water for 25 days (Cade *et al.*, 1966). We further noted that the average amount of water consumed in one draft (one immersion of the beak followed by raising the head) is about 1.5 ml. Out of several hundred observations on individual sandgrouse, 24 drafts was the highest number counted, a figure that equals about 36 ml of water. The average for *P. namaqua* was 9.5 drafts, or 13.3 ml; for *P. burchelli*, 7.2 drafts, or 10.8 ml. These quantities amount to less than 10 per cent of body weight per day, about what one would expect for the body size of these birds (Bartholomew and Cade, 1963). Three adult *P. namaqua* shot by E. J. Willoughby just after they had drunk at a water hole in the Namib Desert on 12 March 1966 gave the following values: a male weighing 183 gm held 30 ml of water in his crop; another male weighing 162 gm held 26 ml; a female weighing 167 gm had 23 ml in her crop. These values range between 16.4 and 13.7 per cent of body weight. Certainly these data do not suggest any unusually high consumptions of water by adult sandgrouse.

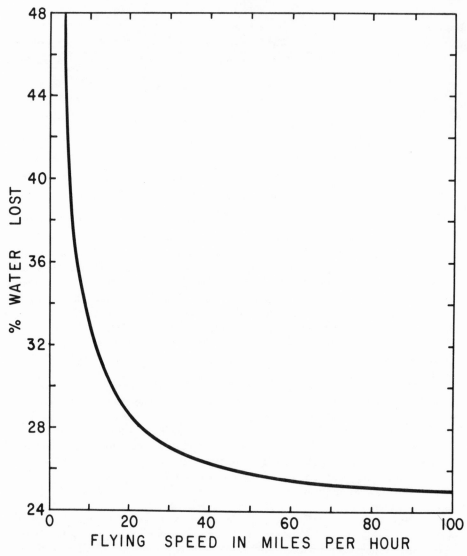

Figure 7. Calculated percentage of water lost by evaporation from a sandgrouse's venter after flying 20 miles at different air speeds, under the following conditions: 20° C air and surface temperature, 30 per cent relative humidity, 50 cm² wet surface area, and 25 g of water in the plumage at beginning.

Moreover, we have noted that many individuals arriving in the flocks at water holes depart without drinking. Some of these undoubtedly return in later flocks, but many do not. For instance, at Gross Tinkas, a natural water hole at the head of an intermittent tributary draining into the Swakop River in Game Reserve No. 3, in the Namib Desert of South West Africa, on 5 December 1965 Cade and E. J. Willoughby watched sandgrouse from 0730 to 1000 hours. An estimated 2000 to 3000 *P. namaqua* assembled in flocks of 10 to 100 on the knolls and hillsides surrounding the water hole, but less than 700 came down to the water, and of these not more than half actually drank. During this time the flocks were continually harassed by a pair of Lanners (*Falco biarmicus*), and once they were disturbed by a jackal (*Canis mesomelas*), which came to drink. Possibly some of the assembled birds drank elsewhere, but we doubt it. We believe that an individual sandgrouse does not need to water every day and that species of sandgrouse are much better adapted to conserve water than are pigeons and doves. When deprived of water in captivity, sandgrouse lose body weight gradually and continue eating seeds (Cade and Willoughby, unpublished), whereas doves reduce food consumption or stop eating entirely and lose body weight precipitously (MacMillen, 1962; McFarland, 1964; Willoughby, 1966).

Thus, it seems likely to us that even if a young sandgrouse eats only air-dried seeds—a point which requires further checking—its water requirement may be proportionately less than it is for the young of many other comparable-sized birds. If so, this would be another permissive characteristic allowing for the evolution of water transport in wet feathers.

CONCLUSIONS

Water transport in feathers as a limiting factor on populations. For the reasons given above, we believe it is possible for a pair of sandgrouse to nest at least 20 miles away from surface water and still allow the male to transport sufficient amounts of water to nourish the young. Field observations on this point are rather meager, as Marchant (1961) pointed out. None of the nests he located was more than 6.5 miles from water, and some were less than two miles away. Similarly, all of the nests we have located in the Kalahari have been close to water, the farthest distance being about nine miles; however, we believe *P. burchelli* must nest considerably farther away. O. P. M. Prozesky (personal communication) found a brood of *P. bicinctus* in the Kruger National Park about 15 miles from the nearest water hole, and Willoughby found a nest of *P. namaqua* in the Namib between 15 and 20 miles from water. Little else has been reported.

It is obvious, however, that the dependence of young sandgrouse on water transport in feathers places a definite limitation on the effective breeding range of these species. Regardless of how far out into the desert away from water adult sandgrouse can forage, the breeding pairs must settle within the range of the male's ability to transport adequate amounts of water to the young. We suspect that only a small portion of the total sandgrouse population of a given region breeds at any one time —even during the peak season—and that the water transport mechanism to the young—greatly restricting the area available for breeding—may be the major limiting factor on population growth.

The present great abundance of sandgrouse in the Kalahari Gemsbok Park and in other arid, settled parts of southern Africa has almost certainly been influenced by the presence of man-made sources of water. Before the bore holes and dams were in existence, sandgrouse must have occurred in far fewer numbers and have been much more restricted in their distribution. In regions such as the Kalahari where natural, permanent sources of surface water do not exist for tens of thousands of square miles, the breeding of sandgrouse must formerly have been entirely dependent upon the infrequent heavy rains that fill pans and other depressions sufficient to last through the period of rearing the young.

Evolutionary considerations. Being seed-eaters, sandgrouse are obligate drinkers. Neither their beaks nor their digestive systems are adapted for the regurgitation of food or water, and the adults do not produce any substance like the "pigeon's milk" of columbiform birds. The sandgrouse have probably evolved from an insectivorous or omnivorous charadriiform ancestor that produced precocial young (Maclean, 1967). How the seed-eating habit evolved we do not know, but it is interesting that there is a South American parallel in the seedsnipes (family Thinocoridae of the Charadriiformes), which also produce precocial young. In any case, the evolution of the seed-eating habit in a line of desert-inhabiting charadriiforms with precocial young, which also eat seeds, necessitated the concurrent evolution of a mode of water transport by the parents to the young. Having no mechanism for regurgitation— itself a highly specialized mode of parental care—soaking a part of the plumage seems to be the only other recourse. Buxton (1923) has suggested that the short legs of sandgrouse, bringing the abdomen in contact or near contact with water when the adults drink, may have been responsible for the evolution of this mode of water transport. It seems more probable to us, however, that water transport in the belly feathers represents a transformation of the bathing response, which is nearly universal among birds.

It would be highly enlightening to know how young seedsnipes obtain water. The problem of water transport to the young in these birds must parallel very closely that of the sandgrouse, and in view of the limited number of possibilities open to birds for watering their young, it would not be surprising to find that seedsnipes have reached the same solution. In this connection, we were most interested to discover that the ventral feathers of seedsnipes have a pubescent area similar to that of sandgrouse, although not so well developed. Furthermore, this pubescence is formed by barbules that are coiled at their bases; but in seedsnipes there appears to be no intertwining of barbules between adjacent barbs as there is in sandgrouse. We have had no opportunity to test the response of seedsnipe feathers to water.

Sandgrouse and the science of ornithology. There is a moral to be read in the literature that has developed about sandgrouse in the last 70 years. We have found it strange—and a little disconcerting—that points of fact, such as how sandgrouse drink and transport water to their young, should so long remain in question. How does one explain the curious acceptance in the literature for more than 30 years of the myth that sandgrouse drink like doves and pigeons, in the complete absence of any published, substantiating data? Yet the fact of water transport in the belly feathers, well documented by direct observations on captives and indirectly corroborated by field observations of adult males wetting their feathers, has been labeled a myth and is categorically stated to be "not so" in *A New Dictionary of Birds* (1964:712), the most recent, "authoritative" compendium of our knowledge about birds.

SUMMARY

Seventy years ago Meade-Waldo described the unusual method by which the male sandgrouse delivers water to his young in his wet abdominal feathers. Although many subsequent authors have doubted the accuracy of Meade-Waldo's statements, our observations on the sandgrouse *Pterocles namaqua* in the Kalahari Gemsbok National Park in the Republic of South Africa confirm them in every detail, and also correspond very closely to descriptions of the water-transporting behavior of *P. alchata* and *P. senegallus* in Iraq.

In short, the adult male flies to a water hole in the morning, soaks his ventral feathers in a special way, and then flies back to the nesting grounds, where he alights, walks to the hiding brood, and presents his abdomen by standing in an upright posture with his feathers fluffed out. The young run from their hiding places, cluster around the male's belly, take the wet feathers in their beaks, and remove the absorbed water by a "stripping" motion.

Although feathers are not very wettable, they do have a good water-holding capacity when saturated. The belly feathers of sandgrouse are superior to all other kinds of feathers tested for this property. Male sandgrouse feathers typically hold from 15 to 20 mg of water per milligram of dry weight, and female sandgrouse feathers hold about 11 to 13 mg per milligram of dry weight. Feathers from seven other species representing six different orders ranged around 5 to 6 mg of water per milligram of dry weight, which is about the water capacity of paper towel or synthetic sponge. Sandgrouse belly feathers retain their structural integrity after repeated wetting and stripping with the fingers to remove water, whereas other kinds of feathers quickly become frayed, matted, and twisted out of shape.

The unusual water-retaining characteristics of sandgrouse belly feathers are correlated with structural peculiarities of the proximal four-fifths of the feather. The structure of the barbules in this specialized zone is unlike that recorded for any other species of bird. The barbules have no hooks or grooves, but are flattened at their base and coiled into helices along both sides of the barbs. The ends of the helices terminate in straight hairlike tips, giving the ventral surface of the feather a peculiar pubescence in the proximal zone. The helices of barbules intertwine, forming a network of coils that is very resistant to mechanical disruption.

When water is dropped on the ventral surface of the feather, the coiled parts of the barbules spring open and project their hairy tips at right angles to the plane of the feather. Water is held in this meshwork by interfacial tension, and possibly also by capillarity. When the water evaporates, the coils spring back, reuniting adjacent barbs into a strong web. These attributes of sandgrouse feathers are interpreted to be adaptations for holding large amounts of water and for resisting structural breakdown from repeated wetting and "stripping" by the beaks of the young.

Measurements made on dead sandgrouse and on pieces of belly skin and plumage indicate that 25 to 40 ml of water can be absorbed and held in the belly plumage of an adult male sandgrouse. Assuming that the wet venter acts as a simple physical evaporating system, and applying realistic values for flying speed, environmental conditions, and water capacity of the plumage, we have estimated that a sandgrouse should be able to deliver from 10 g to 18 g of water for a distance of 20 miles. These quantities should be sufficient to enable young sandgrouse to obtain significant amounts of water from the wet feathers.

The dependence of young sandgrouse on water transport in feathers, by greatly restricting the area available for breeding, may be a major limiting factor on population growth. Before the occurrence of man-made sources of water in the arid parts of southern Africa, sandgrouse probably were less abundant and were more restricted in their distribution.

The evolution of the seed-eating habit in a line of desert-dwelling birds with precocial young has necessitated the concurrent evolution of a mode of water transport by the parents to the young. Lacking a mechanism for regurgitation, soaking a part of the plumage seems to be the only other recourse short of some major anatomical innovation. It seems likely that water transport in the belly feathers represents a selective adaptation of bathing behavior.

The seedsnipes of South America show many parallels with sandgrouse. In view of the limited ways available for adult birds to carry water to their young, it would be most instructive to know how young seedsnipes obtain water.

ACKNOWLEDGMENTS

Our work was supported by a grant from the U.S. Public Health Service, ES 00008 (Environmental Health). We thank the National Parks Board, Republic of South Africa, for permission to do field studies in the Kalahari Gemsbok Park. O. P. M. Prozesky, ornithologist at the Transvaal Museum, kindly provided us with specimens for examination, and E. J. Willoughby and L. I. Greenwald helped with some of the field observations. The latter also performed the calculations involved in the construction of figure 6.

LITERATURE CITED

ARCHER, G. F., and E. M. GODMAN. 1937. The birds of British Somaliland and the Gulf of Aden. Vol. 2. Gurney and Jackson, London.

BARTHOLOMEW, G. A., and T. J. CADE. 1963. The water economy of land birds. Auk, 80:504–539.

BURTON, M. 1959. Phoenix reborn. Hutchinson, London.

BUXTON, P. A. 1923. Animal life in deserts, a study of the fauna in relation to the environment. Arnold, London.

CADE, T. J., E. J. WILLOUGHBY, and G. L. MACLEAN. 1966. Drinking behavior of sandgrouse in the Namib and Kalahari deserts. Auk, 83:124–126.

GOODWIN, D. 1965. Remarks on drinking methods of some birds. Avicult. Mag., 71:76–80.

HEIM DE BALSAC, H. 1936. Biogéographie des mammifères et des oiseaux de l'Afrique du Nord. Bull. Biol. France Belgique, Suppl. 21, 447 pp.

HOESCH, W. 1955. Die Vogelwelt Südwestafrikas. S.W.A. Wissenschaftliche Gesellschaft, Windhoek.

HÜE, F., and ETCHÉCOPAR, R-D. 1957. Les Ptéroclididés. L'Oiseau, 37:35–58.

KLEIBER, M. 1961. The fire of life. John Wiley and Sons, New York.

MACLEAN, G. L. 1967. Die systematische Stellung der Flughühner. J. Ornithol., 108:(in press).

MACMILLEN, R. E. 1962. The minimum water requirements of Mourning Doves. Condor, 64: 165–166.

MARCHANT, S. 1961. Observations on the breeding of the Sandgrouse Pterocles alchata and senegallus. Bull. B.O.C., 81:134–141.

MARCHANT, S. 1962. Watering of young in Pterocles alchata. Bull. B.O.C., 82:123–124.

MCFARLAND, D. J. 1964. Interaction of hunger and thirst in the Barbary Dove. J. Comp. Physiol. Psych., 58:174–179.

MEADE-WALDO, E. G. B. 1896. Sand Grouse breeding in captivity. Zoologist, 1896:298–299.

MEADE-WALDO, E. G. B. 1897. Sandgrouse. Avicult. Mag., 3:177–180.

MEADE-WALDO, E. G. B. 1906. Sandgrouse. Avicult. Mag., new ser., 4:219–222.

MEADE-WALDO, E. G. B. 1921. Observations on the sand-grouse. Bull. B.O.C., 42:69–70.

MEINERTZHAGEN, R. 1954. Birds of Arabia. Oliver and Boyd, London.

MEINERTZHAGEN, R. 1964. Sandgrouse. In A new dictionary of birds. A. Landsborough Thomson, ed. McGraw-Hill, New York. 711–712.

MEKLENBURTSEV, R. N. 1951. The order of Pigeons. In Birds of the Soviet Union. Vol. II. G. P. Dementiev and N. A. Gladkov, eds. Soviet Science, Moscow, In Russian. 3–91.

ROHWER, C. 1931. Evaporation from free water surfaces. U.S. Dept. Agr. Tech. Bull., 271.

SALT, G. W., and ZEUTHEN, E. 1960. The respiratory system. In Biology and comparative physiology of birds. Vol. I. A. J. Marshall. ed. Academic Press, New York. 363–409.

SCHMIDT-NIELSEN, K. 1964. Desert animals. Oxford University Press, London.

ST. QUINTIN, W. H. 1905. The breeding of Pterocles exustus. Avicult. Mag., New Ser., 3:64–66.

WILLOUGHBY, E. J. 1966. Water requirements of the Ground Dove. Condor, 68:243–248.

Division of Biological Sciences, Cornell University, Ithaca, New York 14850, 6 May 1966.

Reprinted from *Proc. Zool. Soc. Lond.* 4:701–706 (1935)

SOME ADAPTATIONS FOR THE DISPOSAL OF FAECES. THE HYGIENE OF THE NEST IN AUSTRALIAN BIRDS

Donald F. Thomson

University of Melbourne

Of the courtship and nesting habits of birds, and of the care that they devote to the building of their nests and the rearing of offspring, much has been written. But of the cleaning of the nest, of the many extraordinary adaptations by which the fæces are removed, little is recorded. Yet the devices for the removal and disposal of fæces in nesting birds are among the most remarkable adaptations found in the entire group; the hygiene of the nest is a matter of some importance, and doubtless has a definite survival value for the species.

Pycraft (2), in his 'History of Birds,' has summarized the available information on this subject as follows : " the sanitation of the nest has also to be attended to, though fortunately nature has provided a method by which this task is simplified, inasmuch as the excrement of the nestling is enclosed within a capsule so that it may be picked up in the beak of the parents, and carried away to a distance from the nest before being dropped . . . Occasionally, whether as a consequence of a depraved appetite, or for some as yet unexplained reason, this excrement is swallowed . . . As the young grow and gather strength, however, they relieve the parents of this work of sanitation by raising the tail above the edge of the nest and expelling the fæces. Among the kingfishers of the Genus *Alcedo*, which are hatched in holes, the cloaca appears to be unusually muscular, since the nestlings are enabled, by turning the hinder end of the body towards the entrance to the nest, to expel the fæcal matter clear of the tunnel."

But the hygiene of the nest has reached a greater development than this, and adaptations for the removal of fæces are often extremely specialized. My attention was first drawn to the methods of fæces disposal among birds when photographing a bird at the nest, and the observations contained in the present paper are the accumulated result of several years of intermittent observation of wild birds.

The nest of the Speckled Warbler (*Chthonicola sagittata*) is dome-shaped, with the entrance at one side, and is situated upon the ground. When watching the birds from a " hide " at a distance of only a few feet from the nest, it was observed that on each occasion, after feeding the young, the parent bird paused at the entrance for a few seconds, and that frequently a whitish body was thrust forward, seized by the waiting bird, carried off, and dropped some distance away. This proved to be the fæcal mass. It was observed that though the young faced the entrance of the nest when being fed, the fæces appeared always to be thrust towards the entrance and within reach of the waiting adult. Moreover, fæces were normally passed only after feeding, and were never seen in the nest.

Better opportunities for observations were afforded at the open, shallow nest of the Grey Butcher Bird (*Cracticus torquatus*). Both birds fed the young, making several visits each hour during the daylight. At each visit, after feeding a nestling, the adult was observed to pause expectantly on the edge of the nest. Frequently the body of the nestling that had received the food would be convulsed by a violent muscular spasm, the cloaca would be thrust upwards, and a mass of fæcal matter, enclosed in a whitish mucilaginous covering, was passed, and as it left the cloaca of the nestling was received into the bill of the waiting bird. Pl. I. is a photograph, taken at the nest, showing the removal of the fæcal mass as it leaves the cloaca of the nestling. On each occasion the fæcal mass was voided by the nestling that was last fed. The head, with mouth open, was raised and waved in the air, but immediately it had received food the bird subsided into the nest, and muscular action accompanied by a violent peristalsis followed. This was observed repeatedly, indicating a definite periodicity in which feeding supplied the stimulus for peristalsis. Subsequent examination of the fæcal mass of nestling birds (Pl. II. figs. 1, 2, & 3) supports this view, for the mucilaginous investment is more than a mere provision to enable the adult to transport the mass, it is a mechanism by which the fæces may be accumulated in the alimentary tract for a considerable period and enveloped in a gelatinous covering, to be expelled by muscular action stimulated by feeding.

Normally, unless the birds were disturbed, the fæces never touched the nest. Similarly, no food was ever permitted to accumulate in the nest ; if the nestling did not take the food at once, or if the adult were disturbed or agitated, it disposed of the food immediately by swallowing it.

Subsequent experience shows that the removal by the adult of the fæcal mass as it leaves the cloaca of the nestling, immediately after feeding, is a phenomenon general among the Passeriformes, although there are exceptions, some of which will be noted below. Among Australian birds in which the removal of the fæces in this way has been observed, the following species may be cited : Yellow-breasted Whistler or Thickhead (*Pachycephala pectoralis*), White-fronted Chat (*Epthianura albifrons*), Coachwhip Bird (*Phosphodes olivaceus*), Variegated Wren (*Malurus lamberti*), Striated Tit (*Acanthiza lineata*), Buff-rumped Tit (*A. reguloides*), and Superb Lyrebird (*Menura novæhollandiæ*).

In each case the fæcal mass is carried away in the bill of the bird and dropped at some distance from the nest ; more rarely it is disposed of by swallowing, as was observed in the Golden-breasted Thickhead. A male of this species was observed to remove the fæcal vesicle in the usual way, and then to swallow the mass. This is not, however, the usual method of disposal, and is probably due in part at least to agitation—under stress of which a bird may frequently swallow anything that it may be carrying, including the food that it has brought to the nest for its brood.

But the scrupulous care that birds devote to the cleaning of their nests is not confined to the removal of fæcal matter, and most birds are intolerant of any foreign object placed in or upon their nests, and will remove it immediately. As soon as the young emerge from the eggs the sitting bird removes the egg shell, not merely throwing it out from the nest, but usually taking it in her bill and dropping it at a considerable distance—frequently at least 50 yards—from the nest, as she does, later, with the fæces. A Striated Tit, brooding on eggs, would take feathers that had been deliberately pulled out, wholly or partially, from the lining of her nest, and would replace them carefully in position, or remove them altogether.   During her absence a feather was removed from the inner lining and placed near the entrance, on the outside of the nest.   On her return the bird took the feather in her bill and replaced it in the inside lining.   This was repeated several times, until finally she carried the feather off and dropped it some distance away.

THE LYREBIRD.

The Lyrebird's nest is always maintained in a scrupulously clean condition, and here the adaptation for the removal by the adult of the nestling's fæces, is even more specialized than in the birds already discussed.   The nest of the Lyrebird is a large dome-shaped structure with side entrance, generally placed on a hillside overlooking a gully with running water.   Only a single egg is laid, and the whole work of incubation, and the feeding and care of the nestling, falls upon the female alone.

The chief food of the Lyrebird consists of a small amphipod that is abundant in the habitat of the bird, as well as earthworms, centipedes, insects, larvæ, etc. The mother may leave the nestling for long periods when she is foraging, and during these excursions she accumulates the food in a temporary gular " pouch " which may readily be seen when she approaches the nest to feed the chick. After the young has been fed, the cloaca is directed to the entrance to the nest and a very large fæcal mass is voided and is received directly from the cloaca into the bill of the mother (Pl. III. fig. 1).

At one nest at Gordon, in New South Wales, the fæcal mass was carried by the bird to the creek twenty or thirty feet away and dropped in the water. Twenty-one similar masses were counted in the same pool, so that it was evidently the habit of the bird to deposit the fæces in the same place on each occasion.   During the greater part of a day that was spent at this nest only one of these masses was voided.

STRUCTURE OF THE FÆCAL MASS.

Examination of the fæcal vesicle of the young Lyrebird showed that it not only represents a special adaptation for removal by the parent bird, but that it also provides a means for storing the fæcal matter over the comparatively lengthy period during which the adult may be absent in search of food. The fæcal mass shown in Pl. II. fig. 1 was two inches in length, contained in a white, opaque, mucilaginous investment, which surrounded a central mass of fæces.   This central mass was spirally coiled and was held in a mucilaginous matrix, and must have represented the contents of the bowel for a distance of more than 12 inches (Pl. II. fig. 2).

Nest hygiene is probably a factor of considerable survival value to the Lyrebird, both on account of its habitat and its nesting habits.   It inhabits dark mountain gullies in regions of high rainfall, and the nest, which is lined warmly with down feathers, would provide a natural incubator for injurious organisms.   Moreover, the bird lays but a single egg and the chick remains in the nest for a long period.

FINCHES.

It was noted above that while in most of the Passeriformes the adult birds removed the fæces of the young from the nests, there were exceptions to this. None of the Australian finches—of the family Ploceidæ—of which I have experience makes any attempt to remove the fæces from its nest. The Red-browed Finch (*Ægintha temporalis*) builds a domed structure of grass etc., and before the brood is able to fly a very large mass of fæcal matter accumulates in the nest.

HERONS.

In addition to these Passerine birds there are a number of other birds, notably the Herons, Ibises, Egrets, and Spoonbills, that make no attempt whatever at the cleaning of their nest. In the White-fronted Heron (*Notophoyx novæ-hollandiæ*), which may be regarded as typical of this group, the fæces of the young are voided over the structure of the nest and its surroundings, which are often completely whitened before the young leave the nest.

KINGFISHERS.

Pycraft (1) has stated that in Kingfishers of the genus *Alcedo*, that nest in holes, the young are able to expel the fæcal matter clear of the tunnel by muscular action.

Several Australian Kingfishers nest in the holes of trees, and in these, too, a special adaptation exists for the expulsion of the fæces from the nesting hollow by the nestling itself, without the intervention of the parents. Of the tree-nesting species of Australian Kingfishers, observations have been carried out chiefly on the Laughing Kingfisher (*Dacelo gigas*). In one nest of this bird containing three fledgelings the nesting cavity was in a hollow limb, the maximum depth to the back wall of the chamber was 12 inches, and the floor of the hollow was almost level with the entrance. The young birds merely turned the cloaca towards the entrance and expelled the fæces through it by muscular action.

One of the birds from this nest was removed and placed on a large sheet of white paper. When voiding the fæces the head was depressed, the cloaca elevated, violent muscular contraction followed, and the fæcal matter was expelled with such force that it reached a maximum distance of almost 32 inches. An actual photograph of this is shown in Pl. V. It will be noticed that very little of the fæcal matter touched the paper for a distance of 10½ inches from the cloaca, and that the greater part fell between 12 inches and 32 inches (Pl. III. fig. 2). As the extreme breadth of the cavity was only 12 inches, this would have been thrown clear of the nest. The process was aided by the consistency of the fæcal matter, which was more fluid than in any of the birds hitherto discussed, and mucilage, which was abundant in the fæces of other birds examined, was conspicuous by its absence. The contrast with the fæces of the young Lyrebird is especially marked, and is instructive, since the birds are fed upon a closely similar diet.

LEPIDOPTEROUS LARVÆ IN PARROTS' NESTS. (Pls. IV. & V.)

Probably the most remarkable of all adaptations for the removal of fæces is that exhibited by of the Golden-shouldered Parrot (*Psephotus chrysopterygius*), which inhabits a restricted area on Cape York Peninsula, North Queensland. The Golden-shouldered Parrot drills a tunnel in a termitarium or "white ant" mound, at the end of which it constructs a nesting chamber. No material is added, but the eggs are laid on the bare ant bed (Pl. VI. figs. 1 & 2). The disposal of the fæces of the nestlings is entirely carried out by a colony of lepidopterous larvæ that establishes itself in the nesting chamber. The presence of these remarkable larvæ was first discovered by Mr. W. McLennan (1) in 1922, and his specimens were subsequently described by Dr. Jeffries Turner (3) as belonging to a new genus and species, *Neossiosynœca scatophaga* (Family Œcophoridæ). In June 1929 I examined a large number of nests of the Golden-shouldered Parrot on Cape York Peninsula, and in many of these found the commensal larvæ described by McLennan. These larvæ were 0·75 of an inch in length. The young had already flown from most of the nests examined, and the larvæ had spun cocoons which were aggregated in masses having a superficial resemblance to honeycomb—the cocoons lying in the horizontal plane, and embedded in the outer wall of the nesting chamber so that the imago would have access to the exterior on emergence. The cocoons measured 1·25 inches in length, and consisted of an outer covering of relatively coarse material and an inner cocoon of finer silk which was separated from the outer covering, and which, in all the specimens that I examined, contained larvæ that had not yet pupated. Pl. V. shows the masses of cocoons, also the individual cocoons, some of which have been opened to show the larvæ enclosed within the inner envelope of fine silk.

In the nests from which the young birds had flown the particles of ant bed on the floor of the cavity were loosely bound together by a tangle of silken threads, which had apparently formed the refuge of the larvæ. In none of the nests in which eggs were still present was any sign of these larvæ noted.

Extracts from McLennan's original field notes on the discovery of these larvæ and their disposal of the fæces of the nestlings of the Golden-shouldered Parrot are quoted below :—

6.5.22. " . . . Reached the place where I found *Psephotus* nest containing well feathered young on 1.5.22 . . . found it contained seven young, one of which was still in the downy state with scarcely a feather showing. Could not make out what was the matter with the bottom of the nest ; it was heaving and undulating in constant movement ; little heads were flickering out and in through trap door-like openings. At first I thought it was alive with maggots ; a few seconds later a number of larvæ came right out of the bottom of the nest and started to eat up the excreta which one of the young *Psephotus* had just voided. I then saw that the larvæ were caterpillars. In a couple of seconds the excreta was eaten up and the caterpillars at once disappeared into the bottom of the nest through the opening from which they emerged. I sat and watched them for some time ; every now and again one or two cater-pillars would come out and go exploring round the bottom and sides of the nest. The young birds were frequently voiding excreta, which at times would get all over their feet and tail-feathers. Instantly the caterpillars would swarm out and devour it, eating up every scrap even off the feet and feathers of the young ; thus the young birds were kept scrupulously clean. The young birds did not take any notice of the caterpillars . . . I now examined the bottom of the nest. The trap door-like openings proved to be the mouths of cocoons in which the caterpillars lived, and these cocoons were lightly bound and matted with web, the interstices being filled with the excreta of the caterpillars and the fine chipped dirt in the bottom of the nest. . . Found another *Psephotus* nest four feet from ground in magnetic termitarium ; it contained half-fledged young with the attendant colony of scavenger caterpillars . . . The next nest . . . contained one fully-fledged young which flew out and away when I went to look into the nest. . . . The caterpillars in this nest were just starting to pupate ; a number of them had clustered their cocoons closely together and bored through to the outside of termitarium at the thinnest part of the side of the nesting-chamber."

7.5.22. " . . . I found a nest of *Psephotus* seven feet from the ground in a big magnetic termitarium ; small young not long hatched, and attendant colony of scavenger caterpillars . . . A couple of miles further on I found a nest of *Psephotus*, from which young had only just flown. The caterpillars were just starting to cluster together at the thinnest wall of the nest."

13.5.22. " . . . several old nests of *Psephotus* with the clusters of empty cocoons of the caterpillars in each. Only one nest from which the young had flown this season, no sign of any caterpillars in this nest ; the bottom of the nest was literally caked with dried excreta of the young birds. Almost every spire-shaped termitarium bore traces of the birds, none of them recent. Nearing camp again a female *Psephotus* flushed from a nest in a spire-shaped termitarium ; on looking in I could plainly see four eggs, which appeared to be slightly incubated . . . Four moths in the chamber, two of them coupling ; evidently the moth responsible for the caterpillars."

It will be evident from these extracts from McLennan's notes that the presence of the larvæ in the nests is not a chance one, but that their association with the Parrots is definitely on a commensal basis. That the Parrots have become dependent upon the larvæ for the removal of the fæces of the young is shown in the finding of a nest, on 13.5.22, from which the young birds had recently flown, but in which no trace of larvæ was observed. This was the only instance in which an accumulation of fæcal matter in a nest of this Parrot was noted either by McLennan or by me. Further, these records show that the breeding period of the caterpillars coincides with that of the Parrot. I have recorded the fact that no larvæ were observed in any nests which still contained eggs ; but McLennan noted the presence of moths *in coitu* in a nest containing eggs of the Golden-shouldered Parrot, in which incubation had already commenced. The period that would necessarily intervene before the eggs of the moth could be deposited and the larvæ emerge might well be expected to coincide with the hatching of the eggs of the Parrot in the same ant bed. In each case in which the young birds had flown, and, in consequence, the food supply of the larvæ terminated, the caterpillars had attained maturity and were pupating, or were about to pupate.

Thanks are due to Mr. W. H. T. Tams, of the Entomological Department, British Museum (Natural History), for kindly supplying the photographs of the male and female *N. scatophaga* reproduced on Pl. V.

REFERENCES.

(1) McLennan, W. 1921–22. Unpublished Field Notes on Expedition to North Queensland.

(2) Pycraft, W. P. 1909. A History of Birds, p. 229.

(3) Turner, A. Jeffries. 1923. Trans. Entom. Soc. pp. 170–175.

32

Reprinted from *Auk* 52:257–273 (1935)

HELPERS AT THE NEST.

BY ALEXANDER F. SKUTCH.

Plate XII.

IN THE great majority of bird species whose nesting has been carefully studied, each pair build their nest and rear their offspring without help from others of their kind. This, indeed, is almost a corollary of the theory of Territory, which teaches that each breeding pair occupy a definite nesting area from which they vigorously expel other individuals of their own species. While the concept of territory in bird life has done much to stimulate and give definite direction to bird study, as so often happens in the first enthusiasm of working out the details suggested by a fertile scientific theory, it has resulted in a tendency to neglect the opposite side of the story. There are many species in which the mated pair are not so exclusive in their territory, and as a result of this, coupled with other peculiar circumstances, receive more or less assistance in the duties of the nest. The number of recorded cases of helpers at the nest which have come to my notice is relatively small, but this appears to be, at least in part, because their discovery requires a more concentrated attention than is commonly devoted to studies of nesting birds. The relatively few species which are known to have helpers at the nest are scattered among the families and orders of birds in a manner which suggests that the custom of giving and receiving aid in the rearing of a family is not restricted to a few unusual groups, but is of widespread if sporadic occurrence.

It is not my intention at the present time to attempt an exhaustive survey of the cases of birds helping at others' nests which have been recorded in the literature, but rather to relate briefly certain instances which I have personally observed. But I am not aware that anyone has classified the various degrees of outside assistance which the mated pair may receive in their breeding operations, and before proceeding to particular cases I should like to attempt such a classification, the better to understand where my own examples fall.

Assistants may be classified as: I, Juvenile Helpers; II, Unmated Helpers; III, Mutual Helpers. These groups, in the order named, represent an increasing degree of sociability during the breeding season.

Juvenile Helpers. In some species the young of the first brood are still unable to shift for themselves and are dependent upon the care of their father while the mother incubates her second set of eggs. Sometimes, in species of which the male takes no part in the duties of the nest, the mother may feed her offspring of the first brood during the intervals of warming

399

her second set of eggs. In the mountains of Guatemala I once watched a female White-eared Hummingbird (*Hylocharis leucotis*) who fed her full-fledged offspring, himself already able to poise before the blossoms, during her recesses from her second nest of the season, in which incubation was in progress. Sometimes, when the young of the first brood can forage for themselves before preparations for the season's second nesting have been completed, the adults prefer to be alone during this period. In the valley of the Rio Motagua I watched a pair of Lichtenstein's Orioles (*Icterus gularis*) drive their full-fledged children of the first brood away from the new nest which the female was engaged in building. In such cases as this it is not likely that the young birds will assist in the care of the later brood. But other species tolerate the presence of their young, after they have become self-supporting, in the vicinity of a later nest. They are more sociable than the Lichtenstein's Orioles, and are often rewarded by receiving the youngsters' aid in the care of subsequent broods of the same season.

Lord Grey[1] gives an attractive picture of the family life of the British Moorhen, a close relative of the Florida Gallinule of America. A family came to enjoy the bread crumbs which he threw to them at Fallodon. The parents picked up the crumbs and passed them to their offspring of the first brood, born in May, and these in turn placed them in the bills of their tiny, downy younger brothers and sisters, hatched in July. There seems to have been more formality than intelligent coöperation in the actions of these birds, but at least the desire to help was there. Those who have read Mills'[2] 'Love Song of Little Blue' may recall that the young Mountain Bluebirds, raised about his cabin in the Rockies, helped the parents satisfy the hunger of their younger brothers and sisters of the second brood. Young Western Bluebirds,[3] and Barn Swallows[4] recently from the nest, have been seen to aid their parents in the care of later broods during the same season. In Honduras I watched a young Groove-billed Ani (*Crotophaga sulcirostris*), slightly over two months old, take a part almost equal to that of his parents in feeding and protecting the latters' younger family. The records of juvenile helpers which I have come upon are not numerous, but from the scattered position of these species among the families of birds, I suspect that the habit is far from uncommon.

Unmated Helpers. Juvenile helpers are of course unmated, but they form a very distinct and easily recognized class, and I prefer to limit the term "Unmated Helpers" to assistants born during the previous nesting season. They may be young birds, outwardly mature but sexually still immature; or sexually mature individuals who, because of the excess of one sex, or from

[1] Viscount Grey of Fallodon, The Charm of Birds. 1927.
[2] Mills, Enos A., Bird Memories of the Rockies. 1931.
[3] Finley, W. L., American Birds. 1907.
[4] Forbush, E. H., Birds of Massachusetts III. 1929.

Plate XII.

Left: Nest of Black-eared Bush-tit with Four Male Nestlings Posed on Outside (They Do Not Naturally Assume this Position). May 23, 1933.

Right: Nest of Banded Cactus Wren. June 14, 1933. Both at Tecpan Guatemala.

other causes, can not find proper mates; or individuals old enough to breed who from accident or disease are sterile. Since in individual cases it is not always possible to decide why a particular bird is unmated during the breeding season, it seems best to include in this class all helpers of approximately a year or more of age. In three species which I have watched, the presence of unmated helpers at the nest seems to be the rule.

Among the Brown Jays (*Psilorhinus mexicanus*) the helpers are largely if not entirely yearling birds, still presumably sexually immature. The status of the helpers among the Banded Cactus Wrens (*Heleodytes zonatus*) is difficult to determine; they are at least a year old, and in appearance indistinguishable from the breeding birds. Among the Black-eared Bush-Tits (*Psaltriparus melanotis*) there is, in certain regions at least, a great excess of males who, not being able to secure mates, help care for the nestlings of other birds. A numerical excess of females seems more likely to give rise to polygamy than to produce helpers at the nests of other pairs. So among Red-winged Blackbirds and Meadowlarks, which normally appear to be monogamous, the occasional presence of an excess of females may result in polygamy. Among species in which the females are normally greatly in the majority, as with Oropéndolas (*Gymnostinops montezuma* and *Zarhynchus wagleri*) and Great-tailed Grackles (*Cassidix mexicanus mexicanus*), polygamy seems to be the rule.

Mutual Helpers. Mutual helpers are breeding birds which coöperate and assist each other in the care of their respective families. Among them we find all degrees of coöperation from casual assistance in repelling a common enemy to the complete sharing of all the duties of the nest. Probably all birds which nest in colonies unite to drive away a Hawk or any other undesirable intruder. Indeed, such service in a common cause is not restricted to members of the same species, for birds of different species will often combine forces in repelling an unwelcome stranger from the neighborhood of their nests. With Oropéndolas and Great-tailed Grackles the situation is a little more advanced, for the males of the colony, who do nothing to help the females directly at their nests, constitute a standing guard, ever ready to give the alarm on the event of danger, and to drive undesirable visitors from the colony. In this last duty the *clarineros*, as the male Grackles are called, are far more active and courageous than the larger male Montezuma Oropéndolas.

Birds which build apartment nests, such as the Green Parrakeet (*Bolborhynchus monachus*) of Argentina and the Sociable Weaver-bird (*Philetærus socialis*) of Africa, show a still higher degree of sociability and group coöperation during the breeding season. Among other species a number of adults take a common interest in the young. The Emperor Penguins (*Aptenodytes forsteri*) undertake in common the task of keeping warm the downy young, hatched during the tremendous cold of the Antarctic winter, and are said to be so eager for the possession of the little birds that the members of a colony engage in desperate struggles to obtain them, and in these engagements the chicks are frequently injured or even killed by their would-be benefactors. The Adelie Penguins (*Pygoscelis adeliae*) incubate their eggs and attend their nestlings in the normal manner (that is, by pairs); but as the youngsters grow older and require more food, satisfying their needs becomes a difficult problem, for the nesting area is often at a considerable distance from the sea, whence comes all food in the inhospitable Antarctic, and the parents must be gone long hours while thay laboriously walk back and forth from the water's edge. In order to release more of their time for the important foraging expeditions, the young, when they can leave the nest, are led to nurseries where a few parents can stand guard over the children of many families, while the rest busy themselves in filling the many hungry mouths. Unmated male Adelie Penguins are an actual menace to the youngsters, and must be driven away from the nurseries by the parents.[1] In the social problem of how to keep idle, unmated birds out of mischief, passerine species like the Brown Jay and the Bush-tit have made great advances over the primitive Penguins.

Complete coöperation or communism, involving all stages of the nesting cycle from the building of the nest to the care of the young, is best exemplified by the Anis. The Groove-billed Anis (*Crotophaga sulcirostris*), which I have watched in Honduras and Guatemala, actually pair at the beginning of the breeding season. The pair may build their own nest and rear their own offspring in the manner of most birds, or two or three pairs may join in the construction of a common nest, in which the eggs, generally four to each female, are laid side by side. Then all the parents, both male and female, take turns in warming the eggs, and later all join in feeding the nestlings. Since I hope later to present a fuller account of the home life of these interesting birds, I shall not at present take space for fuller details.

Helpers at the nest, it may be noted in passing, are not always of the same species as the owners of the nest, and are not always wanted. Forbush records the case of a male Bluebird who busied himself feeding the nestlings of a pair of House Wrens, much to the distress of the latter, until his mate hatched out his own offspring. He also tells of a male Scarlet Tanager who helped feed a nestful of young Chipping Sparrows while his mate incubated. Robins, both the European and the American species, are often ready to place a morsel into the open mouth of any helpless young bird they encounter, not necessarily of their own kind.

With this hasty glance over the field by way of introduction, we may proceed to the more particular object of the present paper, the discussion

---

[1] Levick, G. Murray, Antarctic Penguins, London, 1914.

of certain interesting cases of coöperation which I have had the good fortune to observe in Central America.

THE CENTRAL AMERICAN BROWN JAY

Psilorhinus mexicanus cyanogenys Sharpe.

Two years ago I lived on a banana plantation among the foothills of the Sierra de Merendón, on border territory then claimed by both Guatemala and Honduras, which formed a sort of independent buffer province between the two republics, but has since been awarded to the former. Among the most conspicuous and most interesting of my bird neighbors were the Brown Jays. The family of Jays belongs primarily to cooler regions and is poorly represented in the lowland Tropics, but these dwellers in the hot lands had lost none of the noisy, restless habits which characterize their relatives everywhere. The way they scolded when, in walking through the banana groves, I interrupted their feasts upon the rich nectar of the white banana flowers, left no doubt in my mind that they were near cousins of the northern Blue Jay, although their decidedly larger size and brown, white-vested dress, seemed to belie their blood relationship. Their harsh cries were among the very first notes of purely diurnal birds to greet my ears in the dim light of early dawn; at midday, when all nature drowsed under the rays of a vertical sun, and most other birds were in silent seclusion, they seemed to go out of their way to protest my passage through their haunts. Their lively flocks foraged in all sorts of open and semi-open country; among the banana plantations; along the lagoons; in the bushy pastures of the foothills, where scattered trees remained; in the inextricably entangled second growth which soon takes possession of abandoned clearings of all kinds; but I never met with them in heavy forest.

The Brown Jays possessed one peculiarity which convinced me would make them particularly interesting to watch. Their bills, feet and bare rings surrounding the eyes were not all of the same color, as is the case with most birds, but were so variously marked with yellow and black that it seemed that I should be able to recognize individuals—and the difficulty of recognizing individuals, among other species, is one of the chief handicaps which face the serious bird-watcher. I devoted most attention to the bills, which proved to afford the best recognition marks. Some of the Jays, at the beginning of the breeding season, had bills which were entirely bright yellow; the bills of others, perhaps the majority, were tipped or streaked with black, but hardly two were pied in exactly the same manner. Still other bills were uniformly black. Later I learned that the nestlings' bills, feet and orbital rings are uniformly yellow, and that they turn black with age in an irregular fashion, apparently taking two years or more to become entirely black. Most of the pied-bills which I found at the beginning of the nesting season were probably yearling birds. On breeding birds the parts in question are often entirely black, but the only generalization which it is possible to make is that with breeding individuals these parts *average* far blacker than with the non-breeding birds who help them.

The Brown Jays placed their nests in the crown of a banana plant, just above the ripening bunch of fruit, or else in trees standing in the open or amid the second growth. The former I could reach by lashing together bamboo poles to form a ladder sixteen or eighteen feet long, the latter were

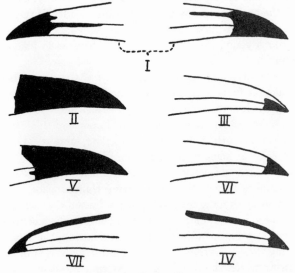

BILLS OF BROWN JAYS.

White areas in the cut were yellow on the bills. I, shows both sides of the bill of the female; an unusual case of variation. II, is the male, evidently her mate. III–VII are bills of the five helpers; IV and VII, so similar that they were taken for the same bird until seen together. (From field sketches, Alsacia Plantation, near Los Amates, Dept. of Izabal, Guatemala, June, 1932).

usually so high, and near the ends of such long, slender branches, that prudence suggested they be left alone. One, in a willow tree beside a lagoon, was against the trunk and could be reached by a hot climb. The nest is a bulky pile of coarse sticks, many of them thorny, which forms the foundation for a shallow cup, neatly constructed of fibrous roots, where repose usually three blue-gray eggs, thickly covered with fine brown speckles. Nesting began in February, but the height of the breeding season was from March to May, while the last nestlings took wing in June.

Male and female share in the construction of the nest. Early in April I watched a pair building a nest high up in a wild fig tree which stood in a hedgerow between two banana plantations. Although both sexes took part in the task, they were unable to coöperate closely, and one, after adding a stick or a root, would sit in the unfinished structure for several minutes and call in a loud, complaining voice, as though berating the negligent mate. As is often the case with mated birds, the bills of both were entirely black and I could not distinguish them, but from the behavior of other pairs, of which one member had some yellow on the bill, that I watched later, I strongly suspect that the noisier of the two was the female. When a pied-billed bird, probably a yearling, alighted on the nest and seemed interested in it, the black-bills did not take offense and drive it away, as other kinds of birds resent the presence of any intruder in the vicinity of their nests. This fact alone promised to have interesting consequences. Later, on a single occasion, I saw a young pied-bill bring a stick to a nest which a mated pair were building.

As the season advanced, I discovered more of the domestic economy of the Jays. Like the female American Goldfinch, the female Brown Jay alone warms the eggs, while the male at intervals brings her food. Like the Goldfinch, too, she calls when she is hungry, but in place of the former's melodious little tinkle she utters a loud, far-carrying, unmusical *pee-ah*, which seems a complaint at being neglected. Once I had learned to recognize this hunger call, it led me to nests a quarter of a mile away, and I had no lack of Brown Jay families to watch. After sitting continuously for an hour, or sometimes as many as four, the female flies off when she has taken the food her mate has brought her, and he remains standing like a sentinel upon the rim of the nest, or on an adjacent branch, guarding the eggs but almost never warming them. Here he awaits until, after a recess of ten or fifteen minutes, during which she has managed to gulp down more food than he has brought her all morning, she returns to settle once more upon her treasures.

One morning in April, while walking along a sharp ridge among the foothills, I discovered a Jay's nest in the top of a small tree projecting above the vine-smothered second growth which covered the steep slope below me. The female was sitting, and even while I paused looking down at her a bird whose bright yellow bill was only tipped with black flew up, protested my presence rather mildly for a Brown Jay, then perched on the rim of the nest and gave her a morsel of food. A few minutes later the yellow-bill brought another offering to the bird on the nest. Still, I doubted greatly whether a Jay with so little black on his bill could be mated, and resolved to await further developments. I seated myself in a grassy clearing near the summit of the slope, whence I could look down over the tops of the vine-entangled bushes, and although I was in full sight, only fifty feet away, my presence seemed to make little difference in the activities of the birds. They have not yet had so many sad experiences with man as have their kin in more densely settled countries.

I had waited nearly an hour before a black-billed Jay appeared. Although there was nothing visible in his bill, his throat was outswollen with the food he carried to the nest. After delivering it he went away, but before long he returned with another offering. At his approach this time the female cried out, rose from the nest to greet him, took the proferred morsel and flew off with it, leaving him standing on a twig beside the nest. Here he remained on guard until the female silently returned, after a quarter of an hour's absence, when he flew silently away. Then I felt convinced that this was the mate of the sitting bird, for the yellow-bill never remained on guard. Later it developed that at least two young birds were bringing food to this female, as much, if not more, than her mate.

After the eggs hatched the parents and helpers left the tender nestlings unguarded while they foraged, with the result that some enemy found them in their exposed position and made an end of them. Then the parents tore apart the ill-fated nest and used its materials in the construction of another a few hundred feet away. In due course the eggs were laid and hatched, and the interesting work of attending the three nestlings began. I erected an umbrella blind for concealment, at a point on the precipitous hillside where I was on a level with the nest in the tree down the slope, and spent hours in its shelter, making a sketch of the bill of every bird who brought food to the nestlings. The father's bill had only a small patch of yellow at the base, the mother's bill was more than half yellow, the bills of the helpers varied from yellow slightly tipped with black to black with a little yellow at the base. When I had spent nearly twelve hours in the blind, and thought I could recognize all six of the birds whose bills I had sketched, no matter which side was turned toward me, two Jays with almost identical markings brought food to the nest, then remained standing side by side on its rim, and I knew there were seven (see figs. p. 262).

The parents and their five helpers continued to attend the young at least until, at the age of twenty-three or twenty-four days, they flew from the nest. While the nestlings were still small and unfeathered they were guarded almost constantly, for each of the attendants, upon delivering the food it had brought, remained standing upon the rim of the nest until another arrived to take its place. The mother alone brooded her offspring, but sometimes, when the sun shone directly into the nest, the helpers stood over it in such a fashion as to shield its tender occupants. At times, when a bird flew up with a bill-ful of food, the Jay standing guard would ask for it in a pleading voice, or possibly even try forcibly to snatch it from the new arrival, and if successful in obtaining a portion of the bounty would pass it to one of the little birds in the nest. This was very much like the behavior of Lord Grey's Moorhens, except that sometimes a parent intermediated between a helper and the nestlings, sometimes a helper took food from a parent and put it in the gaping mouth of a nestling. Rarely a bird abused the privilege of delivering food another had brought, and carried it off for his personal consumption. The helper who most often did this was one of the most faithful attendants, so I felt inclined to forgive him his misconduct. Once, too, he saved one of the nestlings a choking by flying off in this manner with a particle plainly too large, which another bird had brought.

At every one of the five Brown Jay's nests that I watched, I found at least one helper. The nest in the willow tree had three; two aided the parents at a nest in a West Indian birch, growing among the giant canes on a stony flood plain of the Rio Morjá. At the willow tree nest one of the helpers was far more zealous than the parents in defending the young. When I climbed up to look in at them, this pied-bill ventured within a yard of my head, calling excitedly, and finally, lacking the courage to administer the punishment I seemed to deserve, alighted on one of the huge leaves of a nearby banana plant and ripped it into shreds as a substitute. These helpers seem in most cases to be yearling birds who will not have nests of their own until they are two years old. Often, no doubt, they are last year's children of the mated pair in whose duties they assist, but this is not always the case, especially where there are five helpers, for Brown Jays as a rule raise no more than three fledglings each year.

THE BLACK-EARED BUSH-TIT.

Psaltriparus melanotis melanotis (Hartlaub).

The following nesting season found me in the high mountains of Guatemala, only two hundred miles distant from the plantation where I had watched the Jays, a trifle farther to the south, but climatically and biologically in another zone. Here, between eight and nine thousand feet above the sea, where the dawn which follows a clear night from November to April reveals the fields white with frost, the forests are of oaks, pine and alder, the latter not a bush, as in the north, but a tall tree. Here grow violets, buttercups, and the same self-heal we know at home, but as proof that we are well within the Tropics there are begonias, fuchsias which form small trees, and epiphytic orchids which burden the branches of the oak trees. The birds are in part such as one might expect to find among oaks and pines: Bluebirds, Whip-poor-wills, Flickers, Hairy Woodpeckers, Towhees and Blue Jays, living on intimate terms with such distinctly tropical forms as Woodhewers, Trogons, Toucans and Motmots.

There is one bird which seems to belong among oaks and pines for which you may search in vain—no Chickadees live in Guatemala. Their place is occupied by a close relation, the Black-eared Bush-Tit, a tiny gray birdling who, instead of the black crown and throat of the Chickadee, wears his black patches covering the sides of the head. His mate is marked by much smaller areas of black which are confined to her ears, while her cheeks are gray. The male's eyes are black, the female's yellow. During most of the year these lively little birds travel through the more open woods and bushy pastures in flocks of from a dozen to two dozen, maintaining a constant, low, lisping conversation, and exhibiting all of the agility of a Chickadee in clinging to the tips of the twigs in every conceivable position, while they pluck from the foliage the small insects upon which they subsist. It is noteworthy that in these flocks the black-faced males far outnumber the gray-cheeked females, perhaps by four or six to one. In one flock which foraged in the garden under conditions very favorable for observation, I counted eleven males and only one female.

In March, while the nights were still chill and frosty, I watched a pair of the Bush-tits building their cozy nest. The site they had chosen was in the top of a thorny bush (*Solanum mitlense*) just beneath the large purple flowers which clustered at the ends of its branches, high on a bushy mountain-side, where from my place of concealment up the slope I could look out across the high plateau of Chimaltenango, brown and sere after the long dry season, to the three great volcanic cones rising in the east—towering Acatenango, its sister Fuego with a thin wisp of vapor rising from its barren summit, and the perfect cone of Agua.

The nest was a pear-shaped pouch, suspended by its upper end from the twigs, and fashioned of gray foliaceous lichens neatly joined together with cobweb. The top of the pouch was hooded over, leaving only a small circular aperture facing the side, through which the birds were industriously carrying in for the lining bits of such soft and downy material as they could find—tufts of spider cocoons, of the woolly covering of the leaves of the bushes among which the nest was hung, and other kindred substances. Male and female took equal shares in hunting out the down and bringing it to the nest, but the female was far more careful than her mate in arranging this material and in shaping the structure. Their work progressed with much fine twittering, and it was interesting to find that the male was by far the more hesitant in approaching the nest while I stood in plain sight. At a second nest, however, the situation was reversed, and the female was considerably more timid than her mate.

This and the other nests I afterwards found each contained in due time four tiny white eggs, no larger than those of the average Hummingbird. I never dared try to remove them for inspection, but by carefully bending back the hooded top and peering down into the interior with one eye, I could just manage to count them as they lay on their downy bed at the bottom of the pouch. Even before the eggs were laid, male and female slept together in the nest. During the period of incubation, the pair slept together in the nest each night, a habit which, so far as I know, is shared only by the Blue-throated Motmots (*Aspatha gularis*) who nested in the banks along the roads. In the morning, long after the Thrushes had sung their dawn chorus, the Jays had begun to squawk, and all early birds were abroad, the male emerged from his warm shelter. After snatching a bit of breakfast he returned and called, whereupon his mate came out and he re-placed her on the eggs. As the sun dispelled the nocturnal chill, they became most impatient sitters, changing about more frequently than any other birds I have ever watched. Often each remained on the eggs only three or four minutes at a stretch; ten minutes was a long session in the nest while the air was warm. If the bird covering the eggs heard others of its kind chattering close at hand, it answered in fine sibilant twitters from the nest. Often the eggs were neglected while both of the pair sought more down to add to the lining of the pouch, for like the Rose-throated Becard (*Platyp-saris aglaiæ*), the Rufous-breasted Spinetail (*Synallaxis erythrothorax*), and other birds which build very elaborate nests, they continued actively to bring material to it until the eggs hatched.

At one nest, situated in a bushy clearing in the woods, I found that two males were flying into the entrance with small tufts of down in their bills. There was a difference between the two, not in appearance, in which they were indistinguishable, but in behavior. One was careful in tucking into the fabric the bits of down which he brought, while the other often fastened them so carelessly that they were brushed out by the passage of the birds through the entrance. So far as I could determine, only one male, the more careful builder, remained to warm the eggs, and only one slept in the nest with the female. The latter was her mate, the other an unmated helper who, unable to find a partner because of the surplus of his own sex, devoted his time to assisting in the domestic duties of the owners of the nest.

Once the young hatched, after fifteen or sixteen days of incubation, their demands for food left no time for making further improvements to the nest, which had already acquired a soft and ample lining, and the helper joined the parents in bringing minute insects and green larvae to the nestlings. Now he was allowed to sleep in the nest, a privilege which was no mean compensation for his labors in its behalf, for in the thin air a mile and a half above the sea the nights are always chilly. Later at least two more bache-lors joined in the care of the four nestlings, who now, with their parents and three male helpers, had five attendants to satisfy their wants. Possibly there were more, for unlike the Brown Jays one male Bush-Tit appears, to human eyes, exactly like another, and I tried in vain to make them acquire distinguishing marks by rubbing against the red paint which I smeared on a twiglet fastened transversely across the entrance of the nest. Had they fed the nestlings no more frequently than the Brown Jays, I might never have suspected their number, but they brought their offerings with such frequency that I sometimes had all four of the males in sight at one time, and made quite certain that I did not count the same individual twice. After a few days another of the helpers began to sleep in the nest, which then sheltered nightly the parents, two helpers and four nestlings—eight in all.

Each of the other two nests I watched had a single male helper, who fed the nestlings and slept with them and their parents every night. One of these helpers took occasional turns in brooding the nestlings. Although I found only males acting as helpers, once I saw a female take an interest in a nest which was not her own. One morning when I was watching from my blind a nest in which incubation was in progress, I was surprised to see a second female accompany the mother of the nest as she returned from a recess. The stranger followed into the nest, but a minute later climbed up to arrange the down in the top, then emerged and perched in the bush close beside it. Soon the male, returning with a tuft of down in his bill, discovered her and drove her away, but she circled around and, after the departure of the aggressive master of the nest, rejoined the mistress inside. A minute afterwards both females emerged and flew away. When the strange female returned later in the morning, the male again pursued and drove her from the bush, continuing the chase among the branches of a neighboring tree, but the persistent stranger returned a third time despite her two rebuffs. Yet after the eggs hatched she did not appear to assist the parents and their male helper in the care of the nestlings. I do not know why the female stranger was not engaged in the duties of her own nest at the time—certainly not because of the scarcity of potential husbands. Possibly she herself was barren and, unable to raise a brood of her own, might have contributed to the care of the others' nestlings had she been given a more courteous reception.

In the three nests over which I kept watch, all twelve of the fledglings, once they were feathered, turned out to be males, and exactly resembled their fathers. No wonder they would have difficulty in finding mates next year! When the young birds have left the nest, at the age of seventeen to nineteen days, neither they nor their parents return to sleep in it. The downy pouch is an admirable protection against the cold nights of the dry season when the Bush-Tits nest, but it takes up water like a sponge and would make a poor dormitory during the wet season, which begins about the time the fledglings take wing; for they are raised in the brief interval of favorable weather, a scant six or seven weeks, which intervenes between the last frost and the beginning of the rains. From May or early June onward the Bush-tits retire to sleep in the tree tops, with naught but the dripping foliage to shelter their vigorous little bodies from the cold mountain rains.

THE BANDED CACTUS WREN.

Heleodytes zonatus zonatus (Lesson).

The banded Cactus Wrens which I studied in most detail were neighbors of the Bush-tits. They are giants among Wrens; slender, sharp-billed and long-tailed, the plumage of their back, wings and tail is heavily barred with blackish and gray, the white breast is conspicuously spotted with black and the belly is buffy chestnut. They travel through the more open woods in noisy family groups of usually six to a dozen individuals, and neglect to investigate no possible hiding places of their insect food. They search the ground; cling to the bark of trees like Nuthatches; pull the gray lichens from the branches to see what may be lurking beneath, like the Blue-crested Jays; move among the foliage like overgrown Warblers. Although their voices are harsh in the extreme, male and female sing duets in unison, in the manner of Wrens more gifted vocally, and what they lack in sweetness of tone they make up in animated, rollicking tempo.

In the evening the whole family retire into their sleeping nest, a roughly globular structure about a foot in greatest diameter, composed of pine needles, moss, lichens, straws, sheep's wool and the like, with a wide entrance on one side, protected by an overhanging roof. These dormitories are generally high in the trees, at the ends of slender branches where they are difficult to reach. Sometimes a single pair occupy a nest to themselves, but I have seen as many as eleven sleeping together, and all intermediate numbers. From time to time throughout the year they build fresh dormitories, no doubt in the interest of sanitation. It is most amusing to watch them arise on cold or rainy mornings. They lie abed much beyond their usual time for arising, which at best is considerably later than that of most of their bird neighbors, then come out slowly, reluctantly, one by one. Not infrequently a Wren, stepping forth to survey a world of driving cloud mist and drenched foliage amid which he must seek his breakfast, will decide that it is still too early and returns to the snug chamber for a few minutes longer. Having gone through the same painful process of emerging into the wet less than an hour earlier, I know exactly how he feels.

In February and March, the period of courtship, there was much excitement, much pursuit and singing on the wing. The Cactus Wrens built no special nests in which to raise their families, but each female laid five white eggs, either immaculate or faintly speckled with brown, in a structure previously occupied as a dormitory, which appeared old and weathered, although it was still in good repair. So long as it contained eggs and young, the mother alone occupied it at night, while the remainder of the flock retired to another of the several dormitories which were scattered about the territory. In April I devoted much attention to a nest situated forty feet above the ground in an oak tree standing alone in a bushy pasture. The female alone incubated the eggs, but her mate remained close at hand, in company with another Wren, and was always eager to join her in a duet, or follow her, singing, upon the wing, when she came forth to forage. Sometimes he went to the entrance to look in at her as she warmed the eggs, greeting her with queer, harsh notes. After the eggs hatched he brought food to the nestlings. In this work he was joined by the unmated bird who had remained in the vicinity while incubation was in progress, and occasionally looked in to see how things were going. The helper was indistinguishable from the parents, but I stuck a wad of paint-soaked cotton in front of the entrance to the nest, and the bird, brushing against it, acquired some vermilion spots on his breast by which I afterwards recognized him. He was a most faithful attendant, and with the father did most of the work of feeding the nestlings, for the mother rarely brought them food. Her duty was to keep the little ones warm, and she seldom carried an insect to them, as is customary with most birds, when returning to the nest from a recess.

I wanted very much to see whether this interesting division of labor was general among Cactus Wrens, but at the only other nest which was favorably situated for watching there were so many attendants that I could not readily distinguish the mother from her numerous assistants. Five or six birds remained in the vicinity of this nest, and at least four brought food to the five nestlings. Probably the number of attendants was greater than four, but all looked alike, and I never succeeded in keeping a larger number in view at one time, to make sure that they fed the nestlings. At the very least there were two helpers at this nest. After the nestlings could fly they slept in a dormitory nest, not far distant, along with six adults, all of whom not improbably fed them. With the Cactus Wrens it is very difficult to make sure of the number and status of the helpers at any nest, for once the birds have outgrown their fledgling plumage male and female, young and old, are identical in appearance. Possibly the helpers are yearling birds, perhaps older brothers and sisters of the nestlings they attend, who will not themselves breed until the following season.

At the ages of eighteen and nineteen days the fledglings left the nest in the oak tree, and for the next two weeks the whole family, parents, helper, and the three youngsters who survived, went to sleep every night in an old, long-abandoned nest (B) on the opposite side of the tree from the breeding nest. It was the helper who assumed the responsibility of putting the children to bed, and a most interesting time he had of it. The dormitory was difficult of approach for the little birds just out of the cradle and still rather shaky on the wing. Although there were of course twigs all around it, there was none immediately in front of the entrance from which they could easily hop inside. Somewhat before the usual time for the adults to retire, the helper called the three fledglings to the sleeping quarters. A twig about a foot below the entrance seemed at first the most promising mode of approach, but they soon discovered that to reach the entrance from this point required too much of a jump, and they could not yet fly straight upward. Then they tried the alternative of alighting on the roof and climbing down to the doorway. This, too, was no easy matter, for the edge of the roof projected well forward of the entrance, and when they clung to it they found nothing below to which they could drop. While they were trying time and time again these two equally difficult alternatives, the helper was showing them over and over how perfectly simple it was to fly up to the entrance from below, but what was easy for a grown bird was quite a different matter for a fledgling two days out of the nest.

At length one of the three, perhaps the older by a day, succeeded in effecting an entrance by way of the roof, clinging precariously and almost losing its hold as it came over the edge. The others tried in vain to follow the leader. The helper encouraged them and entered at least a score of times, only to come out again at once, teaching them by example how it was done. Several times a fledgling, rising from the lower perch, just managed to grasp by one foot a fibre or stick below the entrance, but found its powers too far spent to raise itself over the sill, and in a moment lost its hold and went fluttering down among the branches, only to return in a minute for another attempt. Several times, too, one flew up while the helper was at the entrance and clung to his back. With more presence of mind the latter might have pulled it into the nest in this manner, but each time he dropped down with the fledgling holding on for dear life, and the struggle began anew. The efforts of the little, short-tailed, pale-breasted fledglings to imitate their long-tailed, patient instructor formed a lovable scene, but their attempts and failures were also very amusing, and at times I shook with silent laughter until I could no longer hold the binoculars steady.

At length, after ten minutes of repeated failures, the other two fledglings managed to gain their bed. There still remained a bit of daylight, so the helper, after looking into the nest to see that all was well, flew off to join the parents and snatch a few more bites before retiring. In ten minutes more the first of the grown-ups came to bed, followed at close intervals by the other two. As each in turn darkened the entrance, the fledglings greeted him with their lisping hunger calls, associating from life-long habit the appearance of a bird at the doorway with the bringing of food. When the last had disappeared into the interior, all remained quiet in the nest, and I longed to be able to peep in and see by the fading light how the six sleepers had arranged themselves for the night.

I kept this family under observation for the remainder of the year, during which they occupied five different dormitories. In the middle of May they moved to a nest (C) in an alder tree, about 200 feet from the oak in which the breeding nest (A) was located. During July the young birds, now in their fourth month, began to molt and acquire their adult plumage, their most conspicuous change being the substitution of the white, black-spotted breast of the adult for the immaculate, light buff breast of the fledgling. They occupied the alder tree nest until the beginning of September, when the whole family moved into a new nest (D) which had just been completed in the same oak tree which already held the breeding nest (A) and an old dormitory nest (B). During October they began a second nest in the alder tree, but never completed it, and continued to sleep in Nest D until the middle of November, when they shifted back to Nest A in the same tree. They did not long remain here, and for a while I lost track of them; but a week later I found them sleeping in a newly constructed nest (E) in the top of a tree hawthorn (*Crataegus stipulosa*) about five hundred feet distant from the breeding nest. The family was now reduced to four members, and I could not determine what had happened to the other two. During December they returned to the oak tree and slept sometimes in the breeding nest (A), sometimes in the nest (D) they had built during August. I do not know why this group moved about so much, for another family of nine continued to sleep in the same nest from September until I left them at the end of the year.

In June, after the close of the breeding season, I watched the construction of a nest in territory which had not been previously occupied by Cactus Wrens. Its builders were a pair of birds who were evidently just establishing themselves. They worked side by side at the task of construction, and began to sleep in their dormitory about the middle of the month, although they continued to add material to it for several weeks longer. The pair continued to sleep alone in this nest until October, when I discovered that they had been joined by a third bird, who remained with them until, at the beginning of December, they changed their residence, or met with some calamity, and I was unable to find them again. I suspect that, if everything went well with these birds until the following breeding season, the third Wren would have turned out to be the helper of the original pair.

While the Brown Jays are restricted to the Tropical and Subtropical Zones, and the Black-eared Bush-tits in Guatemala live entirely in the altitudinal Temperate Zone, between five and nine thousand feet above sea-level, the Banded Cactus Wrens enjoy a remarkably wide altitudinal distribution. I have found the species from near sea level in Costa Rica and Guatemala up to nearly ten thousand feet in a clearing in the cypress forest above Tecpán, Guatemala. In the humid lowlands these birds inhabit the older second growth, riverside groves, and woodlands which have been somewhat thinned and opened by lumbering operations, but, so far as my experience goes, not the heavy virgin forest. In the highlands, where they are far more numerous, they dwell in light woods of oak, alder and pine, or else in bushy pastures where scattered trees remain, and in similar habitats, but avoid the heaviest sorts of forest. The habit of sleeping together in dormitories is not restricted to those individuals who dwell in the cool uplands where nights are frosty, for near Turrialba in Costa Rica, at an altitude of about 2500 feet, I found a family of seven sleeping together, and near sea-level in the Estrella Valley I came upon a dormitory into which two birds retired at nightfall. Probably their custom of seeking shelter from the extremes of nocturnal weather in these commodious nests is one of the factors which enable them to thrive in such a wide range of climates.

To me, the most pleasing aspect of the various associations among birds which we have been considering is that they are entirely voluntary, which puts them on a different and higher plane than those of social insects such as termites, ants and bees. Among these no one pair are complete and able to take care of themselves and raise their families without the aid of other biological forms of their species. With these insects coöperation is obligate, for without it the species would soon become extinct. All birds, save those affected by accident or disease, remain complete in all their faculties; there is no structural or sexual specialization among them, as among the social insects, which makes it impossible for any one pair to live and raise their young without outside help. Perhaps it is necessary to except from this statement only species like Cowbirds and Cuckoos, which place their eggs in the nests of other kinds. Each pair of Groove-billed Anis, as they please, may join in a communal nest with others of their kind, or may build their own nest and raise their young alone. The young Brown Jay who this year helps at the nest of a pair of older birds will next year have a nest of its own; the mated Brown Jays who receive so much voluntary assistance are perfectly capable of raising their family without aid. And so among birds help is given and received entirely in a spirit of good fellowship, neither those who give nor those who receive compromise their independence nor lose their self-sufficiency.

3509 Clark's Lane,
Baltimore, Maryland.

AUTHOR CITATION INDEX

SUBJECT INDEX

INDEX OF BIRDS

About the Editor

RAE SILVER received a B.Sc. with honours in psychology from McGill University, an M.A. from City College, and a Ph.D. from the Institute of Animal Behavior at Rutgers University. Dr. Silver has held posts at Rutgers University College, Hunter College of the City University of New York, and is presently a member of the faculty of Barnard College of Columbia University and a research associate of the American Museum of Natural History. Dr. Silver has done extensive research with birds both in the laboratory and in the field, has written numerous scholarly publications, and has given talks describing the research.